ALSO BY RICHARD CONNIFF

*The Devil's Book of Verse: Masters of the Poison Pen from
Ancient Times to the Present Day* (editor)

Irish Walls (with photographer Alen MacWeeney)

Spineless Wonders: Strange Tales from the Invertebrate World

Every Creeping Thing: True Tales of Faintly Repulsive Wildlife

Rats: The Good, the Bad, and the Ugly

The Natural History of the Rich: A Field Guide

*The Ape in the Corner Office: How to Make Friends,
Win Fights, and Work Smarter by Understanding Human Nature*

*Swimming with Piranhas at Feeding Time: My Life Doing
Dumb Stuff with Animals*

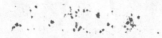

THE
SPECIES
SEEKERS

Heroes, Fools, and the
Mad Pursuit of Life on Earth

RICHARD CONNIFF

W. W. NORTON & COMPANY • NEW YORK • LONDON

For information about permission to reproduce
selections from this book, write to Permissions,
W. W. Norton & Company, Inc.,
500 Fifth Avenue, New York, NY 10110

For information about special discounts for bulk
purchases, please contact W. W. Norton Special Sales
at specialsales@wwnorton.com or 800-233-4830

Manufacturing by RR Donnelley, Harrisonburg
Book design by Dana Sloan
Production manager: Devon Zahn

Library of Congress Cataloging-in-Publication Data

Conniff, Richard.
The species seekers : heroes, fools, and the mad pursuit
of life on Earth / Richard Conniff.
p. cm.
Includes bibliographical references and index.
ISBN 978-0-393-06854-2 (hardcover)
1. Naturalists—Biography. 2. Natural history—History—18th century.
3. Life (Biology)—Research—History—18th century. I. Title.
QH26.C66 2011
508.09'033—dc22

                                2010024532

W. W. Norton & Company, Inc.
500 Fifth Avenue, New York, N.Y. 10110
www.wwnorton.com

W. W. Norton & Company Ltd.
Castle House, 75/76 Wells Street, London W1T 3QT

1 2 3 4 5 6 7 8 9 0

To those who died in the search for species

CONTENTS

THE

SPECIES

SEEKERS

STRANGE THINGS, STRANGE LANDS

What raptures must they have felt to land upon
countries where every thing was new to them!

A T THE HEIGHT of the Battle of Alcañiz on May 23, 1809, as he was about to give the order for a desperate charge by French troops into the center of the Spanish line, Col. P.F.M.A. Dejean happened to glance down. The air around him was thick with gunpowder and blood, but on a flower beside a stream, he saw something unusual. A beetle. Species unknown. He immediately dismounted, collected it, and pinned the specimen to the cork he had glued inside his helmet.

Dejean was a count and a battle-tested leader in the Napoleonic armies. But he was also, above all, a coleopterist, a specialist in beetles. His men knew it because many of them carried glass vials for him and had orders to collect anything on six legs that crawled or flew. His enemies knew it, too, and out of courtesy and respect for the cause of scientific discovery, sent him back vials taken from the dead on the field of battle.

Having collected this latest prize, Dejean swung back up into the saddle and launched the attack. With bayonets fixed, the massed French

1

forces advanced up the slope toward the Spanish artillery. The gap between them slowly closed, everything tense and quiet. Then, at the last moment, the cannons let loose a storm of grapeshot into the faces of the attacking line. Hundreds of French soldiers died. Dejean's helmet was shattered by cannon fire. But he and his specimen survived intact. Years later, he would give his prize from Alcañiz a scientific name, by genus and species, *Cebrio ustulatus.*

To modern readers, Dejean's reckless passion for beetles in the face of enemy fire may well sound insane. There are probably several million beetle species in the world, and even in that species-besotted era, there was little prospect of glory in adding one more to the list. As it turned out, Dejean would not even get credit for his discovery. By the time he got around to describing his prize from Alcañiz years later, some other naturalist had already found the species and recorded it in a scholarly journal. Under the rules of scientific discovery, this reduced Dejean's proposed name to a synonym, an also-ran. In any case, both naturalists were soon forgotten, along with their beetle.

And yet glory and wonder were everywhere in the air then.

Dejean and like-minded naturalists were fanning out across the globe to play their part in a fabulous adventure story. They regarded the hunt for new species as one of the great intellectual quests in human history, and with good reason. At the start, naturalists knew no more than a few thousand species, and often had the basic facts wrong. Even educated people still inhabited a jabberwocky world in which monsters abounded, and one species could slide uncertainly into another. Our own ancestors, just eight or ten generations ago, still thought that dog-headed humans lived in distant lands, probably based on early descriptions of baboons. When the fossil skeleton of a giant salamander turned up, a learned Swiss physician identified it as a sinner drowned in Noah's Flood. Naturalists then could not even clearly distinguish some plants from animals and passionately debated whether one could transform into the other, and back again. (It's a measure of the state of knowledge then that they thought of themselves simply as naturalists or "natural philosophers." The words "scientist" and "biologist" did not yet exist.)

As the great age of discovery began, bestiary creatures still haunted human minds.

That would all change, as a small band of explorers set out to break through the mystery and confusion. The great age of discovery about the natural world was a period of less than 200 years, from the eighteenth century into the twentieth. It got its start in 1735, when the Swedish botanist Carolus Linnaeus invented a system for identifying and classifying species. He was a charismatic teacher, both ribald and full of religious fervor for the wonders of the natural world. His words inspired 19 of his own students to undertake voyages of exploration. Half of these "apostles," as he called them, would die overseas in the service of his mission. Explorers from other nations, also inspired by Linnaeus, soon followed, taking the hunt for new species to the farthest ends of the Earth. They made the discovery of species one of the most important and enduring achievements of the colonial era.

That word "discovery" may stick momentarily in the modern reader's craw. Local people had often known many of these "new" species for thousands of years and in far more intimate detail than any newcomer could hope to achieve. But done properly, the process of collecting a species and describing it in scientific terms made that knowledge available everywhere. Making it available in Europe was, to be sure, the primary objective. But in the process, the species seekers introduced humanity for the first time to our fellow travelers on this planet, from beetles to blue-footed boobies. And gradually we stumbled from the security of a world centered on our species, created for our comfort and salvation, to a world in which we are one among many species.

It would be difficult to overstate how profoundly the species seekers changed the world along the way. Many of us are alive today, for instance, because naturalists identified obscure species that later turned out to cause malaria, yellow fever, typhus, and other epidemic diseases. (This is one of the recurring lessons from the history of species discovery: Useless knowledge has an insidious way of leading people in useful directions. Many mothers would despair, for instance, to have a child make a career out of the study of Chinese horseshoe bats of the genus *Rhinolophus*. But the subject took on global importance when these bats turned out to be the source of SARS, or sudden acute respiratory syndrome, which threatened to become pandemic.)

The discovery of species also shifted the foundations of knowledge and belief. Though early species seekers typically set out to glorify God by celebrating his Creation, the paradoxical outcome of their work was to cast doubt in many minds on the very existence of God. Species that seemed insignificant in themselves would raise disturbing questions about human origins, the age of the planet, the nature of sex, the meaning of races and species, the evolution of social behaviors, and endlessly onward. When we look in the mirror today, we can hardly help but see what the species seekers showed us.

"Human Sacrifices, Strange Currency"

But it began with the adventure.

Naturalists had been prowling beaches and forests, and puzzling over shells and feathers, at least since Aristotle. But it was only in the eighteenth century that a unified cadre of naturalists speaking the same scientific language set out for the farthest corners of the Earth to see the alien world with their own eyes. Armed with the Linnaean system, plus guns, nets, collecting boxes, and an almost missionary sense of purpose,

The splendidly indolent sloth.

they were everywhere—deep in the Namib Desert, up the Rio Japurá, out on the uncharted waters of the Great Barrier Reef. And they were bringing back creatures even the authors of medieval bestiaries could hardly have imagined: Who would have dreamed up the splendidly indolent three-toed sloth, whose genus name, *Bradypus*, means "slow-footed"? Who could have conjured the echidna, an Australian mammal described by a British naval officer as "a kind of sloth about the size of a roasting pig, with a proboscis two or three inches in length" and "short quills like those of the porcupine"? (Closer examination being required, the animal "was roasted and found of a delicate flavor.")

The appetite for new species was almost insatiable, and naturalists went to extraordinary lengths to get them. Sailing in Antarctic waters, one nineteenth-century ornithologist, Titian Peale, attempted to collect seabirds by firing at them as they flew to windward, in the hope that the howling gale would carry their tumbling carcasses back to the deck of his ship. On another expedition in warmer waters, the wealthy eccentric Charles Waterton offered cash to anyone who would swim for a specimen, thereby nearly drowning a sailor to recover a dead bird. In India, Col. Howard Irby, an ornithologist in the British Army, "had a curious soldier-servant whom he had trained as a retriever, and no matter how deep the water was where the duck fell, he quickly brought it to his master." Many explorers also endured the unsettling experience of discovering a species in the dinner pot. Charles Darwin, for instance, had been searching for a small type of rhea in Patagonia when it dawned on him, one Christmas dinner, that he had just eaten it. He was reduced to collecting the bones and feathers from the kitchen scraps.

Then as now, naturalists tended to collect compulsively, even frantically, often pushing themselves to the brink of death. (I have often traveled with modern biological expeditions, and vividly recall being almost catatonic with exhaustion on one trip to Ecuador, lying with my top half inside my tent for a nap while the bottom half stayed outside because my pants were crusted with a body cast of tropical mud. Two naturalists with whom I traveled later died when their reconnaissance plane crashed into a cloud forest. Another survived malaria in Gabon only because a Bakwele pygmy woman carried him 18 miles on her back to get the injection that saved his life.) The normal routine for species seekers was to be out at dawn and again at dusk, or almost any time in between, gathering specimens. Then they often worked late into the night to preserve and identify their booty before insects and rot could reduce it to tatters. The pace took a toll on even the most levelheaded collectors.

In the mid-nineteenth century, for instance, Henry Walter Bates spent 11 happy years collecting in the Amazonian rainforest, undaunted even when robbed and left barefoot ("a great inconvenience in tropical forests," he allowed). He could dwell lovingly on leafcutter ants and their curious behavior of walking in long lines with fragments of leaves carried upright on their backs, like placards. He had the idea that they were using the leaf fragments to construct thatched roofs against rain. (In fact, the ants use leaf fragments in underground chambers to grow the fungus that they eat. They are the Earth's first farmers.) But Bates

Local hunters often did the perilous work of collecting.

was also driven mad by these ants. Waking one night to discover long lines of them raiding his "precious baskets" of *farinha*, or cassava meal, he and a servant resorted to stomping on them with their wooden clogs. When the ants returned the following night, "I was then obliged to lay trains of gunpowder along their line, and blow them up. This, repeated many times, at last seemed to intimidate them."

The American lepidopterist William Doherty, working in the Indo-Pacific in the 1880s and 1890s, was also frequently vexed with tropical afflictions. He wrote home that he couldn't keep his specimen pins from rusting in the rainy season. "Salt and sugar here liquefy every night and have to be dried over the fire every day," he added, "and the boots I take off at night are sometimes covered with mould in the morning."

Doherty was generally too busy to dwell on his misfortunes. He once summed up a year's work in the Indo-Pacific islands in telegraphese, sounding a bit like Fearless Fosdick, the comic book hero who could dismiss it as "merely a flesh wound" even when machine gun bullets made his midsection look like Swiss cheese: "Loss of all my collections, money, journals and scientific notes at Surabaya in Java. Proceed by way of Macassar to the island of Sumba. Dangerous journey in the interior. Discovery of an inland forest region, and many new species of Lepidoptera. King Tunggu, human sacrifices, strange currency. . . . Visit to the Smeru country. Hunted by a tiger when moth-catching. Hunt tigers myself. Leave for Borneo. Ascent of the Martapura River from Banjermasin. Life among the Dyaks in the Pengaron country. Head-hunting. The orang-utan."

Naturalists were often so single-minded in their pursuit of species that great events in world history could seem like a mere distraction, or even at times an opportunity. In 1848, when the blood of revolutionaries ran in the streets of Paris, an American entomologist wrote home from Europe: "Insects are remarkably cheap in Paris at present; now is the time to buy." That same year, a soldier with the U.S. Army waging guerilla war in Mexico wrote home about "leaving in our track blood & fire, burning & plundering . . . sparing none but women & children." Then he added, "You will notice among the Coleoptera I sent home many not duplicated. They were collected on scouts to parts I had no opportunity of revisiting." Small wonder.

Hundreds, or more likely thousands, of naturalists died in the sacred cause of natural history. One modern account, *The Bird Collectors* by Barbara and Richard Mearns, offers a matter-of-fact sampling of hazards: "John C. Cahoon fell off a sea cliff in Newfoundland; William C. Crispin plunged to his death while going after Peregrine eggs; Francis J. Britwell, on honeymoon in the Sierra Nevada, tried to reach a nest in a tall pine, lost his grip and fell, his rope choking him to death while his bride could only stand and watch. Another American, Richard P. Smithwick, smothered to death while digging his way into a soft bank to raid a Belted Kingfisher nest." Appalling injuries were commonplace. Benjamin Walsh, the first state entomologist in Illinois, lost his foot in a train accident and tried to comfort his distraught wife with a joke: "Don't you see what an advantage a cork foot will be to me when I am hunting bugs in the woods: I can make an excellent pincushion of it, and if perchance I lose the cork from one of my bottles, I shall simply have to cut another one out of my foot." Unfortunately, he died of the injury before he could test this idea in the field. When Edward Baker, a British ornithologist in India, died peacefully at home, age 79, his obituary noted that he had lost his left arm by jamming it down the throat of a charging leopard, was tossed twice by bison, and had been trampled by a rhinoceros—but remained (thank God) an excellent shot and a good tennis player.

Disease, while less colorful, was a more efficient killer. In a letter sent from Kenya in February 1901, William Doherty was characteristically dismissive about what he called "the usual adventures. The first were with lions and rhinos. Lately it has been with wild buffalo, a rogue elephant, and a leopard who comes in our boma [a corral or enclosure] every night. . . . Only the other night I had to fight for my life with the marauding Masai." About that time, a friend sent a note to Doherty from England. It came back a few months later stamped "decede" (deceased). The cause, it turned out, was dysentery.

"Odd Fish"

Why did they do it? What drove them to the far corners of the Earth?

The "natives" naturally thought the newcomers were crazy. Along with their passion for butterflies, beetles, and other seemingly worthless species, the ghostly white skin of almost all of these new explorers compounded the impression of abnormal psychology. Traveling to islands in the Pacific, the shell collector Hugh Cuming displayed a genius (and also a pocketbook) for enlisting the help of locals in finding new species. But a friend later wrote that his "apparently unappeasable restlessness" also made them nervous, especially when they saw him through the window of his house, working with his specimens late into the night, "groping and flitting about" by candlelight. In the Philippines, Cuming learned to pretend that he needed the shells for a manufacturing process, the way Filipinos used the ashes of certain shells to make betel nuts easier to chew. Telling them that he wanted shells to stock a natural history collection might have sounded disturbingly ritualistic.

Back home, on the other hand, the pursuit of new species was a

Naturalists were sometimes idealized as the knights-errant of a new heroic age.

topic of ardent public interest at all levels of society, from peasants on up to presidents and kings. Naturalists became the heroic type of the day, like knights-errant in the Middle Ages. "Our perfect naturalist," the English novelist Charles Kingsley wrote in 1855, "should be strong in body; able to haul a dredge, climb a rock, turn a boulder, walk all day, uncertain where he shall eat or rest; ready to face sun and rain, wind and frost, and to eat or drink thankfully anything, however coarse or meagre; he should know how to swim for his life, to pull an oar, sail a boat, and ride the first horse which comes to hand; and, finally, he should be a thoroughly good shot, and a skilful fisherman; and, if he go far abroad, be able on occasion to fight for his life."

Like many other writers, Kingsley also poked fun at naturalists. His popular children's book, *The Water Babies*, featured a daft but kindly Professor Ptthmllnsprts, who liked to collect new species and, exactly as his name suggested, put them all in spirits. At times, naturalists also poked fun at themselves. On meeting John James Audubon, the eccentric biologist Constantine Rafinesque presented a letter of introduction, which declared, "My dear Audubon, I send you an odd fish which you may prove to be undescribed." When Audubon asked the whereabouts of the specimen, Rafinesque replied, "*I* am that odd fish."

But the heroic image of naturalists clearly predominated. One typical young American in Iowa, inspired by an account of Amazonian exploration, became "fired with a longing to ascend" the river and soon lit out for this "romantic land where all the birds and animals were of the museum varieties." But he went bust in New Orleans and had to make do with the river at hand. The experience, recorded in *Life on the Mississippi* and other books, would give him his pen name— Mark Twain. The heroic image of naturalists was powerful enough that nations mounted great biological collecting expeditions to every corner of the planet. At sea with the U.S. Exploring Expedition in 1838, for instance, one young naval officer noted his admiration for these naturalists leaving the comforts of home "to garner up strange things of strange lands." He was also delighted by their staterooms crammed with "dead & living lizards, & fish floating in alcohol, and sharks jaws, & stuffed Turtles, and vertebrates and Animalculae frisking in jars of salt water, and old shells."

Homebodies would eventually share his delight: Specimens from that expedition soon became the basis for the natural history collection of the newly founded Smithsonian Institution in Washington, D.C.

The British, Germans, French, and other nations likewise supported great natural history museums to preserve and display the biological booty their own expeditions carried home with them, and to one-up rival nations. Thus a few naturalists did in fact achieve not just glory but something like immortality: Carolus Linnaeus, Georges-Louis Buffon, Joseph Banks, Alexander von Humboldt, Jean-Baptiste Lamarck, Georges Cuvier, John James Audubon, Charles Darwin, Alfred Russel Wallace, and Patrick Manson all changed the way we think about the world and our place in it. They became names carved into the walls of those natural history museums.

"A Singular Delight"

But public acclaim was hardly the point for most naturalists. Most of them were content merely to inhabit the back rooms of such museums, in dusty little offices during their lifetimes and thereafter as names chicken-scratched on tiny labels attached to the species they had discovered. Vladimir Nabokov, who was a lepidopterist as well as a novelist, once asserted that the glory of naming a new species ("the immortality of this red label on a little butterfly") exceeded even literary acclaim, though he chose to make the point in a poem.

To be a naturalist was to play a part in building a great and permanent body of knowledge. But there was of course more to it than that. The world was full of things that were new and unknown (or "nondescript"), and in their hearts, the species seekers were as desperately acquisitive as any other collector to hold the precious undiscovered thing in their own hands. "The truth is," Charles Kingsley wrote, the pleasure of finding new species is too great. "It is morally dangerous; for it brings with it the temptation to look on the thing found as your own possession, all but your own creation; to pride yourself on it, as if God had not known it for ages since; even to squabble jealously for the right of having it named after you, and of being recorded in the Transactions of I-know-not-what Society as its first discoverer: as if all the angels in heaven had not been admiring it, long before you were born or thought of."

For that sense of private joy in small moments of discovery, naturalists often treated the hunger, loneliness, disease, and other hardships of
field life as an annoying distraction. "I have lost thirty-four pounds, but
feel about as usual and am working as usual," 19-year-old "Willie" Dall
wrote nonchalantly to his mother in 1865, having just completed the
voyage from Boston around to San Francisco.

Things got considerably rougher once he moved on to Alaska.
Among many other adventures, he would endure a long frigid trip in a
sealskin dory across open water, trying to avoid being crushed by waves
loaded with cakes of ice. ("It was a queer feeling," he wrote, "to feel the
bottom and sides of our boat pulsing with the waves, and a still queerer
one when we sprung a leak.") They finally arrived in darkness at a Russian outpost, where the place he was given to bed down was infested
with "millions of cockroaches . . . bent on making our acquaintance, all
night."

Dall was a specialist in marine mollusks, and he gave his family
an eloquent explanation of what motivated him, and by extrapolation
most other naturalists: "There is a singular delight in taking these delicate and almost microscopic animals and putting them under a strong
glass, seeing the tiny heart beat, and blood circulate and gills expand,
counting the muscles and blood vessels and almost the tiny disks that
form the blood and to know that you are the first that has penetrated
these mysteries and are perhaps the only one who ever will, and that all
your notes and drawings and observations are so much solid knowledge
added to the power and grace and beauty of the Infinite."

Limitless Discovery

No one realized how varied the infinite would turn out to be. Linnaeus
had given people the tantalizing belief that they could actually make
sense of the world, and that belief arose partly from the idea that nature
was limited. Naturalists assumed that each species was discrete, a separate creation, no matter how much one might resemble another, and
that God would have put only so many species in the Garden of Eden.
Likewise, only a relatively small number of species could have survived

on Noah's Ark. Thus in the 1770s, when the work of exploration had hardly begun, the wealthy collector Margaret Bentinck, Duchess of Portland, was already declaring her intention to have every unknown species "described and published to the world." She may even have thought she was getting close. When she died in 1785, she possessed thousands of specimens, and the auction to dispose of her collections lasted for 38 days.

A few decades later, the great French anatomist Georges Cuvier asserted that there was no longer much "hope of discovering new species of large quadrupeds" alive in the modern world. He figured that fossils would yield more discoveries. Likewise in mid-nineteenth century, the British anatomist Richard Owen imagined that "nothing remained for naturalists but the business of classification and arrangement"— only to have the largest primate on Earth, the gorilla, discovered by a couple of upstart Americans. (This "diabolical caricature of humanity," as Owen called it, would soon take center stage in the Darwinian debate, and threaten the special standing of *Homo sapiens* in a divinely ordained universe almost as profoundly as had the realization that the sun did not revolve around the Earth.) In truth, whenever anybody had the hubris to declare that we were, in effect, running out of new species, nature always seemed to respond with a spectacular parade of strange new creatures—the okapi, the pygmy hippo, the manatee, the snow monkey, and the panda, to name only a few from the period after Cuvier.

People nonetheless continue to suggest that we have reached the end of discovery today, when scientists have described, catalogued in museums, and often studied in detail about 2 million species. But the current estimate is that roughly another 50 million species remain to be discovered, a number that early naturalists could hardly have imagined.

That is, we still live in the great age of discovery. This book is the story of how it began.

Chapter One

THAT GREAT BEAST OF A TOWN

I have been shown a beetle valued at 20 crowns
and a toad at a hundred . . . whatever appears
trivial and obscene in the common notions of
the world, looks grave and philosophical in the
eye of a virtuoso.

—JOSEPH ADDISON

IN NOVEMBER 1774, a cask of rum arrived in London, spiked with four
dead electric eels, the largest of them almost 4 feet long and up to 14
inches around. They had smooth, snaky bodies, flattened heads, blunt
snouts with a pronounced underbite, and two small fins, resembling
ears, at the sides. Their eyes were small and round, and their dark facial
skin was heavily pockmarked, as if with the point of a knitting needle.

The eels, actually knifefish of the species *Electrophorus electricus*,
had come from Suriname, on the northeastern coast of South America,
and the sensation they caused in England was literally electrifying. A
fifth eel had survived aboard ship all the way to the port of Falmouth
and duly delivered electric shocks to British thrill-seekers, before finally
expiring. But even dead and consigned to rum, a standard preservative
then, the specimens still had the power to excite educated minds.

Among those awaiting the eels in London was John Hunter, who
strictly speaking had no education at all. Having gotten his start as
a carpenter in Scotland, he had managed, with the help of his older

15

brother William, "to lay down the chissel, the rule, and the mallet; and take up the knife," ultimately turning himself into London's leading surgeon and Surgeon-Extraordinary to King George III. His formal medical training had consisted of a few months spent working in a London hospital, and several years' trial-and-error with the British Army at war. A remedial stint at Oxford University had lasted only a month, and with characteristic earthiness Hunter later wrote, "They wanted to make an old woman of me, that I should stuff Latin and Greek at the university, but these schemes I cracked as so many vermin as they came before me."

Hunter cared about only one thing—anatomy—and he cared about it exuberantly. He would dissect "some thousand" human cadavers and more than 500 animal species over the course of his career. It would make him an avid client of both grave robbers and species seekers, and what he learned from the dead he soon applied to the living. His skill with the knife and his willingness to develop radical new procedures, always based on what he learned in his dissections, have earned him a lasting reputation as "the father of modern surgery." Hunter was literally the cutting edge of the developing movement to understand ourselves and our world through the study of other species. His life was also a harbinger of how excitement about this new knowledge could sometimes bridge the vast chasm separating British social classes.

Hunter "danced a jig when he saw [the dead eels], they are so compleat and well preserved," a friend wrote. John Walsh, a Member of Parliament and amateur naturalist who'd gotten rich in the East India Company, promptly paid 60 guineas for the three best specimens— roughly two years' wages for the average London laborer then. Sir Joseph Banks, a young god for his work as a naturalist traveling around the world aboard HMS *Endeavour*, had to rush back to town because Walsh and Hunter were sharpening their scalpels, "bent upon . . . opening one at least at the beginning of next week."

That these prize specimens should have found their way to London and elicited such delight there was unsurprising. Indeed, it was entirely within the city's means to command the collection of addi-

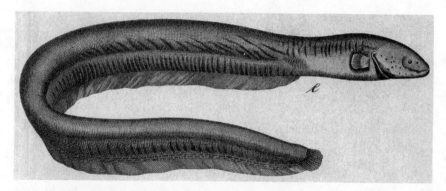

At small gatherings, London's scientific elite joined hands in a ring to experience the shock of an electric eel.

tional electric eels from South America the following year—this time delivered live to Walsh's residence on Chesterfield Street, where an eel soon repeatedly demonstrated its ability to send a 600-volt charge through the joined hands of a ring of up to 27 people at a time. No one seems to have suffered ill effects at these odd gatherings (though fatal electrocutions can occur at even lower voltages). Instead, they came away with a heightened sense of that era's characteristic excitement about the limitless possibilities of the natural world.

Science and giddy spectacle seemed to be everywhere in London then. By the 1770s, the city had become the center of the world, its raucous population of 700,000 people crammed into 7 square miles at a bend of the Thames River, and the Thames itself thick as a forest in winter with the masts of ships serving a global empire. (The crowds did not stop John Hunter, who also loved living animals, from driving his carriage through the streets behind a team of zebu, Asian cattle with humpbacks and long flapping dewlaps.) It was the place and the moment when the modern world of commerce, industry, urban living, and international trade was taking shape. Together with Paris, it was also the nursing ground, if not the birthplace, for the science of natural history. Carolus Linnaeus had published and popularized the first modern system for naming and classifying species. But Linnaeus lived in the backwater of Uppsala, Sweden, a university town of just 5000 people. London was soon to become the world's biggest city, and the British passion for natural history would make it the center for the discovery of life on Earth.

The "Virtuoso Class"

The British interest in exotic animals had deep roots. The royal menagerie already housed a lion and a leopard during the reign of Henry I in the twelfth century. In 1252, it included a Norwegian white bear, which the sheriffs of London had to wait on while it fished in the Thames at the end of an iron chain and a long cord. (Sadly, history doesn't tell us how long a cord was long enough, particularly if the bear came up empty.)

By the start of the eighteenth century, the study of plants and animals had become an obsession of what Alexander Pope called the "virtuoso class," virtuosos being wealthy amateur naturalists vying to add the latest shell or skeleton to their private collections. And by midcentury, the rage for natural history had spread to the rest of the populace, becoming, in the words of one modern historian, "the universal British national pastime, if not the national sport."

Then as now, naturalists with their butterfly nets and killing jars came in for ridicule. The essayist Joseph Addison mocked them for hoarding up the "Refuse of Nature . . . such Creatures as others industriously avoid the Sight of." The critic and wit Samuel Johnson depicted a fictitious landowner named Quisquilius whose zeal for collecting

Little brown lizard.

was such that he let his tenants pay their rent in butterflies and "three species of earthworms not known to the naturalists." But much as he disdained the menial labor of collecting specimens, even Dr. Johnson conceded "there is nothing more worthy of admiration to a philosophical eye than the structure of animals."

The British people plainly agreed. Up to a thousand visitors a day came to ogle the vast natural history collection at the Manchester home of a virtuoso named Ashton Lever. Charging admission would have been beneath his dignity, so Lever, clearly overwhelmed, attempted to discourage the carriageless lower classes by refusing entry to anyone arriving on foot. One spurned but determined visitor promptly returned riding a cow and was duly admitted.

Why this sudden passion for the natural world? For more than a thousand years, nature had been locked away behind moral lessons and mythology. The Church, devoted to knowing the mind of God, had turned its back on nature. Common knowledge about even native European plants and animals had been handed down uncritically for centuries from Roman authors and medieval bestiaries full of imaginary creatures. So when the scientific revolution finally provided the release, people's suppressed curiosity about the real world burst forth with wild abandon.

In England, two great Cambridge University scholars had devised a new way of looking at the world. The physicist and mathematician Isaac Newton had laid out the laws of the physical universe. The philosopher Francis Bacon had promulgated practical rules for investigating natural phenomena based on close observation and careful reasoning, the beginnings of the scientific method. Bacon had also popularized the idea that knowing God meant knowing His Creation, and recovering the mastery of Nature lost at the Garden of Eden. That idea was now in full blossom, as evidenced by a 1765 letter to London from a South Carolina naturalist. With the competitive and proprietary attitude common among naturalists to this day, Alexander Garden first griped about a rival stealing a species that he himself had actually discovered. But then he relented: "Yet, after all, he is an excellent man and I forgive him, because it is a matter of little moment who declares the glories of God, provided only that they are not passed over in silence."

"Rum for the Mole"

Some tradition-bound geographers still populated blank regions on their maps with imaginary monsters, like manticores (a blend of lion, scorpion, and human) or Blemmies (headless humans whose faces were incorporated in their chests). But travelers were now actually visiting the farthest corners of the planet and discovering what really lived there.

This could be alarming, not least for sheer abundance. Arriving in North America in 1748, the Swedish naturalist Pehr Kalm kept a careful expense account ("rum for the mole, 9 pence") and confessed to being "seized with terror at the thought of ranging so many new and unknown parts of natural history." But Kalm's teacher, Linnaeus, had provided the intellectual tools for bringing order to this chaos. The Linnaean system of classification made it possible to take any living thing, give it a two-part name like *Ursus horribilis*, the brown bear, or *Canis lupus*, the wolf, and assign it to a place in the natural order. In time, the Linnaean system would develop into a neat hierarchy by kingdom, phylum, class, order, family, genus, and species.

Not everyone could understand Newtonian physics. But anyone could now approach the natural world, and for a time almost everyone did—"men and women both," writes science historian John R. R. Christie, "doctors and clergy, aristocrats and gardeners' boys, apothecaries, printers, lawyers, military, and naval men." Studying nature took all kinds—explorers willing to face the likelihood of death in some unknown land, as well as homebodies who could sit still and record minute details of morphology and behavior. It called for practical types skilled at "hunting, finding, drying, stuffing, arranging and displaying," as well as philosophical sorts who could make subtle distinctions about what connected different animals and what separated them. (The literary man Dr. Johnson said whales were fish; the naturalist Dr. Hunter was more impressed by their "analogy to land animals.") The new discipline required scholars trained to use "the formidably complex vocabulary" of scientific description for "the increasingly technical and controversial exercises of naming, classifying and describing the natural world." But it also required artists and writers who could bring new species to life for an eager public.

Books of butterflies or birds were perennial best sellers in the rapidly expanding British publishing business. Natural history also rivaled politics as the leading subject of coffeehouse debate. Even the historian Edward Gibbon and the economist Adam Smith turned up at the anatomy demonstrations given in London by John Hunter's brother, William. (William was the eloquent lecturer, while inarticulate John mostly did the bloody work of demonstrating. An enemy later claimed that "he was incapable of putting six lines together grammatically into English." But others described John as kindly, taking his gruffness as merely the price of deep concentration on his work.) Both Hunter brothers also regularly hired the master animal painter George Stubbs, who shared their enthusiasm for dissection (and grave-robbing), to record new species for their collections.

Though he carried on an extensive correspondence with naturalists in other European capitals and around the world, even Linnaeus recognized the extraordinary promise of the forces coming together in London. At Uppsala, his fellow Swede Daniel Solander had been "the wittiest pupil I ever had" and was like a son to him (as well as a likely son-in-law)—and yet Linnaeus dispatched him in 1759 to London, where he soon went to work organizing the collections of the British Museum and helping to propagate the Linnaean system around the world.

Solander later served as a naturalist with Joseph Banks aboard Capt. James Cook's *Endeavour,* and they brought back thousands of plant and animal specimens from their three-year circumnavigation. Among their prizes were skins and bones of a creature with a head like a deer, said to rise up on two legs, in some cases nearly to the height of a man. It was also said to use those legs to go bounding across the grasslands of Australia like a hare, with its long, heavy tail serving for balance. Banks

The "Kanguru" became an eighteenth-century London sensation.

announced it to the outside world with a name borrowed from the Aborigines: "Kanguru."

Wild Beasts of London

Astonishing specimens from distant regions seemed to turn up in London almost daily then—the first zebra, the first moose, the first nilgai—each adding fuel to the raging national passion for the wonders of the natural world. New species arrived not just in the cause of science, but as entertainment. (Either way, most eventually ended up on John Hunter's dissecting table.) For a penny or two, gawkers could see a rhinoceros, a yak, a baboon, or a macaque monkey. Commercial menageries seemed to be everywhere, and taverns and coffeehouses attracted customers with "the living alligator or crocodile, lately arrived from the coast of Guinea" or "an Eel, the largest ever seen in London." It's hard now to imagine all these creatures thrashing around the middle of eighteenth-century London. The human population density then already exceeded that of modern Manhattan, but without high-rise buildings, sewerage systems, or other means of making cities livable. Cartmen and porters pushed through the packed streets yelling, "Make Room there!" or "By your leave!" after having just knocked you to the ground, one writer complained. And yet a 1773 visitor could remark, without too much exaggeration, "Wild Beasts on every Street in Town."

Tame ones, too: One of the performing acts described in Richard Daniel Altick's history *The Shows of London* billed itself as "Breslaw's birds." The birds, of unknown species, arranged themselves in ranks and marched like British soldiers, with miniature grenadiers' caps on their heads and wooden muskets tucked under their wings. For their show-stopper, these feathered soldiers rounded up one bird—designated the deserter—marched him in front of a small brass cannon, and set a match to the priming mechanism. When the cannon fired, the deserter fell down dead—then apparently rose again to accept the applause of his delighted public. Bull- and bear-baiting were common amusements, and in one popular event, the Welsh main, only one fighting cock survived from among 32 combatants. The naturalist Gilbert White, savoring the relatively unmolested wildlife around his rural vicarage in Selborne, disparaged London as "that great beast of a town."

Private collections also catered to the ravenous British appetite for "prodigies," as wonders of the natural world were known then. In 1774, Ashton Lever, the virtuoso whose house in Manchester had been over-run with visitors, moved his collection to London, into 12 rooms of a former royal palace in Leicester Square. This time, dignity be damned, he charged admission. The collection featured preserved specimens, in no particular order, of squirrel monkey, coatimundi, possum, leopard, osprey, bird of paradise, flamingo, saw fish, arctic fox, and thousands of other species. There was also "a Duck with a foot growing out of its head," and a monkey in the pose of the Venus de' Medici. According to

Ashton Lever's museum celebrated odd specimens, notably including himself.

the diarist Frances Burney, Lever himself "pranced about," dressed as a forester "in a green jacket, a round hat, with green feathers, a bundle of arrows under one arm, and a bow in the other."

Lever called his museum *Holophusicon*, "embracing all of nature" and, like Dr. Johnson's Quisquilius, he nearly bankrupted himself in his quest "to possess all nature's wonders." But in the presence of his specimens, he and his public alike seemed to experience something sublime, what one visitor later called "a majestic awe for the power of bones and claws." (And for other parts, too. Among its 300 specimens, one Chelsea pub proudly featured "a whale's pizzle.")

More strictly scientific collectors were less impressed. Sir Joseph Banks apparently despised Lever, perhaps because the virtuoso's higgledy-piggledy displays blatantly flouted the beautiful order the Linnaean system of classification was just bringing to the natural world. Banks may also have resented Lever's knack for beating him to choice specimens. The Holophusicon's natural history collections were superior in variety, if not scientific usefulness, to those of the British Museum itself. Or as Lever immodestly put it, "I am at this Time, SOLE POSSESSOR OF THE FIRST MUSEUM IN THE UNIVERSE."

"Entrails of a Pangolin"

But if the showmen were colorful, bordering on bizarre, so at times were the naturalists. And they sometimes depended intimately on the showmen for their best specimens. One night, for instance, the painter George Stubbs got word that there was a dead tiger at Gilbert Pidcock's traveling menagerie, then performing on the Strand in the heart of London. "His coat was hurried on and he flew towards the well-known place," a friend later wrote, "and presently entered the den where the dead animal lay extended: this was a precious moment."

For three guineas, the carcass was delivered to Stubbs's home, where the artist spent the rest of the night (and probably many weeks thereafter) preparing and dissecting it. Well before human embalming was commonplace, Stubbs was a master at injecting a carcass with wax and other preservatives. As a young man, he'd spent 18 months meticulously stripping horse carcasses down to the skeleton and making detailed drawings at every anatomical level along the way, including five separate layers of muscle. According to a friend, his method was to suspend from the ceiling an iron bar lined with stout hooks, which he fastened between the ribs and under the backbone on the far side of the carcass. Then he winched the animal up, with a platform under its feet, and began methodically peeling back the skin and working his way inward, spending six or seven weeks on each reeking specimen.

The result, apart from some of the greatest animal drawings and paintings the world has ever known, was a dawning sense of all the common elements connecting humans to other animals. Late in life, Stubbs made a drawing of a tiger skeleton running and paired it with a drawing of a human skeleton hunched over like a sprinter leaving the blocks. The one seemed to echo the other almost bone for bone. The quest for the glories of God was already leading some naturalists to think more carefully about the presumed glory of mankind.

His friend John Hunter, the gruff, woolly-headed surgeon, made even more detailed comparisons between humans and other species. He'd spent his youth wandering the family farm in the Clyde Valley south of Glasgow. "I wanted to know about the clouds and the grasses, why the leaves change colour in the autumn," he later wrote. "I watched the ants, bees, birds, tadpoles, and caddis worms. I pestered people with questions about what nobody knew or cared anything about." As an adult, he continued to pester people, including animal dealers and traveling showmen, for the carcasses of recently deceased animals so he could cart them home for dissection.

Naturalists and explorers also routinely supplied him with specimens. "If you will step in at Banks's in Soho Square," his friend Joseph Banks wrote, "you will find the corpse of a fine Sierra Leone Cat, the inside of which is at your service. The skin is to be stuffed for the British Museum." A traveler, perhaps overestimating even Hunter's appetite for gore, offered "the entrails of a Pangolin from Sumatra," but noted that they had been "entirely spoilt from their long detention at the India House."

Hunter did the dirty work of preparing specimens at his country estate in Earls Court, then two miles outside the city. Among other specimens, he handled two dead elephants, an ocelot, a caracal, two hyenas, and an antelope from the royal menagerie. An assortment of live animals helped dispose of excess flesh. It included two leopards, which he once caught by hand when they escaped and started to fight with his dogs. At Earls Court, Hunter also kept a variety of geese to supply eggs for embryological studies, a pond for studying freshwater life, and—apparently for no better reason than the pleasure of their company—a jackal, a zebra, an ostrich, and many other creatures.

Hunter also kept a museum in town to display his more remarkable specimens. A family of "Eskimau" from Labrador visited there in 1772. (The "prodigies" turning up in London then included not just exotic animals, but also human beings of distant cultures, though whether they qualified as fully human was still a common topic of debate.) Horrified by the human skulls and other anatomical displays, the leader of the group supposedly asked, "Are these the bones of Esquimaux whom Mr. Hunter has killed and eaten? Are we to be killed? Will he eat us and put our bones there?" He did not. But the Eskimau family's concern was not so far-fetched. A few years later, Hunter had one of his agents shadow a human sideshow attraction, an Irish giant named Charles Byrne. As he lay dying, Byrne begged his friends to save him from the dissecting room. Instead, the friends took a bribe from Hunter's agent, who spirited the body to Earls Court, where Hunter supposedly removed the flesh by boiling it in a large copper cauldron. Byrne's skeleton is still on display, along with many of Hunter's other specimens, at the Royal College of Surgeons in London.

Hunter's interests made him the likely model for the animal-loving Dr. Dolittle (though Dr. Frankenstein might seem like a better fit). The house he occupied in later years on Leicester Square was probably also the model for the home of Robert Louis Stevenson's Dr. Jekyll and Mr. Hyde. By day, wealthy patients seeking treatment visited the elegant front building, where Hunter lived and had his surgery. By night, the dingy back building lowered its drawbridge to receive cadavers for the dissecting room.

But even if they seemed ghoulish, Hunter's dissections were always highly purposeful, even practical. Most naturalists were only dimly beginning to think about the connection between humans and other

John Hunter's knife work on animals and humans shaped modern surgery. But it also gave him a ghoulish reputation, not least for his pursuit of the "Irish Giant," visible in vivo at left, and in skeletal form at right, above Hunter's desk.

animals. But Hunter was already up to his elbows in detailed flesh-and-blood comparisons. His animal dissections always emphasized where each species fit in the larger biological picture. His method was to focus on some anatomical feature and present dissections from a variety of species side by side, in an orderly series of glass jars. His nervous system series, for instance, included marine worm, medicinal leech, earthworm, sea mouse, centipede, scorpion, lobster, cuttlefish, sheep, ox, ass, porpoise, minke whale, and finally a human brain.

This brand of comparative anatomy yielded practical insights into physiology, to the benefit of his human patients. Among many other contributions to modern science, Hunter demonstrated for the first time how bones grow, how the testicles descend in the fetus, and what course the olfactory nerves travel. The poet William Blake, who lived around the corner, fictionalized him unkindly as the bloodthirsty "Jack Tearguts," but Hunter's work often focused on ways to avoid needless surgery. His influence was such that, even now, physicians still publish articles with titles like "Coronary Artery Disease, Inflammation and the Ghost of John Hunter," or "Popliteal Aneurysms: From John Hunter to the 21st century."

"An Electric Spark"

The idea that natural history might be an obscure or even irrelevant discipline would have struck almost everyone in late eighteenth-century London as woefully misguided. The commonplace modern attitude that if you have seen one tree, or sea slug, or wombat, you have seen them all would have seemed absurd. On the contrary, naturalists then behaved as if they had been blind for generations but could now see the world in all its glory for the first time. Every species was astonishing in its way, and the naturalists who discovered, identified, labeled, and studied them were doing what Linnaeus called "immortal work."

This was true even with humble species like the electric eel. By 1774, the London community of gentlemen naturalists already had an extensive network, bordering on espionage, in some of the more remote corners of the Earth. So they knew well in advance that the electric eels were coming, and what that might mean for them. All five electric eels had been alive early that summer, when an enterprising British mariner carried them to Charlestown, South Carolina, and put them on display. There, a physician and disciple of Linnaeus named Alexander Garden had spied the eels. He promptly wrote to John Ellis, a London linen merchant and naturalist, who maintained an extensive correspondence with collectors in the American colonies. "The person to whom these animals belong, calls them *Electrical Fish*," Garden reported, in the course of a nine-page anatomical description, "and indeed the power they have of giving an electrical shock . . . is their most singular and astonishing property."

Garden naturally tried to buy them. And when the asking price proved too rich, he sized up the likelihood of all five surviving the transatlantic voyage. (As an air-breather, *E. electricus* can endure for months in a barrel of oxygen-depleted water. But not happily.) Then he advised the mariner "to get a small cask of rum with a large bung, into which he may put any of them that may die, and so preserve them for the inspection and examination of the curious when he arrives."

The eels were thus perfectly prepared for London, and London for them. John Hunter and John Walsh had already done careful work a few years earlier on European torpedo fish, which can deliver a comparatively mild electric shock of only about 50 volts. Encouraged by that authority on electricity Benjamin Franklin, Hunter and Walsh had determined that the electrical organs make up half the fish (and also that they tasted like "insipid mucilage"). Hunter had characteristically taken note of the extensive network of nerves running through the electrical organs, and in an offhand phrase about the potential significance of bioelectricity, anticipated much of the future development of neurophysiology: "How far this may be connected with the power of the nerves in general," he wrote, "or how far it may lead to an explanation of their operations . . . future discoveries alone can fully determine."

Not everyone was ready to accept the idea of an animal producing electricity. One skeptic dismissed it as wild nonsense—until he visited Walsh's home and took his place in the ring of witnesses formed to lay hands on a living eel. Others continued to doubt the electrical nature of the shock because they could not actually see or hear it. So Walsh pasted a thin strip of tinfoil on a strip of glass and cut a narrow gap in the foil with the blade of a sharp knife. The idea was to get the electricity from the eels to visibly leap this gap. Soon after, a leading scientific journal reported, "It is with great pleasure that I inform you that they have given me *an electric spark*."

That spark was the beginning of the science of bioelectricity. A few years later, probably inspired by this work, Luigi Galvani undertook his celebrated research demonstrating the electrical nature of neural activity in ordinary animals, using an electrical stimulus to induce a muscle movement in a frog's leg. Soon after, building on animal models, the physicist Alessandro Volta invented the electric battery. In the new century, other researchers would go on to develop the science of neurophysiology and demonstrate the electrical basis of neural activity in the human brain itself.

Discovering new species wasn't about collecting "the refuse of nature" but its wonders—and the gifted community of naturalists gathered in London in the late eighteenth century knew it in their bones.

For them, each new species held the dazzling potential to reveal the secrets of life itself.

Chapter Two

FINDING THE THREAD

Instead of the somnambulism of the previous
ages, naturalists, like men newly risen, went
forth in their morning strength and ardour to
the labour of the day.

—W. H. HARVEY

SOMETIME IN 1798, an amateur naturalist visiting a lake near Australia's
Hawkesbury River watched an Aborigine spear "a small amphibious
animal of the mole kind." What came out of the water, fighting hard
enough to plant a claw in its tormenter, was a taxonomic freak—a venomous mammal that lays eggs and has a bill like a duck.

The naturalist shipped the specimen, packed together with a "wom-
back," in a keg of spirits, to the Literary and Philosophical Society in
Newcastle-on-Tyne, England. There, the story goes, an unfortunate
female servant was carrying the keg on her head up from the docks
when it burst open, dousing her with an early sort of Australian plonk
(oaky flavor, light-to-medium body, nose of dead mammal). The servant was no doubt appalled by the "strange creature, half bird, half
beast, lying at her feet." But the scientific world welcomed this newcomer with delight.

A second specimen soon reached George Shaw, assistant keeper of
the natural history department at the British Museum, who published
a description in June 1799 in *Naturalist's Miscellany,* his own popular

"Half bird, half beast," the platypus baffled early naturalists.

illustrated periodical. (It was a source of income to make up for his dismal salary.) The specimen was 13 inches long, with a "flattish and rather small" head, a "flat, furry" tail, and webbed feet. Shaw called it *Platypus anatinus*, meaning flat-footed and ducklike. Thus the platypus made its debut on the world stage and was officially "discovered."

Or not quite. At about the same time, Joseph Banks passed along a platypus specimen to the German anatomist Johann Blumenbach, who also published a description. Unaware of Shaw's account, he gave the new species a name that reflected its puzzling nature, *Ornithorhynchus paradoxus*, the bird-beaked paradox. The taxonomists who later sorted out these synonymous names found that another naturalist had already preempted the name *Platypus* for a genus of beetles. So Blumenbach's *Ornithorhynchus* became standard, combined with Shaw's *anatinus*. In the annals of science, the full scientific name is now *Ornithorhynchus anatinus* Shaw, meaning that Shaw was the original describer.

But credit for the discovery of a species is often somewhat muddy (like the platypus itself, which uses electrical and mechanical sensors in its peculiar bill to find and feed on invertebrates living on the bottoms of lakes and streams). Why not, for instance, credit the amateur naturalist who actually went to Australia and persuaded the Aborigine hunter to hand over his catch for the benefit of the Literary and Philosophical Society of Newcastle-on-Tyne? He lost out on scientific glory because that specimen only got described in print a year after Shaw's account.

Credit might also have gone to the adventurous soul who supplied Shaw with his specimen, but Shaw did not think to name him.

For that matter, the Aborigines themselves were clearly familiar with this spectacularly odd mammal. So did it really need to be discovered? Did the elephant? The kangaroo? The capybara? Humans had lived with all these species for thousands of years. In the case of the Komodo dragon, for instance, a local rajah was not only aware of its existence, but also took measures to protect it from extinction more than 80 years before a Western expedition early in the twentieth century discovered it and gave it the name *Varanus komodoensis*. So why does it constitute discovery for some scientist, usually Western, to turn up and say, "Eureka!"? What does it actually *mean* to discover a species?

Discovery isn't just a matter of being the first person to lay eyes on some odd duck of an animal. You must also recognize that there's something different about the thing you are eyeing—and explain in print just how and why it's different, so people elsewhere in the world can understand. That means you need a scheme of classification, so you can say how your species fits in with similar species. Because differences among species are sometimes subtle, classification often requires the help of an expert, a taxonomist with an encyclopedic knowledge of a particular animal group—all the *Pheidole* ants, for instance, or all the marine gastropods called *Conus*. Taxonomists have the background to say if a particular group of creatures is distinctive enough to merit classification as a separate species, or if it's just a variety or race of an existing species. So discovery is often a social and collaborative enterprise.

It can also be highly contentious. Taxonomists sometimes disagree, revising one another's work based on closer analysis, a more complete series of specimens, or a competing theory about classification. That can mean shifting a species to a new genus, a new family, or even at times a new order, with the aim of getting closer to the truth about its place in the scheme of life on Earth. In the case of the platypus, for instance, researchers would argue for much of the following century about how it reproduces (sexual and excretory functions occur in both sexes through a single opening, like the cloaca in birds), whether a fur-bearing animal can be an egg-layer, whether it nurses its young (the female lacks nipples but exudes milk through patches on her belly)—and in short whether the platypus ought to be classified as bird, reptile, mammal, or some other category yet unknown.

The discussion has continued into modern times. Scientists recently analyzed the entire platypus genome, resulting in a spate of misleading stories suggesting that the species is part bird, part reptile, and part mammal. In fact, the platypus is merely a species that split off early from the rest of the mammalian line and preserved some traits, like egg-laying, inherited from the reptilelike ancestors of all mammals and birds. It also evolved other traits (like a venomous claw) that are entirely its own. For the record, the platypus is all-mammal, and its official taxonomic classification now looks like this:

Kingdom	Animalia
Phylum	Chordata
Class	Mammalia
Order	Monotremata
Family	Ornithorhynchidae
Genus	*Ornithorhynchus*
Species	*anatinus*

The scheme for describing species in this fashion got started with Linnaeus. Or maybe that's too easy. It started both with Linnaeus and also with his French rival, Georges-Louis Leclerc, Comte de Buffon, whose encyclopedic *Histoire Naturelle* became one of the bestsellers of the eighteenth century. Both men were born in 1707 and came to power in the 1730s. Both struggled with the same fundamental questions, which still trouble scientists today: What exactly is a species? Where does one species end and another begin? What happens when two species hybridize? How do species and habitats affect each other?

Linnaeus regarded himself as anointed by God to solve the species problem and bring order to the chaos of Creation. Buffon, who was in many ways the deeper thinker, questioned the very idea of Creation and provided the first scientific evidence that the Bible was wrong about the age of the Earth. Linnaeus focused his relentless energy on naming species and organizing them into logical groups. Buffon ridiculed the whole idea of imposing order on nature, preferring instead to focus on relationships and behavior.

Each naturally despised the other. But with the questions they asked, Linnaeus and Buffon together launched the great quest to understand life on Earth in all its diversity. In place of the animal mythology that naturalists before them had complacently repeated since Roman times, they demanded actual specimens and eyewitness accounts. Linnaeus gets much of the credit for the effort to classify and make sense of the natural world. But Buffon, whose work is now generally forgotten and who has no societies named for him, deserves equal time.

Nature's Labyrinth

We tend now to take the Linnaean system for granted. Scientific names like *E. coli* and *C. elegans* have become part of our common language. If we think about nomenclature and classification at all, it's mainly to complain about what a mouthful those Greek and Latinate names can be. (For instance, the notorious intestinal bacteria are actually *Escherichia coli*, and the roundworm used as a model species in scientific research is *Caenorhabditis elegans*.) Of Linnaeus himself, even biologists specializing in natural history generally know little or nothing.

But for people struggling to make sense of the world before Linnaeus, the system he invented was a cause for jubilation. Natural history was a mess, and getting messier by the day. The world as Europeans knew it at the start of the eighteenth century did not include Antarctica, or much more than a glimpse of the coast of Australia. But every ship coming home from Africa, Asia, and the Americas seemed to carry some bizarre new creature: a possum on the crowded London quays, an iguana in Antwerp, a chambered nautilus shell fetched up in Paris from the depths of the Pacific, an "animal I believe of the monkey tribe" brought to Philadelphia by a ship captain, who added, "It is called *Mongoose*." The abundance of life-forms was a source of both delight and consternation: How did these creatures live? Where did they fit in the scheme of Creation? How did they affect ideas about our own species?

Anything seemed possible, and one odd effect was the revival of old myths: When they first encountered leaf-mimicking butterflies, people imagined that leaves could somehow transform themselves into insects and take wing. Seeing narwhal tusks, they thought of unicorns. They were thrilled of course with the splendid coloration in bird of paradise specimens shipped from the Moluccas and New Guinea, but even more that these birds seemed to have no feet. People theorized that they must remain perpetually in flight, hence the name bird of paradise. In fact, local collectors had simply been cutting off the legs for easier packing, or to keep as decorations.

Before Linnaeus, naturalists had no language or methodology for discussing the tide of new species. They couldn't agree on how to name the plants and animals in their own backyards. So how could they possibly make sense of species at the opposite ends of the Earth? Finding the answer would take an act of heroic audacity, and Linnaeus saw himself as just such a hero.

He saw himself, in fact, as Theseus, girding to enter the tangled labyrinth of nature. In the Greek legend of King Minos, Theseus volunteered to conceal himself in a sacrificial delegation of young people doomed to enter the labyrinth and be devoured by the half man, half beast called the Minotaur. But Ariadne, daughter of King Minos, became smitten with the Athenian hero and gave him a ball of thread to unwind on the way in. Thus after slaying the Minotaur, Theseus was able to find his way back out again to freedom. For the world of living things, said Linnaeus, his system would be the guiding thread. It would enable the philosopher "to travel alone, and in safety, through the devious meanderings of Nature's labyrinth."

Linnaeus was hardly cut from heroic cloth. He was a provincial, descended from five generations of Lutheran parsons in the Swedish countryside. He'd become immersed in the natural world as a child in his father's flower garden and had gone on to study botany at Uppsala University in Sweden and to qualify as a physician in the Netherlands. He was a careful observer of plants and animals and compulsively organized in his observations. Later in his career, for instance, he proposed a natural clock based on the times of day when different flowers normally bloomed.

But he was also ambitious and spectacularly egotistical ("Nobody has been a greater botanist or zoologist," he later wrote). By the age of 25, he had already completed an expedition to Lapland, sponsored by the Uppsala Science Society. He probably spent no more than a few weeks among the Sami people there, according to historian Lisbet Koerner's 1999 biography *Linnaeus: Nature and Nation*. He also doubled the distance he actually traveled, possibly because he was being paid per mile. But for Linnaeus it was an adventure into uncharted regions, where he endured hunger, thirst, and the risk of death. Presenting himself as a bold explorer proved critical to his success.

In Amsterdam, Paris, and London, he dressed in the colorful reindeer-hide costume of the Sami, complete with tribal drum. Together with his buoyant ego and his deep botanical knowledge, this gained him entrée with the leading naturalists of the day. His contagious enthusiasm and a certain sly humor about the eyes made him good company. He also quickly impressed his new friends with his ideas about the classification of species, which he published in 1735, when he was just 28, in the first edition of his *Systema Naturae*.

Like most naturalists then, Linnaeus believed that God had created all species, with fixed, unchangeable forms, and that "we can count as many species now as were created at the beginning" in the Garden of Paradise. Linnaeus was not merely a creationist, but the "New Adam." He believed that God had chosen him to organize creation and give proper names to its parts.

The system Linnaeus devised for this purpose incorporated three important innovations, none of them completely original. First, he classified flowering plants according to the number of their stamens and

A reindeer-hide tribal costume from his expedition to Lapland helped Linnaeus gain entrée into the scientific world.

pistils, their male and female parts. This sexual system was, he knew, artificial (other naturalists eventually replaced it with reliance on a broader array of traits). But it instantly opened up the botanical world to anyone who could look into a flower and count. Second, he devised precise rules for identifying species, which even beginners could follow. And third, he gradually introduced his binomial system, beginning in 1748. After that, a species that used to suffer under the name *Arum summis labris degustantes mutos reddens* became instead simply *Arum maculata*. So what seems to us like a mouthful is actually a radical simplification.

Linnaeus shrewdly served up this new system of classification with a lyrical dollop of sexual innuendo: He described flower petals as "the bridal bed," perfumed and hung with "precious bed-curtains," awaiting "the time for the bridegroom to embrace his beloved bride." Then he went on to talk blithely about two brides in bed with one husband (two pistils and one stamen), and even "twenty males or more in the same bed together with the female."

Sex undoubtedly attracted newcomers to the charms of botany. But the emphasis on reproduction also had an intellectual basis in the idea that life comes from life, not spontaneous generation. It implied that species had reproduced themselves in "an uninterrupted and unaltered sequence . . . extending from the creation to the present time," according to science historian William Coleman. The reproductive organs also seemed to be "conspicuous, intricate, and more or less constant, and hence formed admirable concentrations of excellent taxonomic characters."

Simplicity gave people confidence about their identifications, and the Linnaean system spread to other countries with astonishing speed, especially given the slow and highly unpredictable means of communication then. By 1737, botanists were already eagerly applying it to discoveries on the American frontier. Testimonials of relief, delight, and gratitude were soon arriving from around the world. The French philosopher Jean-Jacques Rousseau, an early convert, celebrated the Linnaean system as a source of intense pleasure, "une véritable jouissance," because the layman was no longer confined to making isolated observations. The thread Linnaeus had provided made it possible to place almost any specimen in the larger scheme of human knowledge. By 1740, when he was just 33, Linnaeus was already boasting that he was regarded abroad on a par with Newton and Galileo. By the twelfth edition of the *Systema Naturae*, one British shell collector even argued that specimens not included there should be smashed with a hammer because "things not in Linnaeus ought not to exist."

Refuge in Order

Then, as now, Uppsala was a college town, with lots of pink-, and cream-, and ochre-colored buildings arranged around a pretty little river, the Fyrisån. The garden where Linnaeus practiced his craft as a botanist and as a professor at Uppsala University occupies much of a city block in the middle of town, with his house on one corner. The flower beds are lined up in neat rows, as they were in his lifetime, perennials on the left, annuals on the right, all enclosed in hedges of multiple species planted to see which ones might serve best in the Swedish countryside.

From here, Linnaeus used to lead regular collecting excursions into the local countryside, up to 300 students at a time joining in. With his characteristic passion for order, he organized them into platoons. They wore a uniform of loose-fitting clothes for collecting, armed themselves with butterfly nets, and carried their trophies home pinned to their hats. Kettle drums and hunting horns announced their jubilant return at the end of the day, along with cries of "Long live Linnaeus," at least until a university official put a stop to it, possibly at the instigation of envious faculty colleagues.

Linnaeus clearly had the knack for communicating a highly contagious sense of joy, oddly paired with its apparent opposite, an extreme need for orderliness. Unfortunately, his lyrical sensibility gets lost in translation. But according to his Swedish biographer Sten Lindroth, the words with which Linnaeus greeted the short, exuberant Nordic summer "always tremble with a sense of happiness that is his own. . . . To the last he was a priest singing the praises of nature's inexhaustible variety and beauty." Linnaeus himself wrote that he followed behind God "and I became giddy! I tracked His footsteps over nature's fields and found in each one, even in those I could scarcely make out, an endless wisdom and power, an unsearchable perfection."

But it's also clear that Linnaeus felt a powerful sense of alarm at nature's chaotic abundance, perhaps accounting for what Lindroth calls "a compulsion of almost demonic intensity to *arrange* everything." In private, the joy sometimes dropped away to reveal dark, troubled feelings of doubt. One undated journal entry, for instance, sinks into the sexual loathing that apparently lay beneath his delight in the reproductive parts of flowers. It betrays not just misogyny, but also deep misgivings about our God-given nature: "Oh what kind of marvelous animals we are, for whom everything else in the world is created. We are created out of a foaming drop of lust in a disgusting place. We are

born in a canal between shit and piss. We are thrown head first in the
world through the most contemptible triumphal gates. We are thrown
naked and shaking on the earth, more miserable than any other animal.
We grow up in foolishness like apes and guenon monkeys. Our daily
task is to prepare from our food disgusting shit and stinking piss. In the
end we must become the most stinking corpses." Small wonder that
such a mind should compulsively seek refuge in order.

From the start Linnaeus attracted critics along with the army of
his enthusiastic admirers. The German botanist Johann George Siegesbeck protested that the Linnaean emphasis on sex was turning innocent
flower gardens into beds of harlotry. Linnaeus, who suffered criticism
poorly, responded by giving the name *Siegesbeckia* to a small, foul-
smelling weed. Later, he wrote the name *Cuculus ingratus*, "the ungrate-
ful cuckoo," on a packet of seeds, which somehow found its way into
Siegesbeck's own hands. Curious, Siegesbeck planted the seeds, only
to discover that the cuckoo he had been nurturing was his namesake
Siegesbeckia.

Another vocal critic of Linnaeus's, though not on sexual grounds,
was the French naturalist Buffon.

Reprehensible Statements

The Jardin des Plantes on the south bank of the Seine in Paris is today
an enclosed compound of rose gardens, tree-lined alleys, nineteenth-
century greenhouses, a zoo, and museums about the natural world.
Buffon, born Georges-Louis Leclerc, a son of provincial bourgeoisie,
assumed the powerful title of administrator here in 1739 when he was
just 32. Over the next half-century, he more than doubled the size of
the Jardin du Roi, as it was then known, to its present 64 acres. He also
laid the foundations for what was to become the Muséum National
d'Histoire Naturelle, among the finest natural history museums in the
world. Leclerc was a talented administrator, politically adroit, a confi-
dante of everyone from Benjamin Franklin to King Louis XV. But the
key to his reputation was his writing, which made him famous inter-

The French naturalist Buffon doubted that God needed to busy himself "with the way a beetle's wing should fold."

nationally as Buffon—later *Comte*, or Count, of Buffon—a name taken from a small Burgundy village where he owned forests near his country home in Montbard.

From 1740 on, Buffon spent half the year in Montbard ("Paris is hell," he wrote). He built a mansion near the river Brenne and created a private park on the ridge in back by tearing down a castle that had formerly belonged to the dukes of Burgundy. On the far side of the ridge, he built a one-room study, which still looks out today to the hills and valleys of the Burgundy countryside.

Here Buffon set out to produce a catalogue of the king's collection of natural artifacts. His scientific work until then had focused largely on mathematics. But he took up this task with such enthusiasm that he eventually wrote 36 volumes of his encyclopedic *Histoire Naturelle*, published from 1749 on. Instead of a mere catalogue, it became an attempt to synthesize everything then known about the animal and mineral worlds, including the history of the planets. The *Histoire Naturelle* became an immediate bestseller—and remained one of the pillars of French literature until Buffon's lofty manner of writing eventually fell out of favor in the mid-twentieth century.

What made Buffon different was not just his style, but also his scrupulous avoidance of religious or supernatural explanations. Linnaeus and most other contemporaries still rooted their definition of species in the plants and animals created by God to populate Eden. Buffon, by contrast, thought it was absurd to imagine God being "very busy with the way a beetle's wing should fold." He defined a species scientifically, as a group of animals breeding together over time.

Such departures from orthodoxy angered religious authorities, who presented Buffon with a list of 14 "reprehensible statements." Buffon dutifully signed a declaration of his faith in Scripture and published it in subsequent editions of the *Histoire Naturelle*. "It is better to be humble than hung," he commented. But his "reprehensible statements" also continued to appear, unaltered.

Buffon was keenly interested in details of habitat and behavior, often anticipating sciences like ecology and ethology that were still 200 years in the future. Though he had no notion of evolution, he wrote about how species could be transformed by their habitat. He mistakenly believed, for instance, that in the Americas a cold, wet climate caused animals, including humans, to become smaller and degenerate. This "theory of American degeneracy" was intensely annoying to

Americans, who waged a long campaign to refute "Buffonian error." But Buffon was at least right in arguing that species adapt to their environment. And if his examples sometimes missed the mark, his knack for connecting particular observations about animals into general theories about the natural world earned him a reputation as "the Pliny and the Aristotle of France." Given the egos involved, it was inevitable that he would clash with the "Newton and Galileo" of Sweden.

Buffon made his first foray against Linnaeus in the mid-1740s, in a speech to the French Academy of Sciences. He attacked the Linnaean system as an attempt to impose an artificial order on the disorderly natural world, and in his *Histoire Naturelle* he gleefully pointed out absurdities in the groups Linnaeus had proposed. Did tulips really belong with barberries? And elm trees with carrots? Linnaeus had mistakenly grouped these species together because he had no sense of the modern idea that a particular trait, like the number of pistils and stamens, can evolve independently even in the most distantly related species. And if the simplistic reliance on stamens and pistils could produce such odd groupings, it was even worse in zoology, where Linnaeus tried to find other relatively simple traits for organizing species. Based on dental structure, for instance, humans and monkeys both turned up in the order *Anthropomorpha*. But so did two-toed sloths. "One must really be obsessed with classifying to put such different beings together," Buffon wrote.

Linnaeus, no doubt deeply stung, dismissed his antagonist as a "hater of all methods," who delivered "few observations" and much "beautiful ornate French." He quoted the Bible ("And I . . . have cut off all thine enemies out of thy sight"—2 Samuel 7:9) to prophesy that the "Frenchman named Buffon" who "always wrote against Linnaeus" would suffer the wrath of God.

"Imperceptible Nuances"

Buffon's objections to the Linnaean system arose partly from sincere belief. As Linnaeus originally devised it, that system depended on the idea of species as distinctly separate entities, fixed in their God-given form, similar perhaps to other species but not related to them. Buffon knew that the natural world was much fuzzier than this suggested. "Nature moves through unknown gradations and consequently she cannot be a party to these divisions," he wrote, "because she passes from one species to another species, and often from one genus to another genus, by imperceptible nuances."

He was pointing out a problem that bedevils scientists to this day. The Linnaean system, even in its modern form, is far from perfect. New evidence routinely obliges taxonomists to move species to a different genus, or even to an entirely different order. At times, these revised groupings can seem as absurd as the ones Buffon lampooned. For instance, botanists very reasonably used to think the water lotus was related to the water lily, and both were in the same order, Nymphaeales. In fact, molecular analysis recently showed that the water lotus is a cousin of the sycamore tree. Elephant shrews, small African mammals with quivering snouts, used to be classed as Insectivora, like shrews. But genetic evidence now shows that they are more closely related to elephants and belong to their own separate order, the Macroscelidea. Buffon was also correct in arguing that the Linnaean system is often arbitrary. Taxonomic "splitters" tend to recognize new species based on relatively small differences, and "lumpers" group them together based on traits they have in common. Then they fight.

But for all the flaws in the system he invented, the name Linnaeus has endured. Partly this is because binomial identification, by species and genus name, has proved convenient. And partly it's because Linnaeus was extraordinarily lucky. Though he was thinking about God and Creation, he developed a rudimentary hierarchy of classification that would prove congenial, a century later, to the new evolutionary thinking of Darwin. His timing was also perfect. He provided a coherent system of classification just as the age of discovery revealed the overwhelming richness of plant and animal life. Thus Linnaeus prevented a nomenclatural disaster. And even though science would ultimately abandon many of the details of what he proposed, particularly his sexual system, the modern-day permutations of his system continue to be called Linnaean.

Buffon meanwhile proposed no alternative way of coming to grips with the abundance of new species. He also made the mistake, as absurd as anything in Linnaean, of putting man at the center of the animal world, and his *Histoire Naturelle* paid inordinate attention to species that were useful and familiar to us. So his attack on Linnaeus ends up feeling like little more than an attempt to be provocative, or to make his own name by taking on a bigger name in the neighborhood. It was, writes one Buffon scholar, "manifestly unfair and, it appears, ignobly motivated." Perhaps Linnaeus was a mere collector and classifier, as Buffon argued. And no doubt he lacked Buffon's insight into relationships and behaviors. But Buffon somehow missed a point all modern scientists understand: Classification is the necessary first step. You need to know what species you are looking at before you can begin to talk about how they behave.

His attack on Linnaeus mainly hurt Buffon himself. The *Histoire Naturelle* was quickly translated into all the other major European languages, according to science historian Phillip R. Sloan. But it was 25 years before the first translation appeared in England, where the cult of Linnaeus was especially devout. (Even in the eighteenth century he was being celebrated there as "the immortal Linnaeus.") And until 1973, all English editions of Buffon simply omitted the introductory section that includes the attack on Linnaeus.

But does Buffon deserve to be forgotten? His relative obscurity, like the immortality of Linnaeus, turns out to be largely a matter of luck.

From Montbard, it's a short walk out along a canal to a collection of handsome stone buildings with red tile roofs, just outside Buffon's namesake village. It's an old forge where, late in life, Buffon conducted a series of remarkable experiments. He had his workers draw molten balls of iron of different sizes and composition from the smelter and carefully time how long it took them to cool down. His theory was that the Earth originated as a fireball, gradually solidifying as it cooled. By scaling up from iron balls to the size of the entire planet, he hoped to estimate the age of the Earth. The numbers he came up with ranged from 10 million years down to as little as 75,000 years, the estimate he published when his *Époques de la Nature* finally appeared in 1778.

We now know that the Earth is actually billions of years old. But Buffon's work was the beginning of the end for the Biblical belief that all Creation dated back just 6000 years. *Époques de la Nature* was "the most important scientific document ever written in promoting the transition to a fully historical view of nature," according to the paleontologist Stephen Jay Gould. It opened the eyes of educated readers to the vast span of geologic time. It also opened doors for other scientists who would go on to make Buffon's work seem quaint.

The forge is now a museum. The waterwheels, bellows, and other machinery there have been restored to squeal and creak for the amusement of paying visitors, who apparently come out of a strange passion for metallurgy. Amazingly, the exhibits make no reference whatever to the experiments Buffon conducted there. It's as if a museum about Galileo neglected to mention that he had this idea about the Earth orbiting the sun. And this seems to be Buffon's fate in history—his ideas essential in their day for the advancement of science, but consigned thereafter to oblivion.

Thierry Hoquet, a Buffon scholar, credits him with four important ideas in the history of science: the understanding of geologic time; the definition of species on biological terms; the assertion that habitat shapes species; and the conviction that species, far from being immutable, can transform over time. These ideas all stand up to modern scrutiny. But they are also relatively complicated and buried in a prodigious stream of other ideas Buffon wrote about in his lifetime. His reputation also suffered for political reasons.

Buffon died in 1788, a year before the French Revolution, which predictably had little regard for such a close ally of the king. Buffon's son eventually went to the guillotine. (One story says he was sent there by former neighbors Buffon had evicted in the course of expanding the botanical garden.) The revolutionaries at least understood the value of Buffon's work enough to found the Muséum National d'Histoire Naturelle on the collection he had largely created. But one of the early anatomists there, Georges Cuvier, set out to turn natural history into a scientific discipline. And the path to professionalism meant pushing Buffon and the sort of amateur naturalists he had inspired into the dustbin.

Even the Church seems to have taken special satisfaction in diminishing Buffon's legacy. Buffon was buried, as he intended, beside the altar in the Eglise St. Urse, up the hill behind his mansion in Montbard. But for the inscription over the altar, some clever priest pointedly chose the ultimate statement of the Scriptural view of Creation: "And God saw every thing that he had made, and, behold, it was very good."

In contrast to Buffon's complex thinking, the Linnaean system was starkly simple. Linnaeus also relentlessly refined it and campaigned for it. In 10 years, he advised one skeptic, "you will be defending the very ideas that make you vomit now." And he was right. The result is that all names in botany today go back no further than his *Species Plantarum*, published in 1753, and all names in zoology begin with the tenth edition of *Systema Naturae*, published in 1758. Linnaeus is everywhere in modern science.

At the Jardin des Plantes in Paris, a big, handsome plane tree now bears a plaque noting that it was planted in 1785 by Buffon himself. But the plaque also describes the species as *Platanus orientalis* L. That "L" stands of course for Linnaeus, who gave this species the scientific name by which it is now known everywhere in the world.

It is, in truth, not such a bad combination. In a generous moment, even Cuvier later conceded that Linnaeus and Buffon together possessed the essential combination of traits for rapidly advancing the scientific study of nature: "Linnaeus grasped with finesse the distinctive traits of the organisms; Buffon embraced in a glance their most remote relationships." Without both, natural science as we know it would not exist.

Not far from the plane tree, a bronze statue of Buffon presides in casual splendor over the gardens and the natural history museums he helped make great. One recent summer morning, a worker—an unwitting agent of the cult of Linnaeus—set up a sprinkler directly in front of the statue. In the course of its circular ablutions it seemed to be spitting indifferently onto Buffon's ruffled blouse. But then the pressure went off.

And, for a little while, the image of Buffon glistened again under the Paris sun.

Chapter Three

COLLECTING AND CONQUEST

I shall give up the wife for the voyage of discovery.

—LT. MATTHEW FLINDERS

ONE MORNING IN 1773, taking breakfast at the home of a friendly couple in Suriname, the soldier and naturalist John Gabriel Stedman was struck by the appearance of a "beautiful mulatto maid" named Joanna, the daughter of a colonist and a slave, 15 years old and "possessed of the most elegant shape that nature can exhibit."

Stedman retreated to his room "reflecting on the State of Slavery altogether," his ears "being stund with the clang of the whip and the dismal yels of the wretched Negroes on whom it was inflicted *slingslang* from morning till night." He had come to Suriname as an officer in a Dutch military contingent commissioned to defend the plantation owners, he wrote in a memoir published years later. But now he was cursing them for their barbarous ways. Stedman thought that if Joanna's master had only bestowed on her "the education of a Lady," she might be "an ornament to Civilized Society." Instead, he feared, whipping, or worse, "might one day be the fate of the unfortunate Mulatto maid" at the hands of some "tirannical Master or Mistress." To assist readers in sharing these honorable feelings (or as a form of "humanitarian pornography," in the view of one modern critic), his book included

Military naturalists like John Stedman often combined specimen-collecting with the bloody work of soldiering.

an illustration showing Joanna sway-hipped and with one breast bared. Stedman soon set out to "purchase + educate" her.

Stedman's colorful memoir was a bestseller of 1796, under the ponderous title *Narrative of a Five Years' Expedition against the Revolted Negroes of Surinam, in Guiana, on the Wild Coast of South America, from the year 1772 to 1777.* The book was partly a picaresque adventure tale, told on the ribald model of Henry Fielding's *Tom Jones*. It was also an indictment of slavery, though the author was hardly an abolitionist. And, oddly, it was a celebration of South American wildlife.

The mix of elements could be jarring. He recounted, for instance, how a planter's jealous wife had slit a slave girl's throat, stabbed her repeatedly in the breast, and tossed her into a river with hands bound behind. But he also offered his readers loving descriptions of spider monkeys, flying squirrels, cockatoos, and coatimundis. One illustration, by Stedman's friend, the poet and artist William Blake, depicted a slave, still living, hung from the gallows by a hook jammed under his ribs; the next showed "The Toucan and the Fly-catcher." After "Flagellation of a Female Samboe Slave," the reader could contemplate "The Spur winged Water hen" and "the Red Curlew."

Taking delight in the natural world was a way of coping that suited Stedman's "incurable romanticism," according to the historians Richard and Sally Price. His descriptions of the natural world were vivid enough that they may have served as a source for Blake's famous verse "Tyger! Tyger! Burning bright / In the forests of the night." (Stedman wrote of a "Tiger-cat," or jaguar, "its Eyes Emitting flashes of lightning.") It also suited the spirit of the day for an educated officer to adopt the role of scientific observer. Stedman could cite both Linnaeus and Buffon, and he collected specimens and artifacts that later turned up in the Leverian Museum in London and the Rijksmuseum in Leiden. Science gave him a way, say the Prices, "to distance himself from much of what he witnessed."

And perhaps also what he did: The frontispiece illustration of Stedman's book showed him leaning on the barrel of his rifle, with bayonet fixed, beside the bloody corpse of a runaway slave. According to the Prices, Stedman had originally intended to caption the illustration with the characteristically ambiguous phrase, "My hands are guilty, but my heart is free." What actually appeared there was a verse of almost sociopathic self-absorption: "'Twas YOURS to fall, but MINE to feel the wound."

So would Stedman apply that same attitude to his relationship with Joanna? Or could such a thing as real love flourish between soldier and slave?

"Useful Knowledge"

It isn't quite true that everybody traveling abroad in the eighteenth and nineteenth centuries studied natural history, but it can often seem that way. And to a surprising extent—or at least surprising in a time when we associate the subject with conservation—these naturalists were often caught up in the business of conquest and colonization, using natural history as a tool both to advance their own careers and to remake the world on European lines. Most seem to have accepted the belief that they were doing God's work, bringing progress and truth (both Christian and scientific) to an uncivilized world. Stedman was one of the few early explorers to write about the ways appreciation tended to slide down into exploitation. But it didn't stop him from grabbing all the pleasures he could handle along the way.

For most participants in the colonial enterprise, natural history was at the most rudimentary level a source of income. The Dutch East India Company, for instance, not only paid dismally small salaries to employees sent out into the tropics, but also fiddled currency exchange rates in its own favor and held back a portion of pay until the employee returned home, or died. So "everyone from Governor-General to cabin boy traded on the side," according to historian Charles R. Boxer. They stockpiled shells, exotic wildlife specimens (dead or alive), porcelains, lacquer work, and other goods for sale to collectors back home. Ship captains working the trade to the Americas, Africa, and Asia kept an eye out for unusual specimens, and often maintained ties with museums and natural history societies back home.

But apart from cash value, people were also genuinely excited about natural history then, and it was common for even the busiest and most powerful men and women to take time to collect a new species, or to reach across enemy lines in joint pursuit of some precious fossil. The American Revolution had already begun in 1775, for instance, when Benjamin Franklin instructed American warships not to interfere with Capt. James Cook as he was returning on HMS *Resolution* from his second voyage of discovery around the world. And the war had not quite ended in 1782, when Gen. George Washington lent a dozen men with wagons and tools to help an enemy officer excavate a mastodon in the Hudson River Valley.

Thomas Jefferson was so deeply involved in species questions that he assumed the presidency of the American Philosophical Society, devoted to the natural sciences, the day before being sworn in to the somewhat humbler post of vice-president of the United States. He remained president of the Society throughout his term as president of the nation. But it wasn't just polymaths like Jefferson who cared passionately about natural history. Four out of five members of the committee that drafted the Declaration of Independence and 15 of the 56 eventual signers were also members of the American Philosophical Society. Given our experience of modern bureaucrats and politicians, it's a little startling to read that even as late as 1838, the U.S. Secretary of the Navy, Mahlon Dickerson, was a botanist and member of the American Philosophical Society; and the Secretary of War, Joel Poinsett, was a plant collector who introduced his namesake, the poinsettia, from Mexico to the United States. A knowledgeable interest in the natural world was almost a necessity in educated circles.

It was without question "useful knowledge," a phrase Benjamin Franklin helped turn into a watchword of the day. He had dedicated the American Philosophical Society to all studies "that let Light into the Nature of Things, tend to increase the Power of Man over Matter, and multiply the Conveniencies or Pleasures of Life." "Conveniencies or pleasures" may have understated the case. Deciphering the natural world mattered urgently at a time when mysterious epidemic diseases still swept away whole families overnight (Philadelphia was prone to deadly bouts of yellow fever), and when those who dared to travel never knew where in the vast wilderness the foolish myths might suddenly fall away to reveal real monsters.

A career in natural history was often a means of entry to the corridors of power. "Scientific gentlemen" typically hitchhiked on naval expeditions of discovery and conquest, or served in the military themselves. They also frequently used their natural history work as a cover for spying. If they managed to come back alive and with good results, they often moved up into the scientific or colonial hierarchies, which frequently overlapped. Having made his name with Capt. Cook aboard the *Endeavour*, for instance, the botanist Joseph Banks went on to orchestrate future scientific expeditions in his capacity as president of the Royal Society, Britain's national academy of sciences. He supervised Capt. William Bligh's ill-fated voyage on the *Bounty* to transport breadfruit trees from the Pacific to the Caribbean as a potential food stock for slaves, and was an early advocate of developing tea plantations in India to compete with China. He also helped establish the first British colony in Australia and made himself the leading voice on decisions shaping the fate of that continent.

An international reputation as a naturalist likewise turned Linnaeus into an adviser to Queen Lovisa Ulrika, who collected shells and other natural objects at Drottningholm, the summer palace outside Stockholm. (She gossiped to her mother, "He is a most entertaining person, who has the esprit of the high society without carrying its manners, and who for these two reasons amuses me beyond words.") The Queen helped pay to send his students out into distant corners of the world, where Linnaeus wanted them not just to catalogue natural wonders but also to understand their potential usefulness to the Swedish economy. Taking the fir tree as an example, biographer Lisbet Koerner writes, Linnaeus thought a qualified botanist should know how to produce rosin, pitch, tar, charcoal, firewood, and timber; how to bake bark bread; and how to use saps and shoots to cure scurvy. "And in the like manner," he added, "with all other plants."

Linnaeus's concept of useful knowledge was to develop a completely independent Swedish economy by cultivating imported crops at home, rather than depending on international trade. In his youth, he had seen imperial Sweden lose its Baltic colonies to Russia, and in middle age, he survived the great famine of 1756, when so many Swedes wasted away themselves, and also had to listen helplessly, as he wrote to the king, to "their little children's whimpering, suffering and death agonies." So Linnaeus proposed schemes to grow nutmeg, mace, cinnamon, saffron, and tea in Lapland, and he experimented with Swedish rice, coffee, and sugar cane. "Should coconuts chance to come into my hands," he once declared, in a moment of lunatic optimism, "it would be as if fried Birds of Paradise flew into my throat when I opened my mouth."

Most of these agricultural experiments inevitably failed as naturalists came to terms with useful, but daunting, knowledge about the importance of place: You couldn't simply pluck a species out of one climate and reliably expect it to grow in another. Getting to know those other climates—as well as the species they supported—was thus critical, and nations soon began competing to get there first. They competed both out of genuine regard for the value of new scientific knowledge and because they recognized that understanding a habitat was an important step toward colonizing it.

Love and Ambition

On October 19, 1800, two French ships renamed the *Géographe* and the *Naturaliste* set out from Le Havre, France, under the command of Capt. Nicolas Baudin, with 23 scientists aboard in addition to crew. A portrait from the period shows Baudin in uniform, neck cowled in a cotton scarf, with sleepy eyes, a prominent Gallic nose, and bow lips turned up contentedly at the corners. But there was nothing lazy or self-satisfied about him. Planning a voyage to the unknown coast of Australia, he was concerned with potential benefits not just for France, but, remarkably, also for the territories being visited. In a detailed outline to the National Institute of Sciences and Arts in Paris earlier that year, he declared his intent to resolve "certain doubtful points of geography," to "chart unknown coasts," and also to visit the inhabitants and "increase their wealth by exchanging objects with them or by making them gifts of animals or plants that can adapt to their soil." (One scientist, not so *sympathique*, wondered if Baudin could bring back some living human specimens, to enjoy the rest of their days in France—and the afterlife in a museum display case.)

Baudin immediately added the more customary motive that the species he intended to introduce—presumably goats, chickens, pigs, and other livestock—would "subsequently offer resources to navigators." This European practice of provisioning remote islands with livestock was already more than 200 years old. Naturalists then were only beginning to grapple with the heresy that any species in God's Creation could go extinct. So as they contemplated the benefits of these introduced species, they hardly noticed the toll on innocent bystanders. It didn't dawn on them till much later that gifts of the sort Baudin was contemplating would cause the extinction of countless undiscovered species, which had evolved on remote islands in the absence of such competition. The naturalists themselves were inadvertently helping to destroy the very species they were so hungry to discover.

In fact, a grand willy-nilly experiment was already under way around the world, and becoming more intense with each passing year. When Joseph Banks got wind of the Baudin expedition, he lobbied hard to launch a competing British effort. England was at war with France and needed all its resources. But the Admiralty quickly assigned a ship, out of fear that the French might be secretly planning an Australian colony, and they allowed Banks to handpick the captain. He settled on Matthew Flinders, a 26-year-old who had served under Capt. Bligh. Flinders had demonstrated talent as a navigator and scientist and was also evidently a diplomat, applying for the job with a list of priorities calculated to appeal to a naturalist like Banks: "The interests of geography and natural history in general, and of the British nation in particular, seem to require, that this only remaining considerable part of the globe should be thoroughly explored." Flinders set sail in July 1801 in command of the *Xenophon*, a collier, or coal carrier, in doubtful repair,

For Joseph Banks and many others, a passion for nature became a route to power in Europe's colonial empires.

which had been renamed *Investigator*. By then, the French had already gotten a nine-month head start.

Flinders almost didn't sail at all. Sea voyages then provided sailors with a full complement of hardships. It was like being in jail, as Samuel Johnson put it, "with the chance of being drowned." Bad food, scurvy, seasickness, foul odors, crowded living quarters, chronic dampness, and the close company of assorted vermin, not counting crew, all vied for priority as a cause of misery. (On his trip to Suriname, John Stedman had complained that dinner was served from the same bowls recently used to slop out sick-room wastes.) But sailors often suffered loudest from a timeless complaint—the absence of females. This was a sacrifice Banks himself had taken steps to redress in 1772, when he was still planning to join Capt. Cook's *Resolution* on a second voyage of discovery. Not yet 30, wealthy, dashingly handsome, and lionized by London society, Banks had become a little arrogant. His participation in the voyage ended at the last minute, over resistance to his extravagant demands for his entourage. But a traveler had already gone ahead to await his arrival at Madeira. She tried to pass herself off as a man and, more convincingly, as a botanist. Banks's evident intent, to Cook's amusement, had been to have her join him as part of the ship's scientific staff. At home a year later, Banks apparently fathered a child by another mistress.

But by 1801 Banks was 58 years old, long married, gouty, fat, and indifferent to the dull ache of loneliness at sea. Flinders had just wed that April. In July, as he was moving down river and preparing to set sail on a voyage of three years, he smuggled his new wife Ann onto the ship, then promptly ran aground while the two of them were belowdecks. Word got back to Banks, who fired off an angry note to Flinders that the displeasure of the Lords of the Admiralty (no doubt meaning himself) might cause them to withdraw his command. Flinders dutifully replied, "If their Lordships' sentiments should continue the same, whatever may be my disappointment, I shall give up the wife for the voyage of discovery." Ann went ashore, knowing she might never see her husband again.

Despite this rough start, the expedition would prove a great success for British interests. It accomplished the first circumnavigation of Australia, a name Flinders helped popularize, replacing *Terra Australis* and New Holland. The expedition also brought back 150 genera and 1500 species of new plants. The emphasis on plants was evidence not just of Banks's botanical influence, but of the interweaving of scientific and colonial causes: The British wanted to identify circumstances suitable for agriculture, so new colonies would not need to depend on supplies shipped out from England.

For the French, on the other hand, blending scientific and colonial interests proved less successful. Baudin sent home a spectacular assortment of new species, including 144 new birds; he also surveyed parts of the coast of Australia and eventually made peaceful contact with Flinders in what is now Encounter Bay. But his company bickered childishly, according to *Encountering Terra Australis*, a history of the rival expeditions. In the past, such voyages had been under the command of aristocrats, and old attitudes persisted after the French Revolution. Baudin's naval officers were indignant about serving under a captain of humble social background. The scientists meanwhile squabbled about access to resources and about getting proper credit for their work. At one point, zoologist François Péron presented himself to Baudin dripping with blood from a fight with the ship's surgeon over which of them should get the "glory" of dissecting a shark. Péron lost and whined that the surgeon had stolen the shark's heart. Not content to be merely a zoologist, Péron also played at being a spy, making a crude survey of British defenses at Port Jackson (now Sydney) and urging French authorities to destroy the colony as a way of snuffing out British imperial ambitions in Australia.

Citizen Baudin didn't suffer from the common tendency of naval officers to regard scientists as a nuisance (though in this case it would have been understandable); he was a naturalist himself. But instead of capitalizing on their good fortune, one naturalist complained that Baudin "would prefer to discover a new mollusk than a new landmass." And at Encounter Bay, a disgruntled French officer told the British, "if we had not been kept so long picking up shells and catching butterflies at Van Diemen's Land [now Tasmania], you would not have discovered the south coast before us."

Poor beleaguered Baudin died on the trip home to France, of tuberculosis. On his own harrowing trip home, Flinders endured shipwreck, imprisonment, and other ordinary hazards of exploration. From the opposite end of the Earth, he had written to his patient wife Ann, "my heart is with thee, and as soon as I can insure for us a moderate portion of the comforts of life, thou wilt see whether love or ambition have the greatest power over me." But it would be nine long years before husband and wife could share the same bed. A few years later, at the age of 40, Flinders died, apparently from the cumulative physical toll of his travels.

For Nature and Nation

For the modern reader, it is endlessly astonishing how much these early explorers were willing to sacrifice in the conjoined interests of nature and nation. Like Flinders, they were also motivated by personal ambition and appetite for adventure. But adventure was often just a nice word for prolonged hardship followed by painful death. After an expedition into the Suriname rain forest, for instance, John Stedman emerged with a list like the trials of Job: "I have already mentioned the prickly heat-ring-worm-dry-gripes-putrid fevers-boils . . . and bloody-flux, to which one is exposed in this Climate, also the musquitoes, patat and scrapat lice, chigoes, cockroaches, ants, horse-flies wild bees, and bats, besides the thorns, briars, and aligators, and peree [piranhas] in the rivers and to which if added the howling of the Tigers [jaguars], and hissing of serpents . . . the dry sandy-savannahs, unfordable Marshes, burning hot days, cold + damp nights, heavy rains, and short allowance, people may be astonished how one was able to survive it." Here Stedman paused to catch his breath and worry aloud if he was complaining too much, but then could not help adding affliction by snakes, lizards, scorpions, locusts, bush spiders, bush worms, and centipedes, "nay even flying-lice." He omitted other calamities, he said, because he didn't want to be accused of exaggerating.

Whether any person, or more precisely any European, could survive these trials was an entirely reasonable question. Understanding the importance of place wasn't just a matter of whether coconuts would grow in Lapland. In the first half of the nineteenth century, enlisted men in the British Army were seven times more likely to die of disease in India than back home, whereas civil servants doubled their risk of death. In West Africa, notorious as "the white man's grave," 300 to 700 Europeans per thousand died within a year of arrival, according to historian Philip D. Curtin. After that, the survival rate improved because of acquired immunity. But that first wave of death could change the course of history. Laying siege to Cartagena, Colombia, in 1742, for instance, the British lost three-fourths of their 12,000-man army to yellow fever. An 1802 epidemic of the same disease in Haiti decimated a French army there and caused Napoleon to question the value of his American holdings, setting the stage for the American purchase of the Louisiana Territory the following year.

Malaria, supposedly caused by "marsh miasmas," was a constant problem everywhere in the tropics. Jesuit missionaries had learned from Peruvian Indians in the mid-1600s that quinine, from the bark of the South American cinchona tree, was an effective remedy, but it didn't come into general use until the 1850s. Meanwhile standard medical treatments like bleeding just helped patients die faster. Others died, often horribly, from such ubiquitous diseases as cholera, dysentery, typhus, and typhoid fever, for which the causes and effective remedies would remain unknown for another hundred years.

All this gave vital importance to Buffon's idea that species degenerate in foreign climates. Apart from the immediate issue of survival, it would in time become the basis for the pseudoscientific idea that different human races might actually be separate species. "There is no known plant or animal equally fitted for every climate," the Scottish philosopher Lord Kames wrote. Instead, each species seemed to flourish in a particular climate and degenerate elsewhere. "Whence the following question . . . Are not men, like horses or wheat, apt to degenerate in foreign climes?" Sending European naturalists and colonizers out to every corner of the Earth was perhaps no more practical than trying to grow tea in Lapland.

And still they went.

Children of Empire

Not just life and death, but love and sex, parentage and descent became entangled in the business of collection and conquest. John Stedman's book gave an unusually frank account of the process, though not nearly so frank as in his manuscript before the publisher hired a hack to tone things down. The book omitted, for instance, Stedman's calculation, based on what he had seen on the plantations, that 100,000 people a year "are murdered to Provide us with Coffee & Sugar." And where Stedman wrote that a slave gave him "such a hearty kiss—as had made my Nose nearly as flat as her own," the hack bowdlerized it to "imprinted on my lips a most ardent kiss."

Stedman's private journal was even more frank, and it often shows him dining and wenching with the slave owners he later excoriated. In February 1773, the day after arriving in Suriname, for instance, Stedman noted in his diary "sleep at Mr. Lokens . . . I f--k one of his negro maids." And a month later, "Dine at Kennedy's, 3 girls pas the night in me room." However irresistible Stedman may have been, this wasn't just casual sex. Given the coerced relationship between slave and master, we would nowadays call it rape. And perhaps it seemed that way to Stedman, too; he omitted most of these details from his book.

The love story he confected around his relationship with Joanna was far more edifying, and Stedman himself may even have believed the romance. "Her virtue, youth, and beauty gained more and more my esteem," he wrote, "while the lowness of her birth and condition, instead of diminishing, served to increase my affection." Affection led to marriage, or at least what was known as a "Suriname marriage," and the birth of a son.

Other naturalists were themselves the children, legitimate or otherwise, of the colonial enterprise. Stamford Raffles, the British naturalist and empire builder of the Far East, was the child of a slave trader and his wife, and born aboard his father's ship, the West Indiaman *Ann*, shortly after it sailed from Jamaica in 1781. John James Audubon was born in Haiti in 1785, the child of a French plantation owner and a mistress of either French or Creole background (Audubon would become a slave owner himself for a time). Paul Du Chaillu, an explorer in West Africa in the 1850s, was the child of a French plantation owner and a mistress who was most likely a slave on the Indian Ocean island of Réunion, sometime around 1831.

Raffles was a particularly poignant example of how families became caught up, and often sacrificed, to the greater cause of collection and conquest. He was an agent of the British East India Company, a private enterprise that could nonetheless command the full force of the British Empire. At one point, he orchestrated an invasion of Java by a British expedition of 11,000 men. He then served as lieutenant governor there, and among other progressive measures, ended the slave trade and established rules of self-government. But on Java, his first wife, who had survived a previous husband in India, herself died of tropical disease.

Back in England a few years later, Raffles remarried and brought his new wife, Sophia Hull, with him to the Far East, where he became governor-general of Bencoolen in Sumatra. The two of them would have five children together. Raffles meanwhile governed Sumatra, founded the city of Singapore, and somehow also managed to make significant contributions to the study of natural history. He was a passionate student of almost any subject that happened to be before him at the moment. He was also an effective administrator, insisting on hard factual data and deploying subordinates so he could effectively deal with an array of interests simultaneously. Science was always prominent among these interests, and during his years in the Far East, Raffles would manage the discovery of several dozen new species, including the sun bear (*Ursus malayanus*), the crab-eating macaque (*Macaca fascicularis*), the great-billed heron (*Ardea sumatrana*), and the milky stork (*Mycteria cinerea*), as well as the world's largest flower, a genus of plants that parasitize palm trees, now named *Rafflesia* in his honor.

When Raffles first landed on the island of Singapore in 1819, it was, as a colleague later reminded him, "nothing but a vast forest of the largest and most impenetrable kind reaching in every direction to the virgin sea." The landing party could hardly find room to pitch their tents. But Raffles wasn't there to explore. His job was to develop trade, and he recognized Singapore's strategic importance as a chokepoint at the southern end of the Strait of Malacca, commanding the main shipping lane between the Indian Ocean and the Pacific.

The transformation of Singapore into a thriving British colony over the next few years filled him with unabashed mercantile delight: "Here all is life and activity; and it would be difficult to name a place on the face of the globe with brighter prospects or more pleasant satisfaction. In little more than three years it has risen from an insignificant fishing village to a large and prosperous town, containing at least ten thousand inhabitants of all nations actively engaged in commercial pursuits which afford to each and all a handsome livelihood and abundant profit . . . This may be considered as the simple but almost magic result of that perfect freedom of Trade which it has been my good fortune to establish."

His private life, on the other hand, had taken a much darker turn. He and Sophia were deeply in love with their young children, his "dear little rogues." They planned in another few years to head back to the comforts of England "and to time our departure with reference to their health and happiness." He had seen enough of worldly power, he wrote, to know that "there is more real happiness in domestic quiet and repose." But then, within a period of just six devastating months, their first three children died one by one, of dysentery and other common afflictions. A new child arrived in 1823 "to fill the melancholy blank in our domestic circle," but soon also died. They had only a young daughter left and each other, and in Raffles's letters it became an urgent question whether they could find a ship to get them home "in time to save our lives, though we have very little confidence that this will be the case."

Raffles packed up all his notes, maps, books, paintings, musical instruments, and other mementoes. "There were also two or three trunks full of birds in thousands and of various species, and all stuffed," an aide wrote. "There were also several hundred bottles of different sizes filled with snakes, scorpions and worms of different kinds. The bottles were filled with gin to prevent corruption. The animals were thus like life. There were also two boxes filled with coral of a thousand kinds; also shells, mussels, and bivalves of different species. On all these articles stated above he placed a value greater than gold, and he was constantly coming in to see that nothing was hurt or broken." In the hurry of packing, Raffles neglected to obtain insurance. Thus they would have to ship all their property, valued at £25,000, a sizable fortune then, at their own risk.

Two days out to sea, as they were preparing for bed, a fire broke out beneath their cabin. Raffles later described the scene of confusion: "A rope to the side. Lower Lady Raffles. Give her to me, says one; I'll take her, says the Captain. Throw the gunpowder overboard. It cannot be got; it is in the magazine close to the fire. Stand clear of the powder. Scuttle the water-casks. Water! Water! Where's Sir Stamford? Come into the boat, Nilson! Nilson, come into the boat. Push off, push off. Stand clear of the after part of the ship." Everyone aboard was able to get off into lifeboats. "Less than ten minutes afterwards she was one grand mass of fire," Raffles wrote, and with the ship went much of what was left of his own shattered life.

Raffles, Sophia, and their surviving daughter would eventually get back to London, where he would become the driving force behind the creation of the London Zoo and then die at the age of 44. Along with the new species he had brought to the attention of science, Singapore would be his real legacy. Soon after the deaths of his children, he had referred to it as "this, my almost only child" and later he had begged, "Rob me not of this my political child." It was as if, however reluctantly, he had swapped his flesh-and-blood for the glories of empire. And this was a bargain many men implicitly accepted when they went out into the world as naturalists and colonizers—their lives, marriages, children, even perhaps their souls, for the glory of a new city or a new species.

Joanna's End

In Suriname, John Stedman's tour, too, had inevitably come to an end. Then, exactly like the planters—he left his Suriname wife behind. Back home in 1782, he entered into a more conventional union with a Dutch woman. Joanna died later that year, possibly from poisoning. Their son Johnny served and was drowned in the British Navy.

MAD ABOUT SHELLS

> How vast an item is the apparently unimportant
> shell-fish in the wealth and happiness of man!
> —EDGAR ALLAN POE

A RRIVING IN THE mid-seventeenth century after a six-month-long ocean voyage to the East Indies, the colonial administrator who would become known as Rumphius found himself in an improbably dazzling Dutch city, laid out on a neat rectangular grid at the mouth of the Ciliwung River, with step-gabled houses fronting canals lined with palm trees. The Dutch East India Company had built Batavia, now Jakarta, out of glistening white coral, not just "because it was cheap and easily available," historian E. M. Beekman writes, "but also because this material absorbed seventeenth-century bullets, rendering them harmless." The city was the company's trading capital, but still a quasi-military outpost. The company conducted its global enterprise with the help of 40 warships and an army of 10,000 men, in addition to its fleet of 150 merchant vessels. Its underpaid employees worked under the absolute authority of an almost feudal hierarchy.

For all its apparent orderliness, Batavia was a breeding ground for mosquitoes and a tropical death-trap for Europeans. (When Capt. Cook's _Endeavour_ visited more than a century later, a third of his crew died of the combined effects of dysentery and malaria.) But Rumphius somehow flourished in this new world, particularly when he took up his station as a merchant in Ambon, another 1500 miles to the East. The more relaxed customs of the island culture clearly appealed to him after his brutal experience of European civilization.

Georg Eberhard Rumpf had been born in central Germany in 1627, of a Dutch mother and a father who served as an engineer and contractor to various impecunious aristocrats. Europe then was a bloody battleground for the Thirty Years War, and young Rumphius himself soon became a soldier, shipping out for foreign service at the age of 18. "He fled from chaos, from a ruined country that had been pillaged, raped, and murdered into exhaustion," Beekman writes. "He escaped from fraudulent authority, hypocritical religion, and social inequality, a state of affairs which did not warrant allegiance to anything human." Somehow his ship got diverted from Brazil to Portugal, where he spent three years as a soldier before returning briefly to Germany. Then in 1652 he obtained a position with the Dutch East India Company, possibly through his mother's family, and left Europe forever.

His official duties for the Dutch East India Company took up much of his time in his adopted home. But Rumphius devoted every free moment to the search for new species, and the letters he sent back to Europe earned him a reputation as "the Pliny of the Indies," after the great Roman naturalist. The specimens he collected and sketched would eventually make up two great works, a *Herbarium* and his *Ambonese Curiosity Cabinet*, together constituting the first natural history of the East Indies. The burdens and tragedies he faced in the course of his research have helped keep his reputation alive, and the books that he could not get published in his own lifetime continue to appear in new editions today.

In Ambon, he formed a common law marriage with an island woman named Suzanna. She helped him gather the plants and animals of the island, which became, along with their children, his "caritates," or loved ones, according to Beekman. But he lost his heart, in particular, to the ornate seashells of what he happily called "the water Indies."

Conchylomania

The peculiar passion for the exoskeletons of mollusks may be older than the human species itself. Shellfish were of course familiar first as food. According to one scientific theory, mussels, snails, and the like were critical to the brain development that made us human in the first place. But people soon also came to treasure the shells for their delicately sculpted and decorated surfaces. Anthropologists have identified beads made from shells in North Africa and Israel at least 100,000 years ago as among the earliest known evidence of modern human culture.

Different societies also put shells to more practical use, not just as ornaments, but as blades and scrapers, oil lamps, currency, cooking utensils, boat bailers, musical instruments, and buttons, among other products. Marine snails were the source of the precious purple dye, painstakingly collected one drop at a time, which became the symbolic color of royalty. Shells also served as a convenient reminder of sex, with the curves, orifices, and enclosed spaces coming to symbolize a woman's reproductive parts. In Pompeii, for instance, a mural survives of Venus reclining on a shell that represented the vulva, a motif repeated endlessly on up through the more explicit version produced by the French artist Odilon Redon almost 2000 years later.

Shells may also have served as models for the volute on the capital of the Ionic column in classical Greece, and for Leonardo da Vinci's design for a spiral staircase in the Renaissance. In France, shells inspired an entire art movement. Eighteenth-century architects and designers of the rococo favored shell-like curves and other intricate motifs. (The name "rococo" blends the French *rocaille*, the practice of covering walls with shells and rocks, and the Italian *barocco*, or baroque.)

But at the height of the human madness for shells, the Netherlands became the center of the trade, largely because the ships of the Dutch East India Company brought back such spectacularly beautiful specimens from the Indo-Pacific. They became precious items in the private museums of the rich and royal. "Conchylomania," from the Latin *con-*

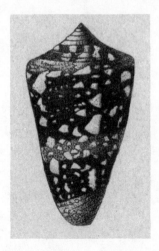

A cone shell species described by Rumphius.

cha, for cockle or mussel, soon became the equal of the Dutch madness for collecting tulip bulbs, and often afflicted the same people.

One Amsterdam collector who died in 1644 had enough tulips to fill a 38-page inventory. But he also had 2389 shells, and considered them so precious that, a few days before his death, he had them put away in a chest with three separate locks, according to *Tulipmania*, a recent history by Anne Goldgar. The executors of his estate each got a single key, so they could show the collection to potential buyers only when all three were present. A Dutch writer mocked both tulip maniacs and "shell-lunatics" in a pair of engravings. Shells on the beach that used to be playthings for children now had the price of jewels, the caption remarked. "It is bizarre what a madman spends his money on."

And he was right: At an eighteenth-century Amsterdam auction of one celebrated collection, some shells sold for more than paintings by Jan Steen and Frans Hals, and slightly less than Vermeer's now priceless "Woman in Blue Reading a Letter," which went for 43 guilders (about $500 in today's money). The collection included a *Conus gloriamaris* shell for which the owner had paid 120 guilders, almost three times the selling price of the Vermeer.

From a financial perspective, valuing shells over Dutch Masters may rank among the dumbest purchases ever. There are only 30-odd Ver-

The Precious Wentletrap.

meer paintings on Earth. But the scarcity that could make a shell seem so precious was almost always an illusion. For instance, C. *gloriamaris*, a 4-inch-long cone covered in a delicate fretwork of gold and black lines, was for centuries among the most coveted species in the world, known from only a few dozen specimens. One shell trade story held that a wealthy collector who already owned a specimen managed to buy another at auction and, in the interest of scarcity, promptly crushed it underfoot. To keep up prices, collectors also spread the rumor that an earthquake had destroyed the habitat in the Philippines and rendered the species extinct. But inevitably, divers finally discovered the mother-lode in the Pacific, north of Guadalcanal Island, in 1970, and the value of *gloriamaris* plummeted. Today you can buy one for roughly the price of dinner for two at a nice restaurant. And paintings by Vermeer? The last time one came on the market, in 2004, it went for $30 million. (And it was a minor and somewhat questionable one, at that.)

But mania can be a matter of perspective. What's common to us was breathtakingly rare to them, and vice versa. Dutch artists produced about 5 million paintings in the seventeenth century. Even Vermeers and Rembrandts could get lost in the glut, or lose value as fashions shifted. Beautiful shells from outside Europe, on the other hand, had to be collected or acquired by trade in distant countries, often at consider-

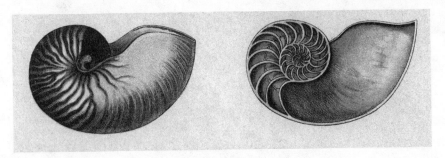

The chambered nautilus.

able risk, and then transported long distances home on crowded ships, which had an alarming tendency to sink or go up in flames en route.

Illustrated guidebooks became essential tools of the trade. Filippo Buonanni, an Italian priest, produced one such volume in 1681 under the title "Recreation for the Eyes and the Mind . . . Through the Study of Shells." At about the same time, Rumphius was writing *The Ambonese Curiosity Cabinet,* his account of the shells and other organisms on the Moluccan archipelago. Because the Linnaean system of classification didn't yet exist, collectors had no common language for identifying specimens. So in their letters back and forth they often relied on a page reference to an illustration in one of these books.

The madness for shells among buyers remained intense throughout the eighteenth century. One French collector then coveted a spiny oyster shell so badly, according to shell historian S. Peter Dance, that he pawned his wife's best silver to buy it. In the heated marital dispute that inevitably ensued, the distraught collector flung himself down on a chair—and snapped off two of the spines that had made the shell so precious. Then husband and wife fell to consoling each other.

To serve this market, some enterprising souls tossed common limpets in hot cinders to strip off their natural coloring, dyed them gold, and then sold them to collectors eager to pay top price for a new species. Linnaeus himself may have fallen for such a ruse: The new species he called *Cypraea amethystea* was really just a species he had already described, *C. arabica,* but with the dorsal surface stripped away to reveal the violet shell layer underneath.

Sailors and colonial administrators also capitalized on the trade, bringing home legitimate new discoveries. "Begging pardon for my Boldness," one such sailor wrote to Joseph Banks in 1775, in a tone of forelock-tugging class deference. "I take this opportunity for acquainting your Honour of our arrival," he wrote, having just returned from Capt. Cook's second trip around the world. "After a long and tedious Voyage . . . from many strange Isles I have procured your Honour a few curiosities as good as could be expected from a person of my capacity. Together with a small assortment of shells. Such as was esteem'd by pretended Judges of Shells."

An officer from the same expedition sold another shell to one of the specimen dealers who regularly met returning voyagers at quayside. It resold immediately to museum proprietor Ashton Lever for more than £10—about seven months' wages to a common seaman then. As *Astraea heliotropium*, it became for a time the most coveted shell in the world. That same voyage also produced the first specimens seen in Europe of the golden cowry, *Cypraea aurantium*. Someone in the crew had noticed the shell as a prized ornament worn on a string by chieftains in Tahiti. Back in Europe, collectors mistook the string hole for a natural feature.

"Stately Mansions"

For many collectors of that era, shells were not just rare, but literally a gift from God. Such natural wonders "declare the skilful hand from which they come" and reveal "the excellent artisan of the Universe," wrote one eighteenth-century French connoisseur. The Precious Wentletrap, a pale white spiral of shell enclosed by slender vertical ribs, first illustrated in *The Ambonese Curiosity Cabinet*, was evidence for another collector that only God could have created such a "work of art."

This religious spin enabled the wealthy to present their lavish collections as a way of glorifying God rather than themselves, according to British historian Emma Spary. The idea of gathering shells on the beach also had spiritual status, not that many wealthy collectors actu-

A European pen shell.

ally did it themselves. It symbolized escape from the workaday world to recover a sense of spiritual repose, a tradition invoked by luminaries from Cicero to Newton.

The pious feelings elicited by shells were also genuine. The shape of many shells suggested the metaphor of climbing a spiral staircase and coming closer with each step to inner knowledge, and to God. The departure of the animal from its shell also came to represent the passage of the human soul into eternal life. The nautilus, for instance, grows in a spiral, chamber upon chamber, each larger than the one before. Oliver Wendell Holmes made it the basis for one of the most popular poems of the nineteenth century: "Build thee more stately mansions, O my soul, / As the swift seasons roll! . . . Till thou at length art free, / Leaving thine outgrown shell by life's unresting sea!"

Oddly, collectors didn't much care about the animals that built the shells. Holmes, for instance, unwittingly blended the characteristics of two separate nautilus species in his poem, according to shell historian Tucker Abbott: "It was as if he had written a poem about a graceful antelope who had the back half of a leopard and the habit of flying over the arctic ice. "Collectors often cared passionately about new species, but mainly for the status of possessing something strange and expensive from a distant land, preferably before anybody else.

The absence of flesh-and-blood animals actually made shells more appealing, for a highly practical reason. Early collectors of birds, fish, and other wildlife had to take elaborate and sometimes gruesome measures to prepare and preserve their precious specimens. (A typical set of instructions to bird collectors included the admonition "open the Bill, take out the Tongue and with a sharp Instrument pierce through the roof of the Mouth to the Brain.") But those specimens inevitably succumbed to insects and decay anyway, or the beautiful colors faded to mere memory.

Shells endured. They were more like jewels than living things. In the 1840s, a British magazine recommended that shell collecting was "particularly suited to ladies" because "there is no cruelty in the pursuit" and the shells are "so brightly clean, so ornamental to a boudoir." Or at least it seemed that way, because dealers and field collectors often went to great lengths to remove any trace of a shell's former inhabitant.

Seeing with His Own Eyes

Rumphius was one of the few conchologists who actually thought and wrote about shellfish as living creatures. Most collectors back in Europe prized pinnae shells, for instance, mainly for their beautiful shape, roughly corresponding to a wide-vaned quill pen. But Rumphius noticed how they behaved. He reported that these "pen shells" tended to live in quiet bays, at a depth of 4 or 5 feet, standing upright with their narrow ends planted in the muddy bottom. He also noted that an ice-colored little shrimp, about 1½ inches long, lived within the shell and stood guard. In the event of danger, "the shrimp pinches the Pinna in order to force it to close its shell," he wrote.

He was describing mutualism—in this case, two species swapping shelter for an early warning system—centuries before a biological term for the phenomenon even existed. That determination to see things

Struck blind, Rumphius still vividly described the natural world of his "water Indies."

originally, with his own eyes, combined with his palpable delight in
what he saw, helped turn Rumphius into one of the great naturalists
of his time. In the text for his *Herbarium*, a botanical guide to the
Indies, he emphasized the medicinal value of the local flora and fauna,
a utilitarian focus calculated to appeal both to his employers and to
other European naturalists. But "he reserved his most poetic texts," historian and translator E. M. Beekman writes, for such "useless" items as
jellyfish.

A *Physalia,* or Portuguese man-of-war, for instance, was "of a transparent color, as if it were a crystal bottle. . . . The little sails are as white as crystal, and the upper seams show some purple or violet, beautiful to see, as if the entire Animal were a precious jewel." The starfish tree similarly came to life under his pen: "The tree resembles a landlubber looking for someone he is in love with out in the sea, because it invariably stands rooted at the edge of the forest, not daring to advance by as much as one step onto the naked beach, and hovers above the same at such an angle as if it desires to fall forward at any moment."

But Rumphius reserved his greatest love for shells. Despite its name, his best-known book, *The Ambonese Curiosity Cabinet*, was neither "a catalogue of lifeless oddities," according to Beekman, nor "an inventory of economic pride." On the contrary, the emphasis was always "on the animal that lives within these handsome shelters." The book was a mild rebuff to what Rumphius saw as "covetousness and pomp" among collectors. He had come to subscribe to the Indonesian ethic that objects could have special power only if one has found them oneself or received them as gifts, but not when "bought with money." (In one of the stranger twists of literary history, Edgar Allan Poe would later get Rumphius's philosophy backward, describing him as "a fool" who once gave "a thousand pounds sterling for one of the first discovered specimens" of the *Venus dione*. Even more strangely, the error occurred in the only commercially successful book Poe published in his lifetime. *The Conchologist's First Book* was a school text Poe edited and improved based on a British volume.)

For Rumphius, getting his books written and published would prove to be a monstrous challenge. Being at the opposite end of the Earth from the centers of scientific discourse put him at an obvious disadvantage. A request sent from Ambon might take weeks to reach Batavia, and then a minimum of 20 months to elicit a reply from Amsterdam. It was like sending a question from a Caribbean island to New York, then waiting for a long-distance runner to bring back an answer from California, with plenty of opportunities for naysaying by bureaucrats or grizzly bears along the way. Rumphius was eventually able to build up a small research library of his own, after the Heeren XVII, the governing board of Dutch East India Company, granted him permission to have scientific books sent out on company ships.

The Blind Seer of Ambon

But then, in 1670, at age 42, Rumphius was struck blind, from unknown causes. Despite its keen focus on profit, the company allowed him to remain in its employ and devote much of his energy to the pursuit of useful knowledge. He carried on with his work, depending on family and staff to read to him and take dictation.

A few years later, an earthquake hit Ambon, and both his wife Susanna and a daughter died beneath a collapsed wall. A witness recorded the terrible spectacle of the blind man sobbing beside their bodies. Rumphius named a white orchid *Flos suzannae* "in memory of her who during her life was my first mate and helpmeet when searching for herbs and plants, and who showed me this flower also." Sadly, the name does not appear to have survived in modern taxonomy.

This was by no means the end of his misfortunes. In 1687, a fire ripped through the island, destroying his library and all the original drawings Rumphius and his assistants had made over more than 30 years. He was at least able to save his notes. Undaunted, he commissioned new drawings and continued to work for another five years, finally sending the first six volumes of his *Herbarium* off to Amsterdam on a ship with the unfortunate name *Waterland*, which sank with its entire cargo. By extraordinary good luck (not something with which Rumphius was otherwise familiar), the governor-general, also a naturalist, had commissioned a copy before sending the *Herbarium* onward from Batavia.

So Rumphius was able to complete his 12-volume manuscript, with lavish illustrations, and ship it to Amsterdam intact in 1696. Even then, he did not get the triumphant reception he surely deserved. Instead, the Heeren XVII, which wielded absolute power over company employees, refused to allow publication, probably because this encyclopedia of the flora of the East Indies seemed as if it might be just a little too useful to rival European powers. *The Herbarium* would remain in the company archives for another 45 years, before a consortium of publishers finally brought it into print beginning in 1741.

But the real tragedy of Rumphius's life lay elsewhere, according to Beekman, and it resulted from the collecting mania he was helping to inspire. In the late 1660s, Cosimo III de Medici, Grand Duke of Tuscany, was touring the cultural centers of Europe, including the Netherlands, where he admired the tropical shells in the curiosity cabinets of the leading collectors. The Dutch conchylomania was highly contagious, and Cosimo set out to develop a shell collection of his own.

Cosimo's guide in Amsterdam was Pieter Blaeu, of the celebrated map-making and bookselling family. The Blaeus often did business with the Dutch East India Company and with Rumphius. (They may even have looked after Rumphius's son when he was a student in Holland.) So it was natural for Cosimo to visit and admire the curiosity cabinets kept by the leading powers within the Dutch East India Company. And it was also natural, Beekman writes, for powerful foreigners like Cosimo to seek help from these influential men in developing their own cabinets of curiosities. The shared interest in collecting shells and other precious objects helped to cement relationships otherwise based only on commerce. "Capitalism subsidized curiosity in a reciprocal relationship that enhanced and dignified both." And Rumphius was the innocent victim.

In 1682, evidently under pressure from the company or the Blaeu family, Rumphius sold Cosimo the best part of his shell collection, 360 specimens gathered over a period of 28 years, "what for him must have been souvenirs in the most literal sense: memorials of the life" that had been snuffed out by blindness and the loss of his wife and daughter. "Giving things away was not the problem," Beekman continues. "What vexed Rumphius was the compulsion, backed by some kind of high-pressure method, to relinquish part of his life to a stranger for money." It was "a clash of Western capitalistic imperatives and a stubborn Indonesian metaphysic," and whenever Rumphius referred to the sale, "the tone is rueful or angry." The note Rumphius sent to Cosimo, along with the shells, began in the required social etiquette of good wishes and then slid inexorably down into regret: "May the Lord . . . favor the passage of these objects and may they arrive safely . . . so that no damage will be done to a treasure which I gathered over many years with much cost and labor and which, in the future, it will be impossible to acquire again, especially since I am now old and blind."

The loss "preyed on his mind for the final two decades of his life," according to Beekman. Rumphius mentions the year 1682 eight times in *The Ambonese Curiosity Cabinet*, twice as often as the year he lost his wife, and four times more than the year he lost his vision. He had been left bereft of his shells, his caritates. And, in the Indonesian scheme of things, the shells themselves had been deprived of their power by being sold.

The experience at least "showed him a way to get his words in print," says Beekman: He would appeal directly to the European collecting appetite and present his caritates as specimens in an *Ambonese Curiosity Cabinet*. It was a Faustian bargain, and it worked. Given the slow and secretive practices of the Dutch East India Company, the book moved relatively quickly into print, appearing in 1705.

In Ambon, Rumphius had died three years earlier, age 74, an unpublished author.

Chapter Five

EXTINCT

It may be asked, why I insert the Mammoth, as
if it still existed? I ask in return, why I should
omit it, as if it did not exist?

–THOMAS JEFFERSON

I N THE SPRING OF 1705, in the Hudson River village of Claverack, New
York, a rainstorm eroded a steep bluff and spalled out a tooth the size
of a man's fist. It rolled downhill and landed like a gift of God at the
feet of a Dutch tenant farmer, who muttered something like "Heilige
koe!" (Holy Cow) and promptly sold it to a local politician for a glass
of rum. Someone (and suspicion naturally falls on the politician) tried
using the tooth itself as a tankard, turning it upside down and filling
the hollow core with liquor. But at almost 5 pounds, the tooth was too
heavy for serious lifting. Deeming it otherwise worthless, the politician
made it a gift to Lord Cornbury, the eccentric governor of New York. A
pioneer in the grand tradition of gubernatorial bad behavior, Cornbury
would soon lose his job for alleged embezzlement and graft. According
to his enemies he also liked to cross-dress as his cousin Queen Anne.

Cornbury did what Queen Anne herself had ordered colonial gov-
ernors to do with a discovery like the tooth—he shipped it off to the
natural philosophers at the Royal Society in London. He labeled it
"tooth of a Giant," after the statement in Genesis that "there were
giants on the earth" in the days before the flood.

Man or beast, this "monstrous creature," as Cornbury called it, would soon become celebrated as the *incognitum*, that is, the unknown and mysterious species. It was the *American Monster*, according to Paul Semonin, author of an authoritative history by that title: The *incognitum* was "the dinosaur of the early American republic," at a time when actual dinosaurs had not yet been discovered. And some inchoate force in the developing American spirit embraced it as "in effect, the nation's first prehistoric monster."

The Massachusetts poet Edward Taylor estimated its height at 60 or 70 feet and wrote a poem celebrating "Ribbs like Rafters" and arms "like limbs of trees," not to mention a nose like "an Hanging Pillar wide." The minister Cotton Mather boasted that the New World possessed Biblical giants to make "Og and GOLIATH, and all the Sons of *Anak*" look like Pygmies. His friend Massachusetts Governor Joseph Dudley wrote that the knobby upper surface of the tooth proved that it couldn't have come from an elephant. He classed the monster as one of the unknown creatures buried in Noah's flood and thought it would not be seen again, God willing, until the Apocalypse.

"Eat or Be Eaten"

People had been puzzling over fossils since Aristotle. But the Earth had now begun to yield up evidence of vanished creatures and lost worlds with discomfiting frequency, producing excitement and consternation in equal measure. Fragments of an ichthyosaur (imagine a crocodile in the skin of a dolphin) had surfaced on the border between England and Wales. And in 1718, in the garden of a parsonage in Nottinghamshire, a "person of curiosity" named Robert Darwin had helped recover bones of a fossil plesiosaur (like the Loch Ness monster, with four powerful fins and an elongated neck for reaching out to seize its prey). Shells and sharks' teeth were turning up on mountaintops and deep within solid ground.

As the Industrial Revolution gathered force, new mines, quarries, and canals sliced open the countryside, exposing past geological ages. Erasmus Darwin, son of Robert and grandfather of Charles, visited one such excavation near Nottingham in 1767 to observe "the Goddess of Minerals naked, as she lay on her inmost bowers," with belemnites, ammonites, and "numerous other petrified shells" wantonly strewn across the fresh-cut banks.

Darwin, a physician, poet, and philosopher, was a likable figure— fat, florid, pockmarked, with a ready smile and a voluptuary's heart. (How else could he have turned a canal excavation into a naked Goddess?) He loved food, and, in a cockeyed foreshadowing of his grandson's ideas about natural selection, his favorite nostrum for patients was "Eat or be eaten." He had a deep faith in human progress (not least in his own family). The sight of fossil shells and other primitive creatures recovered from oblivion made him think about progress in the natural world, too. Years later, in his 1794 book *Zoonomia*, he would venture the "transmutationist" idea that over the course of "perhaps millions of ages . . . all warm-blooded animals have arisen from one living filament," with successive generations acquiring new traits and passing down improvements to posterity. But for the moment, he expressed his belief in the common descent of species more simply, by having a motto painted like a bumper sticker on his carriage door: "*E conchis omnia*" (Everything from shells).

Even that smacked of heresy, to some. A local clergyman mocked Darwin in verse: "Great wizard he! by magic spells / Can all things raise from cockle shells." Long before Charles Darwin was born, the poet Samuel Taylor Coleridge coined the term "darwinizing" to mock this sort of evolutionary theorizing. Fear of offending patients soon caused Erasmus Darwin to display his new motto only on his bookplate, rather than in the high street.

But even in the absence of evolutionary ideas, fossils raised deep, dreadful questions: How could the Earth be just 6000 years old, as the Bible taught, if monstrous creatures had lived in ancient times and then somehow vanished forever, to be replaced by other monsters, and then by still others in turn, till their bones filled up the valleys and formed the limestone of the highest cliffs? How did these lost creatures affect the doctrine that species were a permanent, unchanging heritage from the Garden of Eden? And what did their disappearance say about the supposed perfection of God's work? In Genesis, God had told the animals, "Be fruitful, and multiply, and fill the waters in the seas, and let fowl multiply in the earth." He hadn't told them to go extinct.

The discovery of new species entombed in ancient stone caused people to wonder with dismay if life on Earth had had a history before human beings. The question seems quaint to us now, like a naïve young man in Victorian times crestfallen to discover that his beloved has had *a history* with other men. But it represented a profound shift in thinking. The "new vista of earth history would equal the Copernican revolution in its intellectual implications," writes science historian Nicolaas A. Rupke, "reducing the relative significance of the human world in time just as early modern astronomy had diminished it in space." Long before Charles Darwin brought biological thinking to bear on the issue, astronomical and geological evidence was already driving people to seek desperately for ways to hold the old worldview together.

The Plenitude Problem

At the same time, people doted on these new monsters with horror and fascination. Bones of the *incognitum* turned up again in mid-eighteenth century at Big Bone Lick, a sulphur-smelling patch of marsh near the Ohio River in Kentucky. French and British expeditions separately shipped teeth, bones, and huge tusks back to Europe for further study. In Paris, Georges-Louis Buffon and his sometime-coauthor, the anatomist Louis-Jean-Marie Daubenton, went back and forth on the issue of extinction. They eventually decided, based on a comparison of femurs and tusks, that the *incognitum* must belong to the one living elephant species then known. (Daubenton explained away the peculiar shape of the *incognitum's* molars by arguing that a fossil hippopotamus had gotten jumbled up with the elephant bones.)

In London, the physician William Hunter consulted with his brother the anatomist John Hunter, and then applied the family's characteristic comparative method, putting the knobby upper surface of the *incognitum* molars side by side with the relatively flat, corrugated biting surface of modern elephant molars. He argued that the *incognitum* must be "of another species, a *pseudelephant*," unknown to naturalists, and that it was almost certainly carnivorous. "Though we may as philosophers regret it," he added, "as men we cannot but thank Heaven that its whole generation is probably extinct."

Benjamin Franklin, then on diplomatic duty in London, raised the practical objection that an animal carrying such large tusks was unlikely to be nimble enough "for pursuing and taking prey." He suggested correctly that the knobby teeth might be "as useful to grind the small Branches of Trees as to chaw Flesh." Others objected on philosophical grounds: Primordial killer elephants threatened Biblical orthodoxy that violence and death had come into the world only as a result of human sin in the Garden of Eden.

The disappearance of such creatures also challenged widely held faith in "the Great Chain of Being," the idea that the natural world was a perfect progression from the lowliest matter on up, species by species, jellyfish to worms, worms to insects, culminating in the Earth's most glorious production, *Homo sapiens*. A corollary of the Great Chain, the notion of "plenitude," held that God had created all forms that could be created. There were no gaps in the Chain. So philosophers spent whole careers diagramming and re-diagramming, and seeking intermediate forms, or "missing links," to reveal this innate perfection. Proposing that some of these forms had gone extinct, an anonymous American writer complained, was "an idea injurious to the Deity."

But the process of accommodating religious thinking to scientific fact was already under way. Americans resonated to the idea that the *incognitum* "had been a monstrous carnivore whose disappearance was God's blessing on the human race," Paul Semonin writes in *American Monster*. It was easier to accept extinction if it confirmed "the long-held Christian view of man's dominion over the natural world." As the American Revolution was ending, "patriotism and prehistoric nature became intertwined," with the *incognitum* coming to symbolize "the immense power" of the nation's past—and perhaps also its future.

When a ditchdigger turned up more teeth in New Windsor, New

York, in 1780, Gen. George Washington made the 10-mile trip from the Continental Army's winter encampment and noted that these new finds matched a tooth from Big Bone Lick in his private collection at home. Some of these new specimens ended up in Philadelphia, where Charles Willson Peale, the young nation's most prominent painter, received a commission to make detailed drawings. The bones drew so many curious visitors to his studio that Peale decided to establish his own natural history museum. It would in time play a key role in establishing the national identity—and in the history of the *incognitum*.

More than anyone, though, Thomas Jefferson would make a cause célèbre of the *incognitum*. In his 1781 book *Notes on the State of Virginia*, he wrote, "Such is the economy of Nature, that no instance can be produced of her having permitted any one race of her animals to become extinct; of her having formed any link in her great work so weak as to be broken." But the intellectual problem with extinction—the plenitude problem—was trivial beside Jefferson's powerful attachment to the idea of the *incognitum* as a living, breathing American giant. The quest for the living beast would occupy a corner of his mind for the next 30 years, through his time as minister to France, secretary of state, vice-president, and president of the United States. The mammoth, as he called it, was to be the crowning argument in his case against Georges-Louis Buffon's "theory of American degeneracy." First proposed in 1749, this theory held that the inhospitable climate of the Americas made all species, including humans, smaller and less capable.

Jefferson was determined to prove otherwise. He had originally begun writing his *Notes* as the Revolution was ending, in response to the French government's request for information about the emerging nation. With this audience in mind, he constructed elaborate tables of different American species, setting "the heaviest weights of our animals . . . from the mouse to the mammoth" against puny Old World counterparts. He thought that the mammoth, "this largest of terrestrial beings," should have "stifled in its birth" Buffon's notion "that nature is less active, less energetic on one side of the globe than she is on the other." It set Jefferson's rhetorical wheels spinning: "As if both sides were not

warmed by the same genial sun; as if a soil of the same chemical composition was less capable of elaboration into animal nutriment; as if the
fruits and grains from that soil and sun . . . " When he sailed in 1784
to become American trade minister and then ambassador to Paris, Jefferson carried "an uncommonly large panther skin" in his luggage with
the idea of shaking it under Buffon's nose. Later he followed up with a
moose. Buffon supposedly acknowledged that Jefferson had made his
point. But his theory was another "reprehensible statement" left unre-
tracted when he died a few years later.

Jefferson's concern wasn't just a matter of wounded pride, according to historian Thomas Carl Patterson. For American envoys in the
1770s and 1780s, refuting the idea of innate inferiority was essential
"if they were to obtain sorely needed financial assistance and credit in
Europe." The Americans needed to prove that their national experiment was not doomed by nature to fail. Benjamin Franklin probably got
farther with wit than Jefferson did with his moose. Once, at a dinner in
Paris, a Frenchman (a "mere shrimp," according to Jefferson) was arguing for Buffon's theory. Franklin (5-foot-10) sized up the European and
American guests, seated on opposite sides of the table, and proposed:
"Let us try this question by the facts before us. Let both parties rise,
and we will see on which side nature has degenerated." The Frenchman
muttered something about exceptions proving rules.

Jeffersonian Anatomy

Jefferson believed that the *incognitum* still lived. He gave credence to a
story told him by a visiting delegation of Delaware warriors about how
in ancient times the *incognitum* had come in a great herd to destroy "the
bear, deer, elks, buffaloes and other animals, which had been created
for the use of the Indians." When an angry God struck down this monstrous herd with lightning bolts, one great bull "shook them off as they
fell," and bounded away over the Ohio River to somewhere beyond
the Great Lakes. The Indian tradition, Jefferson wrote, "is that he was
carnivorous, and still exists in the Northern parts of America."

As he was getting his *Notes* ready to send to the French, Jefferson

The statesman-naturalist Thomas Jefferson laid out mastodon bones in the White House, and also described a species of extinct ground sloth, which he mistook for a lion.

also sent off a letter to Kentucky, carried by Daniel Boone, asking his friend George Rogers Clark, commander of the Army of the West, to collect specimens from Big Bone Lick. More than 20 years later, as president, Jefferson would send Clark's brother William, and Meriwether Lewis, to explore the Pacific Northwest, partly in the hope of discovering the living *incognitum*. Later, he laid out a collection of mastodon bones on the floor of the East Room of the White House. (It was a step up from John and Abigail Adams, who had hung their laundry there.)

His Buffonian obsession would lead Jefferson into an embarrassing scientific blunder. He had obtained some leg bones, ending in huge claws, of a fossil dug up in a saltpeter mine in West Virginia. Jefferson had just become vice-president of the United States a few days earlier, but he made time to name this find *Megalonyx*, or "great claw" and to describe it at a meeting of the American Philosophical Society on March 10, 1797. He thought that *Megalonyx* was some type of big cat, that "he was *more* than three times as large as the lion: that he stood as pre-eminently at the head of the column of clawed animals as the mammoth stood at that of the elephant, rhinoceros and hippopotamus: and that he may have been as formidable an antagonist to the mammoth as the lion to the elephant."

Jefferson imagined this pair of American monsters still doing battle in the vast American West: "In the present interior of our continent there is surely space and range enough for elephants and lions . . . and for the mammoth and megalonyxes who may subsist there." Unfortunately for Jefferson, it would soon turn out that his *Megalonyx* wasn't a lion after all, but a giant sloth.

The evidence came from a fossil recently recovered, almost intact, in Argentina. At the Muséum National d'Histoire Naturelle in Paris, an anatomist compared detailed drawings of this fossil with other mammal specimens and determined that the teeth and giant claws were almost identical to those of modern sloths and anteaters. The anatomist named it *Megatherium* (Giant Beast), and at 12 feet in length and 6 feet in height, it attracted considerable public excitement. Jefferson himself happened to see an illustration of *Megatherium* in a magazine shortly before his own *Megalonyx* talk. With what must surely have been horror, he tacked a hasty postscript onto his presentation acknowledging that the two animals might be the same. But he would hold on to his big cat idea, he said, at least until he had teeth to examine.

Happily, another French anatomist would later identify Jefferson's fossil as a different kind of giant ground sloth, preserving the genus name *Megalonyx* and graciously adding the honorific species name *jeffersonii*. Evidence from other fossil species discovered since then has also demonstrated that, despite the misidentification, Jefferson was at least right about his basic premise. Until about 10,000 years ago, an American lion had, in fact, roamed the New World, and it was 25 percent larger than its modern African cousin. Instead of a land of degenerate species, America was a land of giants, home to an armadillo the size of a black bear, a 200-pound giant beaver, the 700-pound saber-toothed tiger, and the largest bear ever known. Even on four legs, the short-faced bear was taller than most men and, at 1800 pounds, it was also heavy enough to crush "Buffonian error" with a casual swipe of one paw.

All of these monsters were of course extinct, a fact we now consider commonplace. "But the questions are simple now only because they have been answered," the twentieth-century paleontologist George Gaylord Simpson once reminded readers. "Every answer was contrary to the accumulated lore of all the millenniums before 1700. They required not only the rejection of some of the fondest beliefs of mankind but also the development of fundamentally new ways of thinking and of an apparatus for scientific interpretation. It was the great achievement of the eighteenth century that it made this revolutionary advance, even more basic than that wrought by the doctrine of evolution in the nineteenth century."

For the world of science, the chain of being shattered and extinction became a reality on January 21, 1796, at the National Institute of Sciences and Arts in Paris. On the podium was the anatomist Georges Cuvier, a handsome, confident 26-year-old, with a full head of reddish hair, a long, thin face, and a strong chin. He had recently joined the staff of the Muséum National d'Histoire Naturelle, which gave him the opportunity to make detailed comparisons among a variety of pachyderm specimens, some freshly looted by the French Army from the Netherlands. In his talk that day, Cuvier presented the first evidence that the African and Asian elephants were separate species, as were the Siberian mammoth and what he called "the Ohio animal." He noted significant differences in bone structure and pointed out that animals the size of mammoths could hardly have escaped notice by "the nomadic peoples that ceaselessly move about the continent in all directions." The evidence he presented made extinction a fact of life.

Cuvier had no political agenda, but his ideas reflected the events of his time. He spoke of extinction as casting new light on "the revolutions of this globe," language that was common enough in the Newtonian scientific world, according to geologist Martin Rudwick. But it now also evoked violent change, "a world anterior to ours, destroyed by some kind of catastrophe," much as the French Revolution had lately swept the Old Régime into the abyss.

Cuvier didn't yet say so. But his work on this lost world would soon lead to a kind of resurrection.

Chapter Six

THE RISING

For the trumpet will sound, the dead shall be
raised imperishable, and we shall be changed.

–1 CORINTHIANS 15:52

TALL, HANDSOME, with keen, deep-set eyes, Charles Willson Peale was a creature of boyish enthusiasms pursued with adult energy and ambition—not just a portrait artist, but also an inventor, a carpenter, a showman, and a family man who fathered 16 children and outlived three wives. His first *incognitum* bones in 1782 had set him on an "irresistibly bewitching" quest for knowledge about the natural world. Characteristically, he wanted the museum he created to include "everything that walks, creeps, swims, or flies, and all things else," with the result that the collection rapidly outgrew the busy Peale family residence at Lombard and Third streets in Philadelphia.

In 1794, Peale arranged to move his museum roughly six blocks north into the heart of the American Republic. The Philadelphia Museum's new home would be in American Philosophical Hall, located almost as an annex to the Pennsylvania State House (now Independence Hall), where national leaders had recently thrashed out the language of the Declaration of Independence and the U.S. Constitution. The U.S. Congress was still meeting just down the block, so nature and nation would be closely allied. With the Smithsonian Institution still

Charles Willson Peale meant his museum to teach the greatness of the American experiment with the help of specimens like the mastodon jaw fragment at lower right.

a half century in the future, what Peale was creating was in effect the national museum.

In Peale's entrepreneurial hands, even the dreary business of moving became a promotional event. He hired men to carry an American buffalo, panthers, and "tyger catts" on their shoulders, and he recruited "all the boys of the neighbourhood" to parade behind bearing "a long string of Animals of smaller size." It brought people to their windows and doors all along the route and provided "fine fun for the boys." The young volunteers also "saved some of the expense of remooval of delicate articles."

Peale had tickets to his museum printed with the slogan, "The Birds & Beasts will teach thee!" (paraphrasing Job 12:7), and he saw to it that they taught lessons in the greatness of the American Republic. American species arranged in dioramas alternated with portraits of the nation's human heroes. The abundance and usefulness of American wildlife foretold the nation's future wealth. Peale even found evidence of innate American decency. Whereas Old World cuckoos were a notorious symbol of infidelity for their practice of planting their eggs in the nests of other birds, American cuckoo pairs built and tended their own nests, he wrote, "and we are proud to believe that they are faithful and constant to each other."

For Peale, as for Jefferson, the massive size of the *incognitum* made it the perfect answer to Buffon's "ridiculous idea," and in 1801 he got word of "an animal of uncommon magnitude" recently discovered in the Hudson River Valley. That June, he traveled three weeks by stagecoach and sloop to Newburgh, New York. A farmer there, John Masten, had unearthed a nearly complete, if fragmented, *incognitum* from a marl pit and was exhibiting it on the floor of his barn for a small admission fee. Peale pretended that he was interested only in making life-size sketches of individual bones, on sheets of paper pasted together, and he managed for a time to disguise his eagerness. The upshot was that he was able to buy the bones for $200—roughly $2500 in today's currency—plus another $100 for the right to do additional digging on his own. On winning this bargain, Peale wrote, "my heart jumpt with joy," causing the farmer to extract deal sweeteners in the form of a gun for his eldest son and a couple of New York City gowns for his daughters. News of the purchase "must have blew like wild fire," and a crowd of the curious came to ogle the monster when Peale stopped in New York City on the way home. Everybody "seemed rejoiced that the bones had fallen into my hands," Peale wrote, because they "would be preserved, and saved to this Country."

By the beginning of August, Peale was back in Newburgh, with a $500 loan from the American Philosophical Society (Thomas Jefferson presiding), and an ambitious plan to excavate more bones from a flooded pond on the Masten farm. Peale later commemorated the scene in a historical painting, with lightning crackling down from a black corner of the sky, and horses panicking in the distance. To drain the pond that dominates the scene, Peale had devised a huge wooden wheel on a high bank, with men treading inside like hamsters in an exercise wheel. The turning of the wheel drove a long conveyor belt of buckets, each carrying a few gallons of water up and over, to spill down a chute into a nearby vale. In the painting, workers on staged platforms pass dirt up from the exposed bottom of the pond, while others sieve for bone fragments, and a shirtless man in the water holds up what appears to be a fibula. Peale gestures to the work with one open hand while holding a long drawing of mastodon leg bones in the other.

Despite the title "Exhuming the First American Mastodon," Peale recovered only a few missing parts from this excavation. He did better from two other, less picturesque, excavations up the road, recovering a second nearly complete skeleton. His showpiece was still going to be the skeleton that he had bought off the barn floor from John Masten. The painting was merely a better story, a shrewd piece of self-promotion, for which Peale made no apologies: "Provided I do not degrade the character of a naturalist by too much puffing—a little seems absolutely necessary to call the attention of the public to such objects."

Back in Philadelphia, it would take three months, and "numberless tryals of puting first one piece, then another, together, and turning them in every direction," to make sense of the scraps. Peale's slave Moses Williams ("my Molatto Man Moses") did much of the work, which seems to have taken place in the family living room. Williams, who would win his freedom soon after, grew up in the Peale household, in something less than a Cinderella role. One of the Peale children, only six months

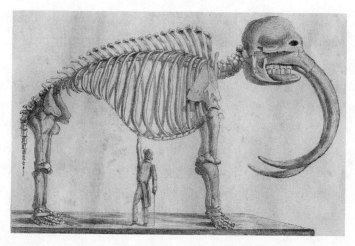

The Peale family thought the incognitum *was "cruel as the bloody Panther" and they displayed it at first with tusks turned down for spearing prey.*

older, dismissed him as "entirely worthless" in his youth. Peale himself managed to praise and insult Williams in the same breath, noting that "the most expert anatomist" could not have done the mastodon work better than "those of the least knowledge," meaning Williams. He "fitted pieces together by trying, [not] the most probable, but by the most improbable position, as the lookers-on believed. Yet he did more good in that way than any one among those employed in the work." Peale filled in missing parts in papier-mâché and wood, scrupulously indicating these substitutions. But the showman or artist caused him to exaggerate the size of his *incognitum* slightly, yielding a skeleton 11 feet high at the shoulder. Later, he corked the joints, adding extra "cartilage" to make it even bigger.

The first complete skeleton was ready before Christmas 1801, and Peale prided himself on having achieved "a second creation." To drum up business for the opening, he had Williams put on an Indian headdress and parade through the city streets on a white horse, with trumpet fanfare, handing out flyers that invoked Indian legend. "TEN THOUSAND MOONS AGO" a creature had roamed "the gloomy forests . . . huge as the frowning Precipice, cruel as the bloody Panther." For 50¢ additional admission to the museum's "Mammoth Room," Philadelphians could now see "the LARGEST of Terrestrial Beings!" with their own wide eyes.

It was only the world's second reconstruction of a fossil species (after Madrid's decidedly less thrilling giant ground sloth) and it became a national sensation, with word spreading "until the masses of the people were more eager than the scientists to see the great American wonder," according to biographer (and Peale descendant) Charles Coleman Sellers. "The mere idea of bigness stirred every heart. Overnight the word 'mammoth' gained a fresh and spectacular currency." In hindsight, "mammoth" was the wrong word. We now know that the specimens from New York were not mammoths, which had lived in the grasslands of the American West; they were mastodons. But for the moment, "a Philadelphia baker offered 'Mammoth Bread' for sale. In Washington a 'Mammoth Eater' dispatched forty-two eggs in ten minutes," and a New Yorker grew a 20-pound "mammoth" radish.

Thomas Jefferson had recently defeated the incumbent John Adams to become president, amid bitter Federalist-Republican squabbling. Given his long interest in the subject, anything "mammoth" could seem like a provocation to his adversaries. When the women of Cheshire, Massachusetts, sent the White House a 1230-pound "Mammoth Cheese," made from the milk of cows not belonging to Federalists, it attracted noisy protesters en route. "From John Adams came a lugubrious warning," Sellers writes, "that one should not talk of natural history when the liberty and morals of the nation stood in peril."

Politics, both national and international, also infected a publicity stunt staged by Peale's son Rembrandt. Thirteen gentlemen, one for each state, sat at a round table beneath the mammoth's monstrous ribcage, while a musician performed at a local inventor's Patent Portable Grand Piano tucked under the pelvis. The diners sang "Jefferson's March" and "Yankee Doodle" and they offered patriotic toasts, being careful not to raise their glasses too high. "The American People: may they be as preeminent among the nations of the earth, as the canopy we sit beneath surpasses the fabric of the mouse!" In June, young Peale boarded a ship with a second skeleton from Newburgh to show off this prodigious sample of American might in Europe.

For the Trumpet Shall Sound

In Paris, Georges Cuvier was rapidly turning natural history into a more scientific discipline. Other naturalists dwelt lovingly on the external appearance of the animals they studied. But Cuvier was a comparative anatomist; he based his classifications on the form and function of internal parts and the ways they varied from species to species. He was also establishing himself as the master not merely at discovering lost species, but also at the almost magical business of putting them back together bone by bone. "In one detailed memoir after another, each founded on the most painstaking comparison with the anatomy of living mammals," writes geologist Martin Rudwick, "he demonstrated that there had been species of elephant, hippopotamus, rhinoceros, armadillo, deer, cattle and so on—all distinct from any living species, many of them much larger in size, and all apparently vanished from the face of the Earth. The resurrection of such a spectacular zoo was a striking accomplishment indeed, and it brought Cuvier fame throughout the world of science." Over discolored bone and monstrous form he would sound the Biblical trumpet, "and the dead shall be raised imperishable."

Cuvier had supposedly gotten his start as a naturalist on a childhood visit to his uncle's house, where he discovered a copy of Buffon's *Histoire Naturelle*. With the obsessive bookish devotion some children display, Cuvier copied the illustrations and then attempted to color them, which sometimes required careful reading of the text. Later, using Buffon's descriptions, he attempted to draw animals that were not actually illustrated. Thus the author whose work he would later disparage as insufficiently scientific provided the key to his own life as a scientist. But this has the ring of a "just so" story. All we know for certain is that Cuvier liked to sketch from nature as a student in Stuttgart and then as a tutor with a family in Normandy, producing precise, detailed drawings of insects, amphibians, and other creatures.

Georges Cuvier persuaded the scientific world that extinction was a fact, not a heresy.

Of his developing fascination with fossils, Cuvier later wrote: "As a new species of antiquarian, I had to learn how to restore these monuments of the past and make sense of them; I had to collect and put into order the fragments of which they were composed, to reconstruct the ancient beings to which these fragments belonged, to reproduce them with their proportions and their characters, to compare them finally with those that live on the surface of the globe today. This was an almost unknown art."

Cuvier alone seemed to possess the necessary combination of vivid imagination tempered by deep anatomical knowledge. A typical excavation was seldom neat or obvious; you did not roll back the rock like a blanket to reveal the intact skeleton of a lost species, preserved as if in sleep. What usually turned up was a meaningless jumble of disassembled bones, often with more than one species sharing the same disorderly grave.

To put comparative anatomy on a rational basis, and avoid the danger of patching together a composite creature, Cuvier relied on two working principles. The "correlation of parts" asserted that if one part of an animal was clearly built for some obvious function, other parts of the animal would also fit that function. If an animal had the teeth of a carnivore, it would have the claws, too. The many subtle correlations in between were as inevitable as mathematics, at least for Cuvier: "the form of the tooth leads to the form of the condyle [the rounded knuckle of any joint], that of the scapula to that of the nails, just as an equation of a curve implies all of its properties . . . similarly, the nails, the scapula, the condyle, the femur, each separately reveal the tooth or each other; and by beginning from each of them the thoughtful professor of the laws of organic economy can reconstruct the entire animal."

In figuring out how to classify the creatures he brought back to life, Cuvier borrowed his second principle, "the subordination of characters," from the botanist Antoine-Laurent de Jussieu. Rather than focusing on a single trait, as Linnaeus had done with the reproductive organs, or treating all traits equally, this principle established a hierarchy of traits and put greater taxonomic emphasis on those that were more fundamental to a particular creature's way of life. Cuvier's rational approach established comparative anatomy as a science. But his own working methods seldom fit his theories, according to biographer William Coleman.

"The Correlation of Parts"

In Montmartre, then a rural neighborhood on the outskirts of Paris, a gypsum quarry had begun to produce impressive fossils. The site now lies buried beneath the gardens leading up to the Cathedral of Sacré Coeur. But its image survives in a later painting by Vincent Van Gogh, depicting the quarry as a gray cleft at the bottom of a sloping field. Its legacy also persists in the white stone of which Paris is largely built, as well as in the dust that became plaster of Paris. Examining an assortment of fossils from the quarry, Cuvier was uncertain that he could bring about the necessary "resurrection in miniature." He worried that he lacked the "omnipotent trumpet." But he was more confident than this implied.

One reconstruction involved the broken bones of a heavyset mammal. Looking at the teeth, a standard starting point because of their abundance and durability, Cuvier thought they resembled the modern tapir in number, but the rhinoceros in the shape of the molars. He ruled out the possibility of a rhinoceros-like horn because the nose bones were too small to support it. And from the shape of the upper jawbone, and the absence of the elephant's huge conduit for the maxillary nerve, Cuvier inferred that the creature possessed only a small tapirlike trunk. The result was a new genus *Paleotherium* and another called *Anoplotherium*, with some dithering about where to assign the available feet. But *Paleotherium* so closely resembled the modern tapir in other regards, Cuvier wrote, that no naturalist could "restrain himself from spontaneously crying out that this foot is made for this head and this head for this foot."

What Cuvier was doing, writes Coleman, "departed significantly from the proclaimed theoretical principles." Instead of relying on the "correlation of parts" within a fossil skeleton, Cuvier depended more often on what he knew of modern animals. In one flamboyant case, a split slab with a fossil exposed on one face had arrived from Montmartre. By examining the fossil's teeth and lower jaw, Cuvier recognized a close resemblance to the modern possum. He not only placed the animal in the same genus, *Didelphis*, but before an audience of learned guests, he boldly predicted that he would find two characteristic bones that project forward from the pelvis of modern marsupials. Then he set to work on the stone and soon unearthed the bones in question. It wasn't quite "a curious monument" to his "zoological laws," as Cuvier maintained. But to his audience, it looked like genius.

Cuvier was not only recovering lost worlds, but populating them with substantial new species almost as fast as living species were being discovered in the modern world. Based on a drawing of a fossil specimen from Bavaria in Germany, he identified the first-known flying reptile, a pterosaur, and gave it the name *Pterodactylus*. He named a giant marine lizard *Mosasaurus* and speculated that reptiles had once dominated the Earth. Another paper gave a new name to the *incognitum*. Cuvier published it in 1806, under the ungainly title, "On the great mastodon, an animal very close to the elephant but whose molars are studded with large bumps, whose bones are found in various places on both continents, and especially near the banks of the Ohio in North America, incorrectly called *Mammoth* by the English and by the inhabitants of the United States."

Cuvier pointed out the critical difference that made mammoths creatures of the grasslands, and mastodons of the forest: The mammoth's molars were relatively smooth and suited for eating grass. The mastodon's had large conical cusps, for the heavier work of browsing on leaves and branches. Those conical cusps reminded Cuvier, a little oddly, of breasts. Hence *mast*, from the Greek for "breast" or "nipple," and *odon*, from "tooth."

The driving question for Cuvier was how all these lost species fit into the history of the world. He came to recognize that fossils found closer to the surface more nearly resembled living species than those found in deeper, and therefore older, strata. So there was direction, if not progress, in the fossil record. Jean-Baptiste Lamarck, his colleague at the Muséum National d'Histoire Naturelle, was already proposing the first cohesive theory of evolution, arguing that species had changed and improved over time in response to environment. But Cuvier dismissed the idea "that during a succession of ages, and by alterations of habitudes, all the species may change into each other, or one of them give birth to all the rest." He noted the absence in the fossil record of intermediate forms showing evidence of "this strange genealogy." He also pointed out that mummified ibises and other creatures from ancient Egypt were identical to their descendants in the modern world. Thus, he concluded, "the ancient and now extinct species were as permanent in their forms and characters as those which exist at present."

Bursting the Limits of Time

Permanent, that is, until one apocalypse or another swept them into oblivion. Cuvier imagined a world in upheaval: "Living organisms without number have been the victims of these catastrophes. Some were destroyed by deluges, others were left dry when the seabed was suddenly raised. Their races are even finished forever, and all they leave in the world is some debris that is hardly recognizable to the naturalist."

This idea of mass extinctions thrilled and terrified the romantic nineteenth-century imagination, as Cuvier had intended. He had chosen his "tragic and deliberately hyperbolic vocabulary," French historian Claudine Cohen writes, "to impress the reader and make of the naturalist a new hero and of natural history a new kind of epic." In recovering this tumultous past, "Cuvier was also building his own myth," presenting himself "like the demiurge of a new creation who could 'burst the limits of time' and 'by observations . . . rediscover the history of the world and the succession of events that preceded the birth of the human species.'" Cohen adds, "And images of catastrophes and deluges would haunt romantic dreams until late in the century."

One or two people managed not to get swept away by Cuvier's grand catastrophic vision. Visiting in 1828, the French-American painter John James Audubon described the great paleontologist with characteristic waggishness, less like the romantic hero of natural history, more like an overstuffed bird (Cuvier was by then sometimes called "the mammoth" and not just for his intellectual greatness): "size corpulent, about five feet five, English measure; head large; face wrinkled and brownish; eyes gray, brilliant and sparkling; nose aquiline, large and red; mouth large, with good lips; teeth few, blunted by age, excepting one on the lower jaw, measuring nearly three-quarters of an inch square. Thus have I described Cuvier almost as if a new species of man."

For almost everybody else, Cuvier was a species of genius. His work describing lost worlds would become the inspiration, paradoxically, both for literary realism and for a sweeping sense of romance—in the same author. Honoré de Balzac, "a novelist trying to look like a scientist," borrowed his close focus on mundane details from Cuvier's *procédé résurrectioniste*, his process for resurrecting lost species, according to literary critic Richard Somerset. "Cuvier was reputedly able to deduce a whole anatomy from a single bone; Balzac thought he was able to deduce a whole personality from a single gesture, from a look, from a profile, from a way of dressing, and so on."

But Balzac also happily borrowed Cuvier's sweep and grandeur. "Have you never launched into the immensity of time and space as you read the geological writing of Cuvier?" he asked at the start of his 1831 novel *La Peau de Chagrin (The Wild Ass's Skin)*. "Carried by his fancy, have you hung as if suspended by a magician's wand over the illimitable abyss? . . . Is not Cuvier the great poet of our era? . . . our immortal naturalist has reconstructed past worlds from a few bleached bones; has rebuilt cities, like Cadmus, with monsters' teeth; has animated forests with all the secrets of zoology gleaned from a piece of coal; has discovered a giant population from the foot of a mammoth."

What had begun with the scattered bones of the *incognitum* was now bearing fruit in the living world. For the United States, the "mammoth" had set loose the characteristic nineteenth-century American delight in things boisterous and big, helping to create a national sense of identity and self-confidence. For science, the extinction of the *incognitum* had created a new view of our world. Not everyone accepted Cuvier's catastrophic history of the Earth. In England, the geologist Charles Lyell was already arguing for a far more gradual process of change. But by shattering the myth of the great chain of being and by making extinction an undeniable fact, Cuvier had destroyed the comforting notion of order and design in nature.

Historians have remembered Cuvier, until recently, mainly as the adamant defender of the fixed nature of species. But by his work resurrecting lost species and placing fossils in their proper sequence in the history of the Earth, he established and populated the scale of geological time. His revolutionary achievements made modern biology possible. He died in May 1832, age 61, of cholera. That month another hunter of species, Charles Darwin, was already making his way down the coast of South America aboard HMS *Beagle*, free to think in new ways about the nature of life.

Chapter Seven

THE RIVER ROLLING WESTWARD

> The time will arrive, when we shall no longer
> be indebted to the men of foreign countries, for
> a knowledge of any of the products of our own
> soil, or for our opinions in science.
>
> —THOMAS SAY

I N THE SPRING OF 1810, an immigrant shopkeeper in Louisville, Kentucky, a prosperous little frontier town on the Ohio River, got an unexpected visit from a wandering naturalist. Exiled from Scotland after a blundering attempt to extort money from a local mill owner, Alexander Wilson had reinvented himself as a birder, an artist, and a zealous American. Now he was on the road collecting species and drawing them. He was also trying to sell subscriptions to the nine-volume *American Ornithology* he had already begun to publish. If Thomas Jefferson was using words to refute Buffon's ideas about "degenerate" American species, and Charles Willson Peale was doing it with his museum, Wilson meant to do it with his sketchpad and his fieldwork.

Because of the feature that gives them their name, for instance, the birds called crossbills can appear to be freaks, particularly to a researcher in Europe seeing only a specimen plucked from its natural environment. The curved top and bottom bills cross at the tip, when the mouth is closed, "an error and defect of nature, and a useless deformity," according to Buffon. But Wilson examined the stomach contents of the birds he collected. He also observed their behavior. It turned out that the oddly shaped bills in fact serve perfectly for prying apart pinecones and extracting the seeds. "Instead of being a defect or monstrosity, as the celebrated French naturalist insinuates," he wrote, the deviation from common form "is a striking proof of the wisdom and kind, superintending care of the great Creator."

Wilson was a lean, somewhat astringent looking figure in his early forties with a high forehead, a "brilliant eye and hair black as an Indian's," according to an acquaintance in Philadelphia. The shopkeeper, who was just 23, depicted him less kindly as hook-nosed and keen-eyed, with a pet parakeet for his only companion, and a tendency to play melancholy Scottish airs on a flute. Wilson made the sales pitch for his *American Ornithology*, which would become the essential guide to the

Exiled after a blundering extortion attempt, Alexander Wilson reinvented himself as America's leading ornithologist.

Wilson saw the crossbill as proof of God's genius.

nation's birds and establish him as the father of American ornithology. And the shopkeeper was about to sign up, according to the account he wrote years later. But just then, his business partner called out in French from the next room that the shopkeeper's own drawings were surely "far better." The partner probably also reminded him that the $120 price could be a crushing burden for a young retailer in a shaky new business.

In any case, Wilson was crestfallen to see the quill pen pause in midair, and again when he inquired, perhaps in the manner of a salesman still hoping to close the deal, whether the young clerk before him might also be a member of the artistic brotherhood. Two years earlier, Wilson had written to a friend, "If I have been mistaken in publishing a work too good for the country, it is a fault *not likely to be soon repeated.*" Now the shopkeeper, whose name was John James Audubon, brought down his own portfolio and spread it open on the counter. It was the beginnings of what would become, years later, the greatest picture book ever published, his *Birds of America.*

Wilson had little choice but to suck in his breath and praise the younger man's work. He took note of two undescribed flycatcher species illustrated there. Audubon, the rare naturalist not obsessed with scoring points for species discoveries, offered them to Wilson, along with use of his paintings. Gloomily, Wilson replied, "I must trudge by myself." He agreed, though, to let Audubon take him to ponds outside Louisville to show him whooping cranes, sandhill cranes, and other species: "We hunted together and obtained birds which he had never before seen," Audubon wrote. Years later, when parts of Wilson's journal were published posthumously, Audubon would be annoyed to see the comment Wilson had jotted down on leaving Louisville without having sold a single subscription: "Science or literature has not one friend in this place."

That the two of them should have met along the Ohio and spent time together in the field pursuing their shared passion for the natural world was hardly unusual. A host of naturalists—or rather, a curiosity of naturalists—swarmed through the Ohio River Valley in those years. Among them, for instance, was the ichthyologist Charles-Alexandre Lesueur, who had previously sailed with Capt. Baudin's French Navy expedition to the coast of Australia, and, somewhat more exotically, Prince Maximilian of Wied-Neuwied on the Rhine.

Philadelphia may have prided itself on being "the Athens of America" and the headquarters of natural history in the New World. But the real work of discovery was taking place on the great river rolling westward. In the forests and marshes there, and up to their necks in the river itself, naturalists were working out the meaning of discovery, down to the nitty-gritty details of how to describe and preserve a specimen. The Ohio River was also the staging ground for expeditions that would give shape to the United States of America, defining the nation beyond mere laws and institutions, as a living thing of creatures and habitats.

Steamboat Sea Monster

In May 1819, a bizarre apparition—the most advanced technology of the day in the body of a fairy-tale monster—passed down the Ohio, bound for the wilderness beyond the Mississippi. The steamboat *Western Engineer* was 75 feet long, with a narrow 13-foot beam for working in tight quarters, and it had its paddlewheel in the stern to minimize damage from snags and floating trees. The official account of the expedition dryly noted that "some peculiarities in the structure of the boat attracted attention" along the river.

A newspaper reporter caught the colorful details: "The bow of this vessel exhibits the form of a huge serpent, black and scaly, rising out of the water from under the boat, his head as high as the deck, darted forward." The gaping mouth poured forth waste steam from the engine, as if breathing fire. At the stern, "a stream of foaming water, dashing violently along" came up from under the boat. "All machinery is hid . . . Neither wind or human hands are seen to help her . . . Her equipment is at once calculated to attract and awe the savage" with the illusion "that a monster of the deep carries" a painted vessel upon its back.

A bit more pragmatically, the *Western Engineer* also carried a painting of a white man and an Indian shaking hands, and another of the calumet, or long pipe, of peace. But it was also "bristling with guns" and the steering house on the quarterdeck was bulletproof. The prospect of traveling among "lawless and predatory bands of savages" caused one member of the party to take stock of the expedition's arsenal: A brass four pounder mounted in the bow, four Howitzers, two smaller artillery pieces, 12 muskets, six rifles, several fowling pieces and an air gun (presumably for collecting birds with minimal damage to feathers), 12 sabers, pistols, assorted private arms, "and a great sufficiency of ammunition of all kinds."

The name of President James Monroe was painted on one side of the steamboat and of Secretary of War John C. Calhoun on the other, as the "propelling powers" behind the expedition. Calhoun had ordered three large expeditions into the American West, two of them military, assigned to build forts in Indian Territory along the Mississippi and Missouri rivers. The third, commanded by Maj. Stephen Harriman Long, was to explore the tributaries of both rivers aboard the *Western Engi-*

neer, and then proceed on horseback and foot to the Rocky Mountains. It would be the first expedition to the American West staffed with trained naturalists.

The spectacle of this dragon-boat, "smoking with fatigue, and lashing the waves with violent exertion," reminded the reporter of the question Hamlet asked on seeing the ghost of his father:

> *Be thy intents wicked or charitable,*
> *Thou com'st in such questionable shape.*

Those on board would no doubt have answered "charitable." Even the Secretary of War urged the expedition to "conciliate Indians by kindness & by presents," while also ascertaining "the number and character of the various tribes," and their territories. Calhoun emphasized "scientifick purposes" and the need to record "everything interesting in relation to soil, face of the country, water courses and productions whether animal vegitable or mineral." It was a portion of "our country, which is daily becoming more interesting," he wrote. As elsewhere, science would be the advance guard for occupation and development.

The Ohio River twisted and rolled through a landscape of low, rumpled hills, and in the first decades of the nineteenth century the shoreline was still blanketed in thick forest. At Gallipolis, roughly halfway between Pittsburgh and Cincinnati, one of those aboard the *Western Engineer* noted "the finest beech woods I ever saw." Below Cincinnati, a companion remarked on the "magnificence of the forests," the "snowy whiteness" of sycamore branches "admirably contrasted with the deep verdure" of other trees. But settlers were quickly cutting back the forests to plant crops and build towns, with 25,000 people already living at Cincinnati by 1819. A traveler remarked that the sudden shifts in vegetation from dense forest to open fields could easily confuse riverboat pilots. Shadows and reflections made it difficult to keep to the middle of the river by night.

Odd Bedfellows

Even before they departed Pittsburgh, the naturalists aboard the *Western Engineer* were getting scooped by a brilliant crackpot named Constantine Rafinesque. Though he did not travel with them, his work would shadow theirs. One of the first species they fished up on a hook-and-line was an odd underwater salamander with lungs as well as gills, now commonly known as the mudpuppy. Chief naturalist Thomas Say gave it the name *Triton lateralis* and, in the notes to the official account of the Long Expedition, he described it in characteristically meticulous detail over four single-spaced pages ("*tail* much compressed, subacutely edged above and beneath, lanceolate"), including notes on his dissection of a comparable species in the collection of the Academy of Natural Sciences of Philadelphia. It turned out later that Rafinesque had described the same species in print a year earlier, in 1818. His report was slapdash, also characteristically. But priority—being the first to publish—counts in scientific discovery. So Rafinesque's name *Necturus maculosus* prevailed.

Though they would prove equally energetic in their devotion to the sacred cause of American natural history, Rafinesque and Say represented the polar opposites of the scientific world. Rafinesque was a species monger, too drunk on the elixir of discovery to take much care with his work. Say was a meticulous observer determined to describe only what was genuinely new and always with the thorough detail that would enable other naturalists to build on his work. Each would spend years working in the Ohio River Valley as well as in Philadelphia. But the resemblance ended there.

A woman who knew him in his Ohio years described Rafinesque as a "lone, friendless, little creature," slight, with dark eyes and silky hair, but so absorbed in his work that he often forgot to clean off the mud from his field trips or wash his face. He had been born in Constantinople, the son of a French merchant and a German mother (in Europe he went by the double-barreled name Rafinesque-Schmaltz). He traveled widely with his parents, and appears to have grown up largely separate

Species-mad Rafinesque got duped by Audubon into describing fictitious fish species, including one with bulletproof scales.

from other children, a possible explanation for his lack of ordinary social skills. He was almost pathologically obsessed with numbers and boasted of having read 1000 books by the age of 12. But he had no formal training in science to discipline what would become his driving passion, the discovery of new species. Rafinesque's zeal for names and categories was such that he once supposedly submitted a scientific paper describing 12 new species of thunder and lightning.

"Nature," he once instructed an audience at Transylvania University in Lexington, Kentucky, "is a beautiful and modest woman, concealed under many Veils." Rafinesque saw himself as one of the lucky few granted the "high favor" to remove these veils and "take a glimpse of her beauty." But to many naturalists, it looked more like he wanted to ravish her, most likely while crying out other women's names—*Rosacia! Peta-lonia! Hippocris!* "His misfortune was his prurient desire for novelties, and his rashness in publishing them," the paleontologist Louis Agassiz later wrote, though he added that Rafinesque "was a better man than he appeared." William Baldwin, the botanist on the Long Expedition, dismissed Rafinesque's work as "the wild effusions of a literary madman."

With his characteristic love of a good yarn, Audubon left behind the most vivid account of what meeting Constantine Rafinesque was like. It was the summer of 1818, and Audubon, 33 and still a shopkeeper, had relocated downriver to Henderson, Kentucky. Rafinesque turned up unannounced, bent beneath a sack of plants and other specimens, and Audubon invited him to stay in his home. "We had all retired to rest," he later recalled. "Every person I imagined was in deep slumber save myself, when of a sudden I heard a great uproar in the naturalist's room."

What followed sounds, in the 1832 recounting, like a demented version of " 'Twas the Night Before Christmas," a poem that had been published just a few years earlier: "I got up, reached the place in a few moments, and opened the door, when, to my astonishment I saw my guest running about the room naked, holding the handle of my favourite violin, the body of which he had battered to pieces against the walls in attempting to kill the bats which had entered by the open window, probably attracted by the insects flying around his candle. I stood amazed, but he continued jumping and running round and round, until he was fairly exhausted: when he begged me to procure one of the animals for him, as he felt convinced they belonged to 'a new species.' Although I was convinced of the contrary, I took up the bow of my demolished Cremona, and administering a smart tap to each of the bats as it came up, soon got specimens enough." Audubon, playing pizzicato with the bats, was still an unknown to his fellow naturalists. But Rafinesque, then 35, was already notorious for his lunatic pursuit of new species. He was the court fool of the scientific world, giving off sparks of genius but always snuffing them out beneath heaps of nonsense.

By contrast, the care and quality of Thomas Say's work would make him one of the unsung heroes of American science, setting a standard for the proper description of a species. Say, age 32 at the start of the Long Expedition, was tall and patrician, with a high forehead and thick, unruly hair. Like Rafinesque, he lacked much formal education. But Say had grown up in Philadelphia, learning scientific methods from some of the best naturalists in the world. He haunted the Peale family's Philadelphia Museum as a boy and then, as an adult, sometimes worked there so late that he ended up sleeping in the only place he could reasonably stretch his 6-foot frame—beneath the mastodon skeleton. (Accommodations were more luxurious over at the Academy of Natural Sciences, which Say helped found in 1812: There, he improvised a tent by stretching a blanket over a horse skeleton.)

Natural history was in his blood. Say's great-grandfather John Bartram had traveled as a naturalist, collecting plants as far south as Florida and west to the Ohio River. Linnaeus, in a momentary lapse of ego, had once called Bartram "the greatest natural botanist in the world." (He was probably giddy from having received one of "Bartram's Boxes" stuffed with new species.) The Bartram family home, a few miles outside Philadelphia on a bluff above the Schuylkill River, had the best botanical garden in North America. There, Say learned natural history from his great uncle, William Bartram, John's son, and from a neighbor, the bird artist Alexander Wilson. From Wilson, the Scotsman, he also acquired a deep belief in the power of America's wildlife to draw the nation's motley citizenry together into a sense of common destiny. By encountering grizzly bears and mule deers, they would become American.

Insects were Say's specialty, and the roughly 1400 species he identified over his lifetime included many of critical economic importance in agriculture. He also discovered and described the mosquito species *Anopheles quadrimaculatus*, the sort of achievement, he ruefully admitted, that could attract "the ridicule of the inconsiderate." But this species turned out, long after his death, to be the cause of the "ague," or malaria, that routinely afflicted and often killed people as far north

Thomas Say was the first trained naturalist to explore the American West—and set a pattern for careful descriptions of new species.

as Boston. Say also specialized in shells, and according to the definitive biography *Thomas Say: New World Naturalist* by Patricia Tyson Stroud, he was among the first to suggest using the fossil record of shell species to date the strata in rock.

On the Long Expedition, Say got capable help from Titian Peale, already making his second major expedition at the age of 19. He had been born in his father's natural history museum and grew up helping preserve and mount specimens from around the world—a bat from Madagascar, an anteater, a llama displayed as if spitting (because, according to the elder Peale, that's "what he used to do when he was alive at the museum").

When Titian was a boy, the family had lived above the shop, and the menagerie downstairs included a pair of "tame" young grizzly bears, which Zebulon Pike had brought east and President Jefferson had donated to the museum. The tameness faded as the bears matured, and when a monkey reached through the bars of the cage, a grizzly casually tore off its arm. The same bear later broke loose and raided the family kitchen. A sign soon appeared advertising the "meat of a remarkably fine Missouri or Grisly bear" for sale to Philadelphia connoisseurs. The hide and bones of the two bears proved more manageable once put back on display as a still life. Titian Peale could hardly have gotten a better preparation for a life of adventure in the cause of natural science.

Audubon, who seems to have been everywhere along the Ohio River then, encountered the Long Expedition in Cincinnati, where he'd gone to work briefly at Drake's Western Museum preparing specimens and painting scenery for the displays. It must not have been Audubon's best work, or maybe artistic rivalries once again tainted the encounter. Titian Peale mentions only that the museum had "a few articles" collected for display, "mostly fossils and animal remains." Audubon seems to have stood in the background, watching in silence: "Messrs. T. Peale, Thomas Say and others stared at my drawings." Then they passed by without further comment. Audubon was biding his time and working steadily on his portfolio. It would be another seven years before he published *Birds of America*, at the age of 41.

How to Describe a Species

Say and Rafinesque must have met, too, and it's probably just as well that we have no record of it. Say had helped found the *Journal of the Academy of Natural Sciences of Philadelphia*, where he pursued his deeply held belief in a clear, coherent system of scientific classification. Almost all naturalists then were amateurs. They were often ignorant of how to describe a species properly, making it available to other naturalists and gaining credit for its discovery. The rules of the game were also still taking shape, and Say was determined to get them right as a vital step toward putting American natural history on a sound professional basis.

Rafinesque likewise touted "proper rules, principles, laws and method" as the key to "complete" scientific knowledge. But the sight of anything remotely like a new species always seemed to make him lose his grip on this idea. Audubon recalled that when he walked his guest down to the river to prove that a plant from one of his drawings actually existed, Rafinesque "danced, hugged me in his arms, and exultingly told me that he had got, not merely a new species, but a new genus."

Among other abominations, Say objected to the endless naming and renaming of species. Certain naturalists, Rafinesque notorious among them, had mastered the trick of racking up "discoveries" by taking one perfectly good species and subdividing it into a half-dozen new species based on trivial differences. These "splitters" drove Say, who was otherwise mild and genial, to fulmination: "posterity will rise up in judgement against all those pirated names & indignantly strike them from the list." (He was right. In the twentieth century, "lumpers" would devote whole careers to the heroic but tedious work of un-discovery, recombining spurious species and ushering them back to their proper places in the taxonomic order.)

Say also objected to the brevity of many scientific descriptions. He wanted a detailed technical description—tediously detailed, if need be—so other naturalists could work through it, compare it to a freshly collected specimen, and determine whether it was the same thing or something new. Rafineque's descriptions often consisted of little more than the species name itself. Some naturalists neglected to publish any description at all. In 1830, when Audubon should have known better, a British naturalist admonished him: "You may not be aware that a new species, deposited in a museum, is of no authority whatsoever, *until its name and its character* are published." By failing to organize specimens properly, "you will, I am fearful, [lose] the credit of discovering nearly all the species you possess."

Other naturalists back to Linnaeus considered a brief description good enough, because in theory every new species name was tied to a "type specimen" preserved in a museum or in the cabinet of a private collector. If you wanted the details about a particular species, all you had to do was go look at the original example. But this wasn't much help when a researcher happened to be working in the Ohio River Valley and the type specimen was locked away in a drawer somewhere in the Netherlands. The technology of preservation was also dismal then, so type specimens often disappeared after a few years.

Some naturalists didn't bother to collect type specimens at all. Rafinesque actually put names on species he had never even seen, much less collected. He sometimes described a new species based on nothing more than a passing reference in somebody else's travel book. (Maybe he thought that's what *type* specimen meant.) He also described nine species of fish based on drawings Audubon had cooked up as a prank, possibly to avenge his smashed violin. Audubon spun him a yarn about how the Devil-Jack Diamond Fish had bulletproof scales, which the woodsman could use with flint to strike a fire. Rafinesque duly recorded it as *Litholepis adamantinus*—meaning roughly "unbreakable stone scales."

Not surprisingly, Rafinesque had trouble keeping track of his own "discoveries." In 1818, he published a paper in the *Journal of the Academy of Natural Sciences of Philadelphia* describing a new fish, only for the editor, Thomas Say, to discover that Rafinesque had already published the same fish under a different species name in another journal. Say was apoplectic at having been drawn into this disreputable business. "Either Rafinesque in his haste to publish had been unaware of this error," writes Patricia Tyson Stroud, "or else he had hoped to be credited with two discoveries." The *Journal* apologized to readers and declared that rejecting Rafinesque's article was "indispensable for the interest of science." Thereafter, Rafinesque had to publish many of his discoveries himself, or place them in obscure European journals in languages most Americans could not read. But since those papers could also count as the first description of a species, it made things even messier.

As if all this were not sufficient cause for loathing, Say also had to endure the indignity of having his first book disparaged in print by Rafinesque. The 1817 notice for Say's *American Entomology* began promisingly: "The United States can at last boast of having a learned and enlightened Entomologist in Mr. Say." But then it degenerated into carping about the cost ($2 for 28 pages) and the handsome illustrations: "It would be well," Rafinesque sniffed, "if this style was left for the use of the princes and lords of Europe." It was like Ralph Waldo Emerson being taken to task by P. T. Barnum.

The criticism must have been especially galling because natural history books with engravings, often hand-colored at considerable expense, weren't just tea table fodder. They were guides for identifying species. And they were essential because the likes of Rafinesque repeatedly failed to provide type specimens, or illustrations, or even vague descriptions—leaving other naturalists to muddle through a maddening tangle of scientific names that had priority, but were otherwise meaningless. Illustrations would become especially important for Say, who spent the last nine years of his life in the Utopian community at New Harmony, Indiana. He lamented that this remote location gave him "not even a loophole through which to peep at the scientific world."

A Few Kind Words for Rafinesque

The sad thing is that Rafinesque had the makings of a great naturalist. A twentieth-century ichthyologist said he was "in many respects the most gifted man who ever stood in our ranks." A modern commentator calls him "beyond question the best field botanist of his time." Rafinesque scholar Charles Boewe credits him with "very acceptable descriptions of nine Kentucky bats," and with "no fewer than five bat genera," as well as with the prairie dog and many other mammals. Wading below the Ohio River Falls and feeling around with his bare feet in the mud, Rafinesque also brought up an astonishing number of freshwater mussel species. Other naturalists were irritated by what looked like another attack of his mania for splitting. But many of those mussels have turned out to be true species, representing an animal group where America has the richest biodiversity in the world.

Rafinesque also championed advanced scientific ideas. When other naturalists still regarded species as permanent and unchanging creations, he clearly saw that they could evolve. There may have been an element of wishful thinking in this insight: Criticized for naming three species of the lopseed plant where most botanists saw only one, Rafinesque replied in effect that if they weren't species yet, they soon would be, affording "a fine illustration of incipient species forming under our eyes in our woods." Charles Darwin knew a good line when he saw it, even if he deemed Rafinesque a "poor naturalist." So in the third edition of *On the Origin of Species*, he cited Rafinesque among his many antecedents, quoting his remark that "all species might have been varieties once, and many varieties are gradually becoming species." But Rafinesque was merely piggybacking on the work of more serious scholars, according to biographer Charles Boewe. In any case he had no clue about what caused evolution, Darwin's key insight.

Rafinesque was also an early proponent of the move away from Linnaeus's artificial system, which classified species based on a few obvious traits. Instead, Rafinesque adopted the natural system of classification introduced by the French naturalists Michel Adanson and Antoine Laurent de Jussieu, which would soon become standard. To designate a species under this system, a naturalist had to consider a variety of different traits, many of them obscure, and take the time to make detailed comparisons with similar species. But since detail was never a strong point for Rafinesque, his conversion to the new system was mostly theoretical. Thomas Say was also an early advocate of the natural system, and no doubt a far more effective one.

The new system was the beginning of a major change. Linnaeus had opened up the natural world to ordinary people because his system was so simple and easy. Unfortunately, the biological world was neither. The new system helped naturalists come to terms with the astounding complexities they were discovering, by enabling them to use all available traits for making a distinction between one species and another. But it would also inevitably balkanize the natural world into esoteric specialties presided over by narrowly focused experts. One old-fashioned botanist complained in midcentury that the natural system "takes botany from the multitude and confines it to the learned." The old "natural philosophers" were fading away and being supplanted by professionals. In 1834, probably at a meeting of the British Association for the Advancement of Science, someone took the model of such words as "artist," "economist," and "atheist" and introduced the new name "scientist." For all their differences, both Rafinesque and Say would ultimately fall victim to this change.

Chapter Eight

"IF THEY LOST THEIR SKALPS"

> The object of the expedition is to acquire as
> thorough and accurate knowledge as may be
> practicable of a portion of our country which is
> daily becoming more interesting.
>
> —SECRETARY OF WAR J. C. CALHOUN

SHORTLY BEFORE the Long Expedition got under way, Charles Willson
Peale painted portraits of Maj. Stephen Harriman Long and other
members of the party, resplendent in their high-collared uniforms. He
also had the strategic sense to paint President Monroe and Secretary
of War Calhoun, and persuaded them to deposit specimens from the
expedition in his museum. Thomas Say and the assistant naturalist,
Peale's youngest son Titian, sat for portraits, too. "If they lost their
skalps," the elder Peale remarked, with a father's false bravado, "their
friends would be glad to have their portraits."

In fact, the Long Expedition would enjoy remarkably friendly rela-
tions with the Indians, apart from a few minor incidents. From the
Ohio River, the *Western Engineer* pushed on up the Mississippi and into
the Missouri, before stopping for the winter just north of what is now
Omaha, Nebraska. The following spring, in June 1820, the explorers
would continue on horseback and foot along the Platte River to the
Rocky Mountains, returning to the East later that year by way of the
Arkansas River and New Orleans.

Say would prove himself a careful observer of Indian customs, and often surprisingly modern. He recounted without judgment the worldly attitude some Indian husbands displayed toward infidelity by their wives while they were absent hunting. (An Oto chief admitted, "I am not so silly as to believe that a woman would reject a timely offer. Even this squaw of mine, who sits by my side.") The official account of the expedition also reported that, in some tribes, sodomy was "not uncommonly committed; many of the subjects of it are publicly known, and do not appear to be despised, or to excite disgust."

On the other hand, Say was plainly disgusted by white newcomers to the West: "It is common for hunters to attack large herds of [bison], and having slaughtered as many as they are able, from mere wantonness and love of this barbarous sport, to leave the carcasses to be devoured by the wolves and birds of prey; thousands are slaughtered yearly, of which no part is saved except the tongues. This inconsiderate and cruel practice is undoubtedly the principal reason why the bison flies so far and so soon from the neighbourhood of our frontier settlements."

This was 30 years before the first vague stirrings of the American conservation movement. Naturalists elsewhere were earnestly praying for the wilderness to be tamed and Christianized. But Say was already advocating "some law for the preservation of game . . . rigidly enforced" to prevent "the wanton destruction of these valuable animals, by the white hunters."

"The Great American Desert"

Along with equipment failures, illness, and hunger, the major hazard faced by members of the Long Expedition seems to have been their own company. Maj. Thomas Biddle Jr., from a prominent Philadelphia family, quit at the end of the first season because he thought the expedition "chimerical & impossible"— and also because he was "mortified" by an episode in which he and the naturalists under his protection had been robbed of their horses by Pawnee renegades. Biddle and Long held the same rank, so his subordinate status may also have rankled. He described Long as "I daresay respectable in his own department" (probably meaning steamboats and other machinery) but "entirely unqualified for an expedition of this description." Biddle would die years later in a duel with a Congressman.

His replacement, Capt. John R. Bell was also resentful, and with a histrionic manner fit for bad melodrama. At one point, he snarled at Long, "We are out of the U.S., enforce your orders if you can," and "By God, we both wear pistols." He was later court-martialed in an unrelated case. Geologist Edwin James, also a second-year replacement, complained about the haphazard salary arrangements, accusing Long of "contemptible knavery" for trying to pay wages in a Kentucky currency "now at 37½ cents below par and on the decline." Congress had cut funding because of failures in the military, but not the scientific, expeditions. Now Long was under orders to explore a huge swath of the American West while also bringing the expedition to "as speedy a termination as possible." Thus the first scientific exploration of the American West often proceeded at a half-starved trot, routinely covering 20 miles a day. "I have been allowed neither time to examine and collect nor means to transport plants or minerals," James fumed, adding that he had also gone for weeks without bread or salt, and eaten "tainted horse flesh, owls, hawks, prairie dogs, and other uncleanly things," hardships he seemed to think Long imposed for the subsequent "amusement of the publick."

Caught between the "detestable parsimony" of the government, as the *North American Review* soon put it, and the grumbling of malcontent subordinates, Long seems to have performed his mission with skill and even grace. His critics charged that he failed at two of his assigned tasks, to discover the sources of the Red and Platte rivers. But their East Coast perspective clearly led them to underestimate the vast scale of the Western landscape. Historians would later condemn the expedition for branding the territory east of the Rockies "the Great American Desert," a label that stuck and discouraged settlement into the 1880s. But the Long Expedition provided a good basic map of the American West and a remarkably intelligent guide to its human inhabitants and wildlife. Long also led his fractious party of more than 20 men over thousands of miles of wilderness, through the territories of numerous Indian nations, thickly populated with grizzly bears and other dangerous animals, with only a single death: Botanist William Baldwin had thought travel in open country would help his tuberculosis. Instead, it soon killed him.

The naturalists Say and Peale also seem to have displayed grace under extraordinarily difficult circumstances, especially near the end when they suffered what Say called "the greatest of all privations." As they were coming down the Arkansas River in August 1820, the explorers found themselves in a race for survival across an endless succession of "bends & hills & hollows & ravines," in Capt. Bell's words, driven forward by swarms of green-headed flies and held back by the constant danger of rattlesnakes. The heat was so debilitating that one of the expedition's dogs threw himself down in front of the horses to force a halt. The explorers were reduced to a single meal a day, consisting at one point of "a few mouldy biscuit crumbs, boiled in a large quantity of water, with the nutritious addition of some grease." A skunk got tossed into the pot when one of the hunters got lucky and it "tasted skunkish enough." Then, on August 31, three soldiers deserted in the night, taking the best horses with them.

Worse, they stole saddlebags containing five notebooks Say had filled over that season's 2000 miles of travel. In his journal, Say wrote, "All these, being utterly useless to the wretches who now possessed them, were probably thrown away upon the ocean of prairie, and consequently the labour of months were consigned to oblivion by these uneducated vandals." The deserters were never heard from again. The rest of the party reached the relative safety and comfort of Fort Smith, on what is now the Oklahoma-Arkansas border, just nine days later.

Back in Philadelphia, after a bout of malaria, Say reconstructed his notes from memory and from the specimens that had survived. Thus as a result of "the trials, difficulties and dangers" of the Long Expedition, he was able to introduce America to some of its most iconic animals, among them the coyote, the swift fox, the great plains wolf, the lazuli bunting, and the orange-crowned larkspur. In all, Say brought back 13 new mammals, 13 birds, 12 reptiles and amphibians, and 4 arachnids and crustaceans, according to *The Natural History of the Long Expedition* by Howard Ensign Evans. Insects, too, of course—more than 150 species. At one point, Say recounted how he was seated with a Kansa chieftain, "in the presence of several hundred of his people assembled to view the arms, equipment, and appearance of the party," when a darkling beetle came scurrying out from among the feet of the crowd. Diplomatic dignity wrestled briefly with the passion for species. Then Say went plunging after the beetle and impaled it on a pin, for which the astonished Kansa admiringly dubbed him a medicine man. The beetle is now known as *Eleodes suturalis* Say.

Old Disharmony

Say would return to the Ohio River Valley in 1826, joining other scholars on the "boatload of knowledge" bound for the utopian community at New Harmony, Indiana. On board, he met Lucy Sistare, the daughter of a Spanish sea captain who had settled in New London, Connecticut. She would become the illustrator for Say's *American Conchology*, a landmark in the study of shells. She also became his wife and, apart from an expedition Say made to Mexico, they remained together at New Harmony, happily married, until his death in 1834.

The little community on the banks of the Wabash was hardly an ideal environment, keeping them isolated from the scientific mainstream and often without adequate tools or books for their work. The social atmosphere was Old Disharmony. A local schoolteacher directed the full blast of snout-faced early American anti-intellectualism at the naturalists and their specimens: "tell me what benefit will arise from their work to the present and even the future generations," she demanded in a letter. "That is the case with all Scientific people. Their knowledge is not only useless (because there is no application to it) but hurtful; it carries the mind astray, in fact it is false knowledge."

Say would also struggle for the rest of his life with that genuine font of false knowledge, Constantine Rafinesque. As with the mudpuppy, it would turn out that several discoveries from the Long Expedition had already been described sketchily, but first, by Rafinesque, among them the prairie rattlesnake and the pocket gopher. Say would have been bothered less by the loss of his discoveries than by having inadvertently cluttered the scientific landscape with synonyms. He believed passionately in the taxonomic system. So he was "tireless in ferreting out all previous references to particular species," writes biographer Patricia Tyson Stroud, "in order not to name as new those that had been previously described in published articles."

At one point, Say wrote in distress to a friend that he couldn't obtain a key essay by Rafinesque describing new species of freshwater mussels: "I made a narrow escape with [*Unio*] *monodonta*, & great danger is to be apprehended in publishing any species as new without that essay." When it came to bad taxonomy, Say's language betrayed a sense of personal peril that never seems to have troubled him in the presence of rattlesnakes, wolves, "savage" Indians, and other conventional hazards.

Rafinesque's work would prove particularly exasperating in the case of the mule deer. Near the end of the Long Expedition, Say and Peale had offered a reward for the perfect specimen of mule deer, which they thought was "without doubt, a new species." But when one of the expedition's hunters finally carried a trophy buck into camp one evening after dark, the rest of the party was too famished to have much patience with science. Peale had to set up the deer by the light of the fire to make a quick sketch, with half-starved soldiers sharpening their knives nearby. Say managed to salvage the skin and the head, with its large, handsome antlers and characteristic mule ears, to be preserved and carried several thousand miles home to Philadelphia.

There it later turned out that Rafinesque had already named the species without ever actually having laid eyes on it. His source was a published account by an alleged captive of the Sioux that is now considered a hoax. Decades later, even more bizarrely, a taxonomic revision would move all North American deer, including the mule deer, into the genus *Odocoileus*, originally proposed by Rafinesque based on what he took to be the fossil tooth of a dwarf oxen. As for Say, the mule deer

This mule deer seemed like a precious discovery, but quickly became dinner for the starving explorers.

specimen he had gone to such trouble to collect and carry home was ravaged by insects. But back at the Philadelphia Museum, Titian Peale was able to work some taxidermy magic, along with a little unintended poetic justice: When the head of the mule deer (*Odocoileus hemionus* Rafinesque) went on display, it was under the foot of a coyote (*Canis latrans* Say).

Lost in the Wilderness

The idea that one man could provide the first scientific description for species from grizzly bears and songbirds to beetles was already fading by the 1830s. Field naturalists found themselves increasingly at odds with "closet naturalists," specialists who stayed home and presided over the specimens, books, and journals needed to identify a species correctly. The closet naturalists generally controlled the money, in the

In Philadelphia, the closet naturalist George Ord seemed to spend most of his time under-cutting his fellow naturalists in the field.

form of museum jobs or personal wealth, to support their own work. They wielded their institutional influence to promulgate new standards aimed at shutting out the Rafinesques of the world. They also sometimes enhanced their own professional standing as scientists by depicting field naturalists as eccentrics wandering in the bush.

The field naturalists had little means of defending themselves. Their travels isolated them from the latest intellectual fashions and also from the cliques being formed at places like the Academy of Natural Sciences of Philadelphia. "Science here is in the hands of a few Aristocrats, who wish to be alone before the public, and keep off all competitors by management," Rafinesque complained. "I keep aloof from all their strifes and petty intrigues." But his idea of keeping aloof involved raging at his critics in print as a "host of revilers and croaking frogs . . . the most contemptible sophisters, aristarchs and moles . . . oh, Wisacres and silly Sanchos."

George Ord, from a prosperous rope manufacturing family, was one such "aristarch." He had sailed with Say and Peale on an 1818 expedition to Florida, and as a member of the Academy he seemed in the early years to encourage their work. At one point, after arranging the purchase of a valuable book for the Academy's library, he exulted, "Say's heart will throb with delight at the sight of it." He also contributed important work to finish *American Ornithology* after Alexander Wilson's death. But as his standing in the Academy rose, his malevolent nature slowly unfolded and poisoned everything around him.

Ord is best known for his hostility to Audubon, who had finally published his *Birds of America* in England, beginning in 1827. It contained 435 hand-colored plates, depicting more than 500 species in spectacularly lifelike poses, and at life size. (Perhaps, in the anti-Buffonian spirit, Audubon wanted to make American wildlife look big.) One French critic hailed it as "a real and palpable vision of the New World." With his rough clothes, shoulder-length hair, and gregarious manner, Audubon himself became an instantaneous celebrity in Europe, as a backwoods New World type. James Fenimore Cooper had just made an international hit with his new novel *The Last of the Mohicans*, and Audubon, a shrewd promoter, capitalized on the sensation by presenting himself as a real-life version of Cooper's hero, the capable frontiersman Leatherstocking.

Ord was furious at him not just because Alexander Wilson's *American Ornithology* all but vanished beneath Audubon's colorful dust, but also because Audubon's writings "basely traduced the character of . . . the illustrious Alexander Wilson" in recounting Wilson's unhappy visit to Louisville. Ord complained that Audubon's "wretched collection of engraved birds" looked as if it could have been done better with a pitchfork. He also noted that Audubon had been blackballed for membership in the Academy, but neglected to mention that he himself had done the blackballing.

Ord became almost as angry with Say for having had the temerity to criticize Ord's taxonomy in print, and also for having abandoned the Academy by moving to New Harmony. He made it difficult for Say to get his papers published in the very journal Say used to edit. After Say's death, Ord also delivered a slanderous eulogy that would permanently undermine Say's reputation. He called Say's *American Conchology* "a disgrace" and described Say as suffering from "deficiency in elementary learning" and want of "technical precision." The Academy found his remarks so offensive that Ord had to resign the vice-presidency. But when Lucy Say shipped her husband's insect collections back to Philadelphia after his death, the closet naturalists of the Academy left them to rot in their packing boxes. And in time, Ord regained his status sufficiently to become the Academy's president.

Titian Peale would also come under Ord's dark shadow. From 1838 to 1842, he served as a naturalist with the U.S. Exploring Expedition. It meant spending four years at sea, separated from his family, in some of the most hostile environments on Earth. The expedition mapped vast areas of the Pacific and discovered the frozen continent of Antarctica. It also brought back more than 60,000 plant and animal specimens, which would become the basis for a new national museum, the Smithsonian Institution. But on his return, Peale was reduced to complaining to Ord that the Academy staff wouldn't even grant him access to basic reference materials he needed to describe the specimens he had collected. Ord replied disingenuously that there must be a "plot" afoot to defraud field naturalists of the fruit of their labors. No doubt he was shocked.

As for Rafinesque, he would die penniless, alone, and probably insane in 1840. One of the last books he read was Darwin's *Journal of*

Researches (better known now as *The Voyage of the Beagle*) and in its pages he found and described, sight unseen, a new genus of fungus. This time, fortunately, professional standards prevailed. A British researcher had already described the species, based on the actual specimen, and that name, *Cyttaria darwinii*, had priority. At the sale of his estate, Rafinesque's material goods amounted to little more than "trash," disposed of by the bushel "for trifling sums." His specimens "were absolutely worthless."

City of the Dead

There is a sorry postscript to this tale. In 1825, James Fenimore Cooper was working to complete the third novel in his Leatherstocking series, a sequel to his best-selling *The Last of the Mohicans*. For *The Prairie*, Cooper created a new character named Obed Bat, or as he liked to hear himself called, on the Latinate model of Carolus Linnaeus, Dr. Battius. It was a comic caricature of the naturalist, as one scholar has put it, "absently endangering himself and others in an addled quest for new species, and spouting unintelligible Latin phrases from the new Linnaean taxonomic system." It was a caricature, in fact, of Constantine Rafinesque.

At one point, having spotted some strange monster bounding across the countryside, Battius describes it aloud as a new species: "*Quadruped*; seen by star-light, and by the aid of a pocket-lamp, in the prairie of North America—see Journal for latitude and meridian. *Genus*—unknown; therefore named after the discoverer, and from the happy coincidence of having been seen in the evening—*Vespertilio Horribilis, Americanus . . . Greatest length*, eleven feet; *height*, six feet . . . *horns*, elongated, diverging, and formidable; *color*, plumbeous-ashy with fiery spots; *voice*, sonorous, martial, and appalling . . . an animal . . . likely to dispute with the lion his title to be called the king of beasts." A young woman of no scientific background points out to him that the "new species" he has seen is in fact his own donkey.

Vespertilio was the genus into which Rafinesque had put many of the bats he had discovered, possibly including one of those Audubon

Audubon cut a swaggering backwoods figure on the international stage.

caught for him that night of the smashed violin in Henderson, Kentucky. Audubon had been back East in 1824, about the time Cooper was writing *The Prairie*. As a natural storyteller, Audubon may well have dined out on his comical encounter with the "eccentric naturalist." In any case, it seems likely that the story made the rounds and reached Cooper's ears as he was inventing Dr. Battius.

But another naturalist also informed this caricature of the naturalist. Cooper's background reading for *The Prairie* included the official *Account of an Expedition from Pittsburgh to the Rocky Mountains*, which the Long Expedition had published in 1823, to considerable popular interest. There, in keeping with his "scientifick" marching orders, Thomas Say had included many of the detailed descriptions he considered essential to good natural history. Of the coyote, *Canis latrans*, for instance, Say wrote, "*head* between the ears intermixed with gray, and dull cinnamon, hairs dusky plumbeous at base . . . *tail* bushy, fusiform, straight, varied with gray and cinnamon, a spot near the base above, and tip black."

It was easy pickings for Cooper's satire.

So while Audubon was capitalizing on Cooper's work by presenting himself as a real-life Leatherstocking, the equally capable field naturalist Thomas Say was being forgotten. Or worse. To what must have been Say's everlasting horror, Cooper's fiction had wrapped him up in the same shroud with his dismal opposite, Constantine Rafinesque, to moulder together for eternity.

Chapter Nine

THE BURDEN OF SPECIMENS

Natural History is far too much a science of dead
things; a *necrology*. It is mainly conversant with
dry skins furred or feathered, blackened, shriveled, and hay-stuffed; with objects . . . impaled
on pins, and arranged in rows in cork drawers.

—PHILIP HENRY GOSSE

W HEN HE WAS KEEPER of zoology at the British Museum around 1820,
William Elford Leach made a reputation "for certain peculiar
eccentricities and crotchets," as a history of the museum later put it,
"mixed up in close union with undoubted learning and skill." One such
crotchet was a tendency, common among naturalists, "to undervalue
the achievements of past days, and to exaggerate those of the day that
is passing." In Leach's case, this disdain took a particularly incendiary
form.

The British Museum had been founded in 1753, largely based on
the collections of the virtuoso par excellence, Sir Hans Sloane. Sloane, a
London physician and naturalist, had acquired thousands of specimens,
from tiny invertebrates on up to elephant tusks, often describing them
in meticulous scientific detail. In death, he gave his carefully chosen
trustees the task of making the nation take charge of the collection—
and pay a high price for the privilege. "He valued it at fourscore thousand," Horace Walpole, who was one of the trustees, carped to a friend,

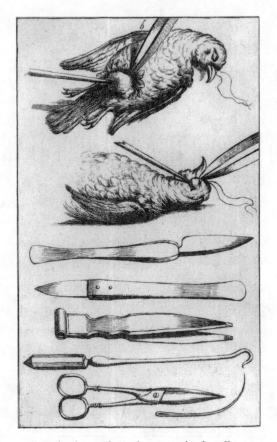

Tools of an eighteenth-century bird-stuffer.

"and so would anybody who loves hippopotamuses, sharks with one ear, and spiders as big as geese! It is a rent-charge to keep the foetuses in spirit! You may believe that those who think money the most valuable of all curiosities will not be purchasers." But the trustees shrewdly threatened to sell the collection abroad, and nature mattered enough to national pride that Parliament soon agreed to a £20,000 purchase price, establishing the museum in Montagu House, an elegant seventeenth-century mansion in the heart of London. Sloane's bequest would in time give rise not just to the British Museum, but to the British Library and the Natural History Museum, London. It was a handsome legacy, though perhaps secondary to the result of a collecting trip in Jamaica, on which he also invented chocolate milk.

But time is harsh and memory mostly fickle. In the first part of the nineteenth century, 70 years after Sloane's death, Bloomsbury was one of London's more fashionable neighborhoods. The Bloomsburians liked to promenade in the handsome gardens on the north side of Montagu House. And William Leach seems to have taken equal pleasure in annoying them. "He despised the taxidermy of Sir Hans Sloane's age, and made periodical bonfires of Sloanian specimens," Edward Edwards wrote, in his 1870 *Lives of the Founders of the British Museum.* "These he was wont to call his 'cremations.'" Unfortunately for his neighbors, "the attraction of the terraces and the fragrance of the shrubberies were sadly lessened when a pungent odour of burning snakes was their accompaniment."

The taxidermy of the Sloane bequest was probably no worse than in other natural history collections of its day. Rot, insects, careless handling, and other hazards were the common fate of specimens everywhere. Even as Parliament was voting to purchase new specimens in 1823, a writer complained in the *Edinburgh Review* that those the British Museum already owned were "mouldering or blackening in the crypts of Montagu House, the tomb or charnel-house of unknown treasures," where moths and beetles were "busily employed amid the splendours of exotic plumage, or roaring through the fur of animals."

The species seekers often talked as if they could confer immortality on the specimens they collected. But the discovery of new species raced far ahead of the technologies for preserving them. Desperate questions of flesh and corruption invariably preoccupied naturalists—how to kill animals without destroying their beauty; how to carry eggs and other fragile specimens while trudging through jungles and deserts; how to transport them safely aboard damp, crowded, sea-tossed ships; and, once home, how to keep them intact, if not forever, then at least long enough to study them and elicit their secrets, or even more wishfully, to bring them back to life in some characteristic pose from the moment of flight, predation, or parenting.

Like the naturalists themselves, the ordinary public wanted to fondle and possess these strangely beautiful creatures. The species seekers were simultaneously driving this long-suppressed appetite for nature into frenzy and, until they could figure out how to preserve their treasures, also inevitably frustrating it. At his Philadelphia Museum, Charles Willson Peale had set out in the 1780s to bring into "one view a world in miniature," arranging birds and beasts in a naturalistic landscape, with trees, shrubs, and a pond. He put his artistic family to work painting scenic backdrops. It was the forerunner of the museum diorama, and visitors were plainly beguiled by these "most romantic and amusing" displays. They idly fingered the birds even while reading a sign that warned: DO NOT TOUCH THE BIRDS. THEY ARE COVERED WITH ARSNIC POISON.

The Art of Killing

Many naturalists were ambivalent about killing wildlife. In the autumn of 1807, for instance, the residents of Philadelphia were slaughtering and eating robins in such vast numbers that ornithologist Alexander Wilson resorted to conservation by false rumor. He planted a story in local newspapers advising cautious readers that the pokeberries the birds were then gorging on made their flesh unwholesome, and he was "righteously elated," according to one historian, when his story knocked the bottom out of the robin market.

But killing was also essential to the discovery of species. Most field guides then assumed that naturalists already knew how to handle a gun. Matching the shot to the size of the specimen was critical for collecting specimens without blasting a hummingbird, say, into a cloud of mist. One common technique was to carry a double-barreled gun, loaded with large shot on one side and dust on the other, according to the science historian Anne Larsen Hollerbach, "so that the zoologist could shoot whatever came into his sights first, whether buzzard or bullfinch." All the better if lightweight shot merely stunned the bird. Then the naturalist could quickly kill it "by introducing the thumb and fore-finger under the wings and pressing with as much force as the size of the bird requires," *The Naturalist's Pocket-Book, or Tourist's Compan-*

ion advised in 1818. Thus "the feathers are not soiled and the animal is immediately freed from pain."

For tourists of a certain persuasion, the same guidebook also advised "steeping seeds, beries, or grain, in some deleterious infusion" so that birds consuming this feed would "only be stupefied," to be finished off gently by hand. (John James Audubon wrote that hummingbirds "are easily caught by pouring sweetened wine in the [chalices] of flowers—they fall intoxicated.") Baiting a likely spot could also bring birds in range of "common folding or clap nets" or foot snares "formed of stout horse hair." For bats, the oddly ritualistic method was to walk along in the evening with one person carrying a lantern to attract insects, and two others following behind holding up a large net to entangle the bats eagerly pursuing the insects, a technique called "bat fowling."

This new variety of hunting, for species far removed from traditional game, and with an emphasis on avoiding visible damage, often necessitated a fatal intimacy between man and animal. When Audubon obtained a live Golden Eagle in 1833, he spent three days observing its behavior. In the field, Audubon could bring down a hundred birds a day without qualm, but killing this bird to get the pose he required turned into a prolonged nightmare. A friend had recommended suffocating the eagle with charcoal fumes, and Audubon duly covered the cage with blankets and placed it in a closet over a pan of burning coals. He listened "for *hours*" (the italics are his) expecting "every moment to hear him fall down from his perch." But when he peeked inside through the "mass of suffocating fumes," the eagle was still sitting there "with his bright, unflinching eye turned toward me." The air in the closet was "insupportable" to Audubon and his son, and the adjoining rooms were beginning to feel "unpleasant," when Audubon finally gave up and went to bed at midnight.

Incredibly, he tried again in the morning, adding sulphur to the toxic mix. But the bird "continued to stand erect, and look defiance at us whenever we approached his post of martyrdom." In the end, resorting to a technique "always used as a last expedient," Audubon "thrust

a long pointed piece of steel through his heart, when my proud prisoner instantly fell dead, without even ruffling a feather." After working through the night to draw the pose, Audubon spent the next few
days laid flat with "a spasmodic affection," though whether from carbon
monoxide poisoning or emotional turmoil he does not say.

Killing insects could also be surprisingly difficult and disturbing,
according to Hollerbach; "the familiar stomp and swat techniques
would not do." It was easy enough to impale an insect on a pin set
in cork. But some insects could wriggle for days like that before they
finally died. One naturalist was shocked to see that a dragonfly he had
just pinned through the thorax still clutched a struggling fly in its forelegs and proceeded to eat it. When the humanitarian and antislavery
activist John Coakley Lettsom published a field guide for naturalists in
1799, he was moved to quote Shakespeare: "And the poor beetle, that
we tread upon, / In corporal sufferance finds a pang as great / as when a
giant dies." Then he went on to recommend techniques for killing each
type of insect as quickly as possible: For beetles, pinning and dipping in
hot water. For hemipterans like aphids or planthoppers, a combination
of pinning and anointing with "a drop of the etherial oil of turpentine
applied to the head." For small moths and butterflies, heat applied to a
jar to make the sulfur within give off fumes—"the insect instantly dies,
without injuring its colours or plumage."

For the larger moths, Lettsom, who was a physician, devised a 6-inch-
long steel killing needle with an ivory handle. First, he squeezed the
thorax to "deprive the insect of motion, and probably of sensation too for
some minutes." Then he passed the needle through the thorax, mounted
a thin brass or tin shield on the far side, and placed the needlepoint in
the flame of a candle. The heat passing through the needle into the core
of the insect killed it in 30 seconds, while the metal baffle prevented
scorching of its plumage.

Entomologists experimented with the wide-mouthed "killing jar"
through the first half of the nineteenth century. One such experiment, using ammonia in the form of smelling salts, proved unsatisfac-
tory, according to Hollerbach, "because it bleached bright colors and

because the jars containing the poison had an annoying tendency to explode when exposed to hot sun." Finally in the late 1840s, entomologists settled on a layer of plaster soaked with potassium cyanide (later ether) in the bottom of the jar as the best way to dispatch most types of insects with minimal pain or damage.

Field Prep

"Never put away a bird unlabelled, not even for an hour," a nineteenth century field guide advised, "you may forget it or die." Attaching a label to the leg, with the chicken-scratched details of place, date and habitat, was the only way to make sense of a specimen months later, especially since naturalists then sometimes took dozens of specimens in a day. Collecting multiple individuals within a single species was a way to compile a thorough scientific record of normal variation—the little differences between juveniles and adults, or males and females, or separate populations on neighboring islands. Even the ordinary differences among individuals in the same population could be crucial for sorting out where one species ended and another began. Taking multiple specimens also mattered to some naturalists because selling duplicates to collectors back home was their only means of support.

But specimens needed to be preserved almost immediately to avoid what one ornithologist called the "Mortification to see them every Day destroyed by ravenous Insects." While an expert might prepare 10 bird skins an hour, 4 was a more likely average, including all the business of skinning, cleaning, plugging, and packing. So working late into the night after a long day in the field was routine.

Ideally, if he was staying in one place long enough, the naturalist could train local help and set up a field station. "In the room are three large tables, one of them against the window, at which a negro youth is sitting," the British naturalist Philip Henry Gosse wrote, on a trip to Jamaica. "Before him lie half a dozen *birds*, one of which he is skinning; beside him lie scissors, knives, nippers, forceps, a pepper-box of pounded chalk, a jar of arsenical soap, needles and thread, cotton-wool, and other apparatus, with several cones of paper ready to drop each skin into, when finished. Across the room are strung lines in various directions, from which are suspended some hundreds of similar paper cones, each tenanted by a birdskin; they are thus placed in order that they may dry out of the reach of rats, which nevertheless sometimes manage ingeniously to scramble along the slender lines and gnaw the feet and wingtips of the specimens."

At another table, Gosse wrote, someone was working with boiling water to kill the animals in land shells, extract them with a needle, and clean out the residue with a toothbrush. Someone else pressed botanical specimens between sheets of coarse brown paper and weighed down each stack with stones. Meanwhile, the naturalist himself might be dipping beetles in boiling water, making field notes, or "taking sketches of forms and colours that death would destroy." Gosse presented it as an idyllic scene, with freshly gathered orchids heaped on the floor, and hummingbirds flitting about the corners of the room.

But the pace could also be lethal. Late at night, having finally completed his labeling and packing, the birder "throws himself on his bed if he has one," the Australian ornithologist Samuel White wrote, during an expedition to the Aru Islands, "to rise again at dawn and begin afresh with not time to wash the dirt from his hands and face, scarcely time to feed himself. This goes on for weeks or months till frequently hard work and scanty living brings on some climatic fever when his labours suddenly cease for the time." A few weeks later, at the age of 45, White himself died in just that fashion, of pneumonia or fever, with arsenic-damaged hands from his late-night work.

Preparing specimens in the field was, however, merely a step in the process of discovering a species. The field naturalist also needed to get his discoveries home. One guidebook recommended specially built protective boxes, with cork lining soaked in a solution of corrosive sublimate (mercury chloride), turpentine, and "camphorated spirit of wine." Sealing the joints with paper strips and a sizing made with corrosive sublimate or arsenic was also customary. "If the least opening presents itself," one field guide warned, ants and other "rapacious vermin" will commence "their work of devastation. A box of 200 or

300 insects will be destroyed in this way during one night, and even before some specimens are quite dead." To kill the moths that often hatched even within sealed boxes, one guidebook recommended keeping preserved skins in a glass-topped box, then setting it in front of a fire for eight hours so heat would cause moths to flutter about and die. If being put on ship, the whole box also needed to be covered with a thick coat of oil paint.

The burden of specimens could quickly become overwhelming. "Our progress was often held up by having to drag after us for five and six months at a time from twelve to twenty" mules loaded with specimen boxes, Alexander von Humboldt remarked during his travels in South America. At times he had to throw away old specimens to make room for new ones. Prudent collectors took every opportunity to ship their treasures back home, carefully distributing them and sending duplicates as insurance against the usual vicissitudes of sea travel.

But that wasn't always possible. The French ornithologist and botanist Jules Verreaux, for instance, spent 13 years collecting species in Africa, finally loading his specimens aboard the *Lucullus* and sailing for home in 1838. Within sight of France, off the coast of La Rochelle, the ship got caught in a storm and sank. Verreaux managed to swim ashore, empty-handed, but also the sole survivor. Evidently undaunted, he set out a few years later on a new expedition, this time to the Pacific.

ARSENIC AND IMMORTALITY

> When a visitor shall gaze upon it afterwards, he
> will exclaim, "That animal is alive!"
>
> —CHARLES WATERTON

A TROUBLING QUESTION for every collector was what to do with specimens if you actually managed to get them home—how to unpack
them, preserve and display them, and recapture a hint of their living selves. Some of the most innovative answers came from Charles
Waterton, now considered "the father of modern scientific and artistic
taxidermy." Waterton's favorite preservative was corrosive sublimate—
a variant on the same mercury that caused madness in hatters and
other industrial users. This preference may explain his reputation, preeminent even by the colorful standards of nineteenth-century natural
history, as an eccentric—or as one biographer put it, a "unique dodo."

Waterton was the fourteenth generation of a family of landed gentry to live at Walton Hall, on an island in a lake in West Yorkshire.
Among other genuine accomplishments, he turned the 260-acre estate
into what is widely regarded as the world's first nature preserve. He
also launched the earliest successful action to stop pollution, managing
by 1853 to drive out a soap manufacturer who had set up shop next
door. (The manufacturer eventually extracted his vengeance, buying
Walton Hall out from under the fading gentry when Waterton's son
went bust.)

Charles Waterton's adventures in South America, including a ride on a caiman, thrilled many young readers who later became naturalists themselves.

Waterton's own way of life was decidedly austere. He made four expeditions to South America and urged his imitators to "leave behind you your high-seasoned dishes, your wines, and your delicacies; carry nothing but what is necessary for your own comfort." At the age of 47, he married a 17-year-old girl, part Arawak Indian, part Scottish royalty, who had been his ward from childhood. But after she died in childbirth a year later, he lived like a monk even at home, sleeping on a floor mat with a wooden block for a pillow and rising at 3:30 a.m. He apparently worked in the same room where he slept, amid the perfume of corrosive sublimate and dead flesh. "Dressing table and taxidermy bench were one and the same," writes biographer Brian Edginton. "The eviscerated baboon hanging from the bare rafters served as inducement to profitable contemplation and sweet dreams." His only extravagance was the family's Catholic religion, and the chapel on the first floor at Walton Hall was lavish.

Waterton never pretended to be a scientist and he regarded Latinate species names as a nuisance. As a result, it wasn't always clear just what animal he was talking about, though his firsthand descriptions of animal behaviors were otherwise vivid. His 1825 book *Wanderings in South America* was full of colorful adventures—wrestling with boas and riding on a caiman to capture it without damaging the skin. Skeptics dismissed such tales as fiction. But Waterton inspired many budding naturalists, including Charles Darwin and Alfred Russel Wallace. His travels into little-known regions were motivated by the search for perfect specimens. So where other naturalists brought back hundreds or thousands of discoveries, Waterton somehow won credit for only a single species new to science, the long-tailed woodnymph hummingbird, *Thalurania watertonii*.

Waterton's real interest had nothing to do with taxonomy, and everything to do with taxidermy and the thin line between life and death. Among other achievements, he discovered curare, the plant-derived toxin used by some South American tribes to poison their arrows. He also tested how rapidly it causes asphyxiation by paralysis in a barnyard fowl (5 minutes to death), a three-toed sloth (11 minutes), and a 900-pound ox (25 minutes). But his real purpose became evident when he tried the drug back in England on an ass named Wouralia (from the Indian name for the drug). It seemed to die in 10 minutes. Then Waterton revived it by artificial respiration using bellows to inflate the lungs through an incision in the windpipe. The subtext of resurrection ran through all his work.

Waterton complained that other taxidermists took what once was a bird and made it look as if "stretched, stuffed, stiffened, and wired by the hand of a common clown." His own method of skinning a specimen in the field was painstaking almost to the point of being sacramental: "find out the shot holes, by dividing the feathers with your fingers and blowing on them, and then, with a penknife or the leaf of a tree, carefully remove the clotted blood, and put a little cotton on the hole . . . wash the part in water without soap, and keep gently agitating the feathers, with your fingers, till they are quite dry."

To bring an animal back to life, Waterton also devoted hours to watching how it moved in its native habitat: "You must pay close attention to the form and attitude of the bird, and know exactly the proportion each curve, or extension, or contraction, or expansion of any particular part bears to the rest of the body." Thus, for instance, "your sparrow will retain its wonted pertness, by means of placing his tail a little elevated, and giving a moderate arch to the neck."

Waterton was confident that a specimen he had prepared would "retain its pristine form and colours for years after the hand that stuffed it has mouldered into dust." And in fact, his taxidermy remains in excellent condition today, on display at the Wakefield Museum, not far from Walton Hall, which also survives, as a hotel. He was confident enough in the preservative value of corrosive sublimate that he applied it not just to his specimens, but also, according to biographer Clare Lloyd, to "top hats, coats, pants, the interior of his carriage, and the furs and ostrich feathers on the bonnets of his sisters-in-law." It did not kill him. He remained active into his eighties, still climbing the trees on the grounds to count the eggs in bird nests. His peculiar—indeed, mercurial—wit also survived, and he liked to amuse friends by scratching the back of his head with his big toe, or by hiding under a table and lunging out to bite unwary visitors on the leg.

His quirkiness also turned up in the taxidermy specimens he used to express his many prejudices, and he earned permanent notoriety among scientists as a result. His masterpiece was the "Nondescript," a distinctly human bust mounted on a wooden stand, frowning, with large brown eyes, anxious upraised brow, and dark skin fringed with reddish hair. He displayed it on the grand staircase at Walton Hall, and also published it as the frontispiece for *Wanderings in South America*. The rumor went round that he had "shot a native human being" and set it up as an ordinary natural history specimen.

In fact, the "Nondescript" was a howler monkey skillfully manipulated to give "the brute a human formation of face, and to this face an expression of intellectuality." Waterton was doing something that, in the 1820s, would have been more disturbing than either shooting natives or making a mockery of natural science. "Are you not aware," he told an inquisitive friend, "that it is the present fashionable theory that monkeys are shortly to step into our shoes? I have merely foreshadowed future ages or signalised the coming event."

Kind Friends

Other naturalists remained mired in the past, and paralyzed by the generally dismal state of preservation. The British naturalist William J. Burchell had become a natural history collector in the aftermath of an unhappy love affair. He was working as a teacher on the island of St. Helena in 1805 and sent for his fiancée to join him from England. Instead, she married the captain of the ship that brought her out to him. Their presence on the island, a 5-by-10 mile volcanic rock in the middle of the South Atlantic, must have been torture for Burchell. He remained a bachelor for the rest of his life, and threw himself into his work.

During his travels in southern Africa from 1811 to 1815, he collected 265 bird species, many of them new to science. He traveled in a canvas-topped ox-wagon packed with collecting gear and was aided by expert Khoi Khoi hunters. Unlike many naturalists then, he credited them by name, though with a note of colonial possessiveness: Speelman, he wrote, "added to my ornithological collection more than any other of my Hottentots. Juli . . . was little inferior to him, either in the number, or in the value and rarity, of the objects which his zeal and industry procured for me. I ranked myself only third."

But Burchell, like many other collectors, came unstuck over the challenge of dealing with his treasures once he got home. "The collection of birds which I made during my travels in Africa has always remained packed up in the chests," he confessed to his friend William Swainson in 1819, four years after his return, "nor have I yet been able to get any of them stuffed. The chief reason for which is that they would take up much more room than I have at present in my power to allow them. They have indeed never yet been all seen by any person since my arrival in England." He was immobilized by the fear of decay: "I am obliged, in order to secure them from moths and other insects, to keep them so pasted up that I have not myself any access to them, but intend to keep them in this state till I begin to work at the ornithological part of my collection."

Burchell published an account of his African travels in the early 1820s. Then in 1825, with the African specimens still untouched in their boxes, he grabbed a sudden opportunity to make another collecting expedition to Brazil. But he still entertained the hope of being able to describe his discoveries. From Rio de Janeiro, he wrote Swainson of a rumor he had heard "that you are in possession of all my birds and are employed by me to describe them. Do contradict so groundless a story; for I hope at my return to have the long-looked-for gratification of publishing *all* my African discoveries myself, and which I reserve as one of the most pleasing results of my toils in that country."

By 1831, it had begun to dawn on him how futile this hope was. Getting into the field to collect species was far easier than obtaining a salary for the unglamorous business of cataloguing them. Government would not "take me by the hand in the affair of rendering all my collections & observations useful to science & the public." When Swainson offered to work with him on a joint publication, Burchell now replied that the proposal "exactly meets with my present ideas on the subject." The specimens from Brazil, on top of those from Africa, had "increased my collection in every department beyond all hope of being able single handed to get through such enormous a mass of materials." But it wasn't exactly an order to proceed at full speed. "We must talk over this subject when we meet," he wrote.

Then in 1834, another African traveler, Eduard Rüppell, visited the British Museum and persuaded one of the young keepers there, John E. Gray, to take him to see Burchell's collection. Gray, who would play a part in some of the great biological controversies of the mid-nineteenth century, was a curious character, sharp-tongued, quarrelsome, a closet naturalist who was always ready with an invidious comment about field collectors. All these traits were evident in his account of the visit with "my kind friend" Burchell, who was deeply reluctant, and ultimately unable, to open his carefully sealed boxes.

Not taking the hint, Gray and Rüppell came back a few days later, "provided with a hammer and chisel to prevent a recurrence of the same difficulty. Mr. Burchell laughed at our persistence and agreed to our opening the box containing the Vultures which was most carefully packed," Gray wrote. But when the lid came off, the box "contained nothing but the naked skull, arm and leg bones, all the rest had been eaten up, and this was unfortunately the state of all the boxes of African birds which we examined much to our grief and disgust." Some species that Burchell had collected would only be "discovered" when Rüppell collected and described them on his own decades later.

Burchell was hardly alone among the intrepid travelers who found themselves paralyzed by the twin perils of cataloguing and preservation. Anders Sparrman, one of Linnaeus's apostles, made expeditions to southern Africa and West Africa and also accompanied Capt. Cook on his second voyage around the world. But his collections remained in their shipping crates for years afterward. Likewise, specimens from the Lewis & Clark expedition to the American West went largely uncatalogued. Even Charles Darwin would have trouble finding experts to take responsibility for all the species he brought home from his years on HMS *Beagle*. The geologists were "willing to render all assistance," he wrote to his sister Caroline in October 1836, "but the Zoologists seem to think a number of undescribed creatures rather a nuisance." And with good reason: They were already buried in new species, which often decayed faster than the experts could describe and catalogue them.

For Burchell, the burden of specimens remained overwhelming until his death in 1863. "What have I travelled and laboured so hard for!" he had written at one point to Swainson, "only to sit still in years of inactivity to grieve at the sight of my unpacked boxes, and patiently submit to those circumstances which have so long rendered all my hard labours abortive." But Gray's assessment of the dire condition of Burchell's specimens—and his own fears—were overstated. In fact, Burchell seems to have done the fieldwork of preservation with care. The bulk of his collection survives today at the Oxford University Museum of Natural History, where his southern Africa skins were finally catalogued and named in 1953, 138 years after the adventurer's return home. Moreover, the disappointment and grieving that afflicted Burchell and other naturalists had come to an end during his lifetime, as new technologies helped perfect the preservation of specimens.

"Jackos" and "Bloated Piñatas"

It had been a long, perilous struggle. A typical eighteenth-century taxidermy technique recommended by one British writer involved a dry mixture of corrosive sublimate and a varnish made from turpentine and camphor, the one "a dreadful poison," according to a 1773 critique from a French physician, and the other highly explosive. Making varnishes "requires the experience and skill of a trained artist," the physician warned. Heating turpentine over a bed of coals produced a "thick, fetid, black vapor" that could "catch fire suddenly, ignited by a light incautiously brought near, or igniting itself by a draft from an open window or door which pulls the smoke down on to the coals." Recommending the technique "to people ignorant of the details of a new and dangerous art" exposed them to the "almost certain danger of harming themselves" and immolating their neighbors. Besides, it didn't work. The physician had specimens prepared using all known preservation methods. Then he placed them in containers and added the familiar ravenous insects, which generally got fat and happy as the specimens twitched into nothingness.

Techniques for mounting specimens realistically were almost as bad. Early taxidermists tried wire supports, wrapped in stuffing, to return a bird's skin to something like its natural form. But the results were often goggle-eyed and stretch-necked, and many such specimens sagged. Some taxidermists combined wires with a solid core wrapped in stuffing and produced "overpadded, barrel-chested birds" and creatures that looked like "bloated piñatas." Other specimens "shrunk and twisted to become grotesque caricatures of their former selves." Fine details of anatomy often got lost or distorted in the course of mounting. The Dutch anatomist Petrus Camper complained that stuffed orangutans and "Jackos," or chimpanzees, in eighteenth-century European museums had human knees and other features tending "to keep alive

A subtext of resurrection ran weirdly through Charles Waterton's work as a naturalist, perhaps influenced by his background as a member of the English Catholic gentry.

the silly idea" that these species "very nearly resemble man." (He was also disturbed to note that "in the Jacko of Buffon the virile member is entirely like human.")

Waterton regarded the lot with disdain: "It may be said with great truth that, from Rome to Russia, and from Orkney to Africa, there is not to be found, in any cabinet of natural history, one single quadruped which has been stuffed, or prepared, or mounted (as the French term it) upon scientific principles. Hence, every specimen throughout the whole of them must be wrong at every point."

But in the early nineteenth century, taxidermy underwent a transformation. The naturalistic style of mounting that the Peale family had pioneered at their Philadelphia Museum in the 1780s now became widespread. By 1819, one Liverpool "bird-stuffer" was proudly offering to mount a bird, thrush-size or under, for three shillings six pence, promising "the very mind of the bird—(wildly speaking) brought back again with action in the Limbs and in attitude easy." This was something new and thrilling, as the same taxidermist made clear when he allowed that the customer on a budget could still get birds stuffed for half that price—"if you would like them in the old goggle eye'd Rear'd up way."

Witchcraft and Enchantment

The key to the dramatic improvement in taxidermy was arsenic. From the French town of Metz near the German border, an apothecary and naturalist, Jean-Baptiste Bécoeur, was already sending arsenic-preserved specimens to the *Cabinet du Roi* in 1755, as a way of advertising his taxidermic skills. Buffon himself wrote back, calling the samples "very well preserved" and noting that the work of the best taxidermists then lasted only three or four years, even when kept in a vacuum flask. Three years later, Buffon wrote again, marveling that the samples were still "in excellent condition." This was apparently an understatement, to judge by the effusive praise from other naturalists.

"Had Bécoeur lived in ages past," a contemporary later wrote, "he would have been accused of witchcraft and enchantment . . . These are truly immortal animals."

Bécoeur managed to keep his recipe secret in his lifetime. "If my discovery concerned the health or welfare of men, I would hear only the cry of humanity and sacrifice to the public the fruit of my work," he wrote, a little defensively. "But as my secret cannot be placed in that class, I believe that I can wait without qualm while I am repaid for the trouble and expense that it has caused me." Oddly, Bécoeur claimed that his recipe contained no poisons. He may have been aiming to mislead competitors, science historian Paul Lawrence Farber suggests, or perhaps he meant only that it did not give off poisonous vapors. After his death in 1777, the recipe passed to the Muséum National d'Histoire Naturelle and it became public knowledge after 1800. By 1830 arsenic, or arsenical soap, had become such a standard tool almost everywhere, according to Farber, that naturalists "no longer regarded taxidermy as a problem but considered it a technique." The like of Leach's cremations in the back gardens of the British Museum would no longer be a routine necessity, and Bloomsburians could promenade without fear of burning snakes.

The taxidermy revolution mattered not just for collectors (and their sorely put-upon neighbors), but also for science. For the first time, the idea of a type specimen as a voucher permanently tied to the original description of a species became a practical reality, helping to standardize nomenclature and classification. Taxidermy also made it possible to build up the large collections naturalists needed to make sense of the world. These collections "determined the character of the nascent discipline of ornithology," Farber writes, and of zoology at large. The style of zoological writing progressed from descriptions of individual species to detailed studies across families, orders, and classes, requiring "meticulous comparison of thousands of specimens often from distant areas," an impossible task unless specimens could be kept intact. Biological thinkers, among them Charles Darwin and Alfred Russel Wallace, could now focus on how species varied and how these variations reflected their geographical distribution—and thus their evolution.

Taxidermy also encouraged museums to think more about the behavior of animals in their natural environments, according to Farber. Instead of displaying species like suspects in a police lineup, they began to present them in habitat groups. The development of lithography at about the same time also made zoological illustrations more accurate and lifelike "because naturalists could draw directly onto the stone instead of relying on engravers, who often had little appreciation for scientific illustration," Farber writes. A few years earlier, naturalists had had to make those drawings from live or fresh-killed specimens. Zoological artists like George Stubbs and John James Audubon had often worked late into the night trying to capture the essence of animals that were rotting and stinking before their eyes. But now proper taxidermy made it possible for artists to work through thousands of specimens at leisure, opening up a great era of natural history illustration.

Charles Waterton was a peculiar exception to the arsenical enthusiasm. He described arsenic as "very dangerous" and its use "sometimes accompanied with lamentable consequences," citing the case of a naturalist he'd met in French Guyana in 1816, who showed him a box containing 16 human teeth. "These fine teeth once belonged to my jaws," the naturalist remarked. "They all dropped out by my making use of the *savon arsênetique* for preserving the skins of animals."

Other enthusiasts tended to deny or downplay the hazards. "Since arsenic is said to be so dangerous," one natural history magazine airily ventured in the 1880s, "we will ask for a little information on the point from anyone who has felt any deleterious effects from its use." It apparently did not occur to the editors that potential informants on the subject might remain silent, on account of being dead.

Raphaelle Peale, the artist and son of Charles Willson Peale, had died of arsenic and mercury poisoning from his taxidermic work in the family museum, in 1825, age 51, after years of physical and mental instability. Also from Philadelphia, the physician and pharmacist John Kirk Townsend undoubtedly recognized the hazards of his work discovering and preserving bird species in the American West, but he died of arsenic poisoning anyway, in 1851, age 42. The Philadelphia ornithologist John Cassin, who constantly handled preserved bird skins in the course of describing 198 new species, worried about "mortgaging myself by perpetual lease to Arsenic and Liver complaint," and died of apparent arsenic poisoning in 1869, age 55. In California, Henry Hemphill, a bricklayer-turned-naturalist, somehow managed to poison himself with arsenic while curing shells, in 1914. Arsenic probably also sent countless local assistants—unnamed in the annals of science—to their graves.

Eternity's Sunrise

But for armchair travelers in mid-nineteenth century, perfectly preserved animal specimens, displayed under a bell jar or in a glass-fronted case, were among the wonders of the age. People became "caught up in taxidermy's magical ability to offset rot and decay," Judith Pascoe writes, in *The Hummingbird Cabinet*. Where the jaded modern eye sees only "death gussied up," they saw hints of immortality in the way the sparkle and shine had been caught on the wing.

"It is impossible to imagine anything so lovely as these little Humming Birds, their variety and the extraordinary brilliancy of their colours," Queen Victoria remarked, when the ornithological entrepreneur John Gould put on a spectacular display of hummingbirds at the London Zoological Gardens in 1851. Gould had set up 24 hummingbird display cases in a specially built room, with the sun from a skylight filtered through awnings and with stage lighting to catch the metallic glint of the feathers. He arranged the birds in lifelike poses, hovering over a bed of bromeliads, or darting their beaks deep into flowers. "They poise themselves in the air," Charles Dickens wrote, "we hear not the humming of the wings, but we can almost fancy there is a voice in that beauty." The art critic John Ruskin, whose "rational and disciplined delight in the forms and laws of Creation" had led him to collect minerals as a hobby, was moved to lament: "I have wasted my life with mineralogy, which has led to nothing. Had I devoted myself to birds, their life and plumage, I might have produced something myself worth doing. If I could only have seen a humming-bird fly, it would have been an epoch in my life."

Remarkably, Gould himself had never seen a living hummingbird. But he was a deeply knowledgeable ornithologist, aided by the expert guidance in taxidermy provided by Charles Waterton in his book *Wanderings in South America*. Waterton had detailed the latest preservation techniques and urged taxidermists to show "Promethean boldness, and bring down fire and animation as it were, into your preserved specimen." William Blake's famous poem had declared

> *He who binds to himself a joy,*
> *Does the winged life destroy.*
> *But he who kisses the joy as it flies*
> *Lives in eternity's sunrise.*

Now, by capturing the winged life seemingly forever (and sometimes in species that later went extinct), taxidermy threatened to turn Blake's meaning on its head.

Chapter Eleven

"AM I NOT A MAN AND A BROTHER?"

I am haunted by the human chimpanzees I saw
[in Ireland]. . . . To see white chimpanzees is
dreadful; if they were black, one would not feel
it so much.

–CHARLES KINGSLEY

T HE 22-YEAR-OLD naturalist who went aboard HMS *Beagle* just after
Christmas 1831 was the son of a privileged family with a long history in the antislavery movement. His grandfather Josiah Wedgwood, a
pottery manufacturer, had mass-produced the jasperware cameo that
became the symbol of opposition to the slave trade, often worn as a
bracelet or hair ornament, the way people might wear a pink ribbon
or peace symbol today. It depicted a black man in chains pleading for
his freedom on one knee, framed by the slogan, "Am I not a man and
a brother?"

But when Charles Darwin experienced his first horrifying encounter with human beings in an uncivilized state 12 months later at the
southern tip of South America, brotherhood was the furthest thing
from his mind. The people he met in Tierra del Fuego struck his
genteel eyes as "stunted in their growth . . . hideous faces bedaubed
with white paint, skins filthy and greasy . . . hair entangled . . . voices
discordant . . . gestures violent and without dignity." In times of hunger,
he reported credulously, the Fuegians would sooner kill and roast their

old women for food than sacrifice their dogs. ("Doggies catch otter," they explained, "old women no.")

The visit to Tierra del Fuego still troubled him years later, at the conclusion of *The Voyage of the Beagle*, where he asked with evident anguish, "Could our progenitors have been men like these?—men, whose very signs and expressions are less intelligible to us than those of the domesticated animals." The shock of the aboriginal also leapt to mind when he encountered a great ape for the first time at the London Zoo in 1838. "Let man visit Ourang-outang in domestication," Darwin wrote in his notebook, and he would find this creature expressive, intelligent, and affectionate. By contrast, seeing a human in the "savage" state "roasting his parent, naked, artless, not improvable," would make it impossible for any man "to boast of his proud preeminence." The notion of descent from apes was almost preferable to kinship with certain humans.

The people visited by early European explorers no doubt reciprocated the sense of horror, though we have only the explorers' bemused word for it. In Gambia, for instance, young Africans questioned the eighteenth-century explorer Mungo Park about the whiteness of his skin and the prominence of his nose: "They insisted that both were artificial. The first, they said, was produced when I was an infant, by dipping me in milk; and they insisted that my nose had been pinched every day, till it had acquired its present unsightly and unnatural conformation." Explorers sometimes invited their readers to laugh at the naïve failure to recognize Europeans as the human archetype. Park reported that when he praised "the lovely depression of their noses," his Gambian friends told him that "honey-mouth," or flattery, would get him nowhere. But doubt resonated beneath the joke.

Coming to terms with other races was only one among a host of discomfiting ideas and experiences resulting from the pursuit of new species. Naturalists were also grappling with the sprawl of geologic time; the evidence that a vast herd of species created by God had somehow gone extinct; the queasy suspicion that nature was not a great chain of being, perfect and orderly, but a sprawling hodgepodge; the paradox that species thought to result from separate acts of creation should nonetheless share so many similarities, as if God had needed to plagiarize His own work; and always the ominously rising tide of evolutionary ideas.

In an angry letter to a colleague sent in 1831, when he was a zoological assistant at the British Museum, John G. Children briefly slipped by writing about "the connexions" between species, but instantly caught himself and corrected it to mere "*similarities* which subsist between created beings." He went on: "I do protest, and ever will, against the abominable trash, vomited forth by Lamarck and his disciples, who have rashly, and almost blasphemously imputed a period of comparative imbecility to Omnipotence! when they babbled out their juvenile crudities about a progression in nature—that is in God—from abortive imperfect first attempts, to more and more perfect efforts and results!" Children asked to have his remarks kept "sub silentio" and finished with a hapless plea that the entire scientific community might well have echoed: "I beg I may not be entangled in controversy."

But nothing matched the anxiety inspired by simultaneously encountering other primate species and other human races for the first time. That the two things—apes and human races—so often came together in intellectual debate seems outrageous to the modern mind. We tend to feel clear about where we draw the line between humans and other primate species. It's also clear that early European thinkers often drew that line in a calculated effort to preserve Europe's grand delusion about its own special nature—and consign "lesser" races to submission, enslavement, and even extermination. We now know, moreover, that it's impossible to identify a boundary separating one human race from another; racial categories have no scientific meaning.

Even so, it's revealing to set aside present knowledge and imagine just how confused early thinkers were by the entire question of species: Why did there suddenly seem to be so many kinds of animals—and of humans? Why did they vary so much from place to place? When were the differences just the normal variation that occurs from one individual to the next, and when did these differences add up to dis-

Though most eighteenth-century naturalists had never seen one, they worried that "scarcely any mark" distinguished apes from humans.

tinguish one species from another? These questions took on particular point where they touched on our own nature not so much as children of God, but as primates—a term coined by Linnaeus that was itself fraught with the subtext of hierarchy: It means "of the first rank."

Apart perhaps from a few creatures in royal menageries or the figures in medieval bestiaries and travelers' tales, people in Europe had lived largely apart from apes and monkeys, as well as other human groups, for a period extending roughly from the migration out of Africa 50,000 years ago and the extinction of Neanderthals 30,000 years ago to the start of the age of discovery not quite 700 years ago. It was long enough to develop a considerable sense of separateness from the rest of nature, as well as a splendidly puffed-up notion of their own special place in the world.

One of the earliest mentions of an ape species in European literature appeared in 1641, and the reference was characteristically confused. The animal in question was a juvenile chimpanzee, shipped to Europe by a slave trader in West Africa. But the word "chimpanzee" itself would remain unknown outside Africa for another century. Probably because of the extensive Dutch trade in Malaysia and Borneo, Amsterdam anatomist Nicolaes Tulp instead used the more familiar name "orangutan," from the Malay for "person of the forest." The London physician Edward Tyson also said *orangutan* (or *orang-outang*), as well as *pygmy* and *Homo silvestris* (man of the woods) when he dissected a juvenile chimpanzee in 1699. Orangutan remained the catch-all for all great apes until mid-nineteenth century.

Both Tulp and Tyson overstated the similarities between apes and humans, no doubt because, at first viewing, they were so striking compared with anything Europeans had seen up to that point. Tulp's essay included an illustration of an ape with modestly downcast eyes and hands concealing its genitals. Tyson similarly illustrated his treatise with a picture of his "pygmy" walking upright, with the help of a cane. These apes seemed to be almost perfectly human in part because both subjects were juveniles, which have a flat face much like ours. (The chimpanzee jaw only begins to jut forward in adolescence.) Belief in the great chain of being also encouraged naturalists to minimize the gap between humans and apes as a way of filling in the missing links. Like Aristotle, they thought that "nature makes no sudden leaps."

And yet they also wanted to keep apes at a distance. The idea of a finely graded progression, from the lowliest amoeba on up, extended into the human species, helping to reinforce traditional thinking about rank and social hierarchy. Thus Tyson's essay introduced a rhetorical device that would become standard in the literature about primates. In a "dedicatory epistle" to his aristocratic patron, he emphasized the similarities not so much between men and animals at large, as "between the lowest Rank of Men, and the highest kind of animals." In just this fashion, later thinkers would repeatedly deflect the full impact of our similarity to apes onto people of the wrong social class, nationality, or skin color.

Firsthand knowledge for making careful distinctions about apes was often woefully lacking. "In many cases, naturalists never set eyes on the animals they described," Londa Schiebinger writes, in *Nature's Body*, "but drew their ideas about these creatures from the rather fanciful

Nineteenth-century illustrators were still making chimpanzees look like humans gone slightly askew.

teachings of the ancients combined with the untrained observations of voyagers." The Roman naturalist Pliny had described subhuman cave-dwellers in the Atlas Mountains, and Plato's *Republic* had included an allegory of African cave-dwellers. Linnaeus made this the basis for a proposed second human species, *Homo troglodytes*, taking the species name from the Greek words meaning literally "one who creeps into holes." (Chimpanzees are known to this day as *Pan troglodytes*, though they live in trees, not caves.) Satyrs, pygmies, the occasional wild child (supposedly raised by wolves), and unusually hairy humans all attracted serious consideration.

Linnaeus put apes in the same taxonomic order with humans, to the consternation of many critics. "I know full well what great differences exist between man and beast when viewed from a moral point of view: man is the only creature with a rational and immortal soul," he wrote. "If viewed, however, from the point of view of natural history and considering only the body, I can discover scarcely any mark by which man can be distinguished from the apes." He based this broad assertion on having seen only a single juvenile chimpanzee specimen, in 1760, but he continued, "Neither in the face nor in the feet, nor in the upright gait, nor in any other aspect of his external structure does man differ from the apes."

Apes even seemed to view the world as we do. In the tenth edition of his *Systema Naturae*, Linnaeus projected his own obsession with hierarchy onto the orangutan, also sight-unseen: "By day hides; by night it sees, goes out, forages. Speaks in a hiss. Thinks, believes that the earth was made for it, and that sometime it will be master again."

"An Universal Freckle"

The European debate about where human races fit into this bewildering picture usually started from the Biblical tradition that all humanity had descended from Adam and Eve. In the late eighteenth century, Johann Blumenbach, a German physician often regarded as the father of anthropology, described five separate human varieties—black, white, yellow, red, and brown, the last referring to Southeast Asian and Pacific Islanders, and he argued adamantly that all of them belonged to a single human species. Along with most other naturalists then, he believed that the ancestral human form was white and European, probably originating around the Caucasus Mountains. Hence he coined what historian Winthrop D. Jordan has called "the inept but remarkably adhesive term *Caucasian*."

Other races had diverged or degenerated from this original type, according to conventional wisdom, because of environmental factors like diet, mode of living, and climate. Darker skin, for instance, resulted from exposure to the tropical sun, much as sunlight caused white people to develop freckles, except that black skin was what one thinker called "an universal freckle." The environment seemed to draw out latent powers in human nature, making some races better suited to particular circumstances. For instance, the German philosopher Immanuel Kant attributed both skin color and "the strong body odor of the Negroes, not to be avoided by any degree of cleanliness" to their special ability to survive in the "phlogisticated air" of swamps and thick forests. According to a nonsense theory dating back to the Ancient Greeks, phlogiston was a combustible essence produced by respiration and decomposition and it abounded in moist, tropical habitats. So Kant thought Africans had developed the special ability to "dephlogisticate" the blood through the skin, in their sweat. The theory suggested practical implications for the European colonial enterprise, in that whites, lacking this special ability (or, as we now know, lacking acquired immunity to tropical diseases), often died on visiting the same dank regions.

Proponents of the single species idea believed that people could acquire traits like skin color and pass them on to their offspring. Some thinkers argued that these traits could also *change back* in the right circumstances. In an influential 1787 essay on human varieties, Samuel Stanhope Smith, president of what is now Princeton University, argued that transplantation to temperate America, combined with freedom from slavery, would gradually cause blacks to revert to the original type. He reasoned that foxes, bears, hares, and other animals in cold regions turn snow white. Likewise blacks would become the equal of whites by losing "the peculiar deformities of the African race" and literally becoming white. Smith noted that domestic slaves often seemed to become less African over three or four generations, "a striking example of the influence of the state of society upon the features." Perhaps because he was a Presbyterian minister, it did not occur to him that some factor might be at work other than the gracious company of civilized white people.

Smith regarded Henry Moss, born a slave in Virginia, as living proof of the potential for racial change. At the age of 38, Moss's "natural colour began to rub off," according to an ad published in Philadelphia in 1796. Now "his body has become as white and as fair as any white person, except some small parts, which are changing very fast." Moss's ad, for an appearance at the Black Horse tavern on Market Street, promised that the sight would open "a wide field of amusement for the philosophic genius." Making the trip over from Princeton, Smith was among those paying a quarter for the privilege. The physician Benjamin Rush also took in the spectacle and concluded that Moss had been "cured" of the disease of blackness. And in a sense this was true. Moss, who suffered from the skin depigmentation called vitiligo, used his freak-show income to purchase his freedom from slavery.

The supposed superiority of whites went almost unquestioned, even when one iconoclastic thinker argued that humanity had originally been black. James Cowles Prichard, a physician and ethnologist in Bristol, England, recognized the power of artificial selection to alter the appearance of animals in domestication. He concluded that achieving civilization was a process of self-domestication, which had resulted in the development of whites as the new, improved model of humanity.

Equality and Facial Angles

Within the limits of his time, the Dutch anatomist Petrus Camper was, like Rev. Smith of Princeton, a forthright proponent of equality among races. By background, he was a physician and a thorough naturalist, who discovered the lightweight, air-filled character of bird bones, identified a rhino from Java as a new species, and discredited the myth of the unicorn. He used his connections in the East Indies trade, which had made him rich, to obtain and dissect orangutan specimens, demonstrating that they were a separate species, not a degenerate human variety.

He also set out to disprove the myth that black Africans "descended from the mingling in olden times of white people with great Apes or Orang-Outangs." On dissecting a black Angolese boy in Amsterdam in 1758, he reported that he found "his blood very much like ours and his brains as white, if not whiter." The dissection included a part-by-part comparison "with the famous description of the Bush-man or 'Orang-Outang' of the renowned Tyson. I must confess that I found nothing that had more in common with this animal than with a white man; on the contrary, everything was the same as for a white man." Camper invited his listeners to join him "in holding out the hand of brotherhood to Negroes and Blacks, and in recognizing them as true descendants of the first man whom we all recognize as our Father."

The horrible irony, given this egalitarian outlook, is that Camper went on to develop what would become the chief visual icon for belittling and enslaving blacks. Science historian Miriam Claude Meijer has recently argued that later polemicists distorted and misused Camper's work. He started, she says, from an honest study of differences in skull shape. Whereas Buffon thought Africans acquired a flattened nose from being carried in a sling pouch after birth, pressed up against their mothers' backs (the flip side of Mungo Park's "pinched" nose), Camper attributed the facial characteristics of different races to the underlying structure of the skull.

In the 1770s, he began to theorize about facial angle, meaning how far a line from the front teeth to the forehead deviated from the vertical. He had noticed how the straight, perpendicular profiles in classical sculpture contrasted with the more sloping, prognathous faces depicted in Dutch and Flemish art. Camper didn't attribute greater intelligence to one shape of skull over another. But out of a sense of aesthetics, like someone arranging the books on a shelf by the color of their bindings, he lined up the skulls in his anatomical collection according to facial angle. The result, he wrote, was "amusing to contemplate." First came monkeys and apes with a facial angle of 40 to 50 degrees. Then came Negroes at 70 degrees. A succession of intermediary races progressed on up to Europeans, who were, at a facial angle of 80 degrees or better, just short of Greek gods and archangels. Camper didn't draw conclusions. But by the simple trick of pairing Africans with apes, his lineup of facial angles would become the visual shorthand for racial inferiority well into the twentieth century.

The notion of innate white superiority predominated even in the most progressive intellectual circles. Thomas Jefferson regarded blacks as irredeemably debased. He lamented the absence of a proper natural history of the race and wrote, "I advance it therefore as a suspicion only, that blacks, whether originally a distinct race, or made distinct by time and circumstances, are inferior to the whites in the endowments of both mind and body." (This was years before the slave Sally Hemings would become his likely mistress and mother of several of his children.)

David Hume, one of the great liberal philosophers of the Enlightenment, did not stop at mere suspicion. "There never was a civilized nation of any other complexion than white," he declared. Other races did not even belong to the same species. "In JAMAICA indeed they talk of one negroe as a man of parts and learning; but 'tis likely he is admired for very slender accomplishments, like a parrot, who speaks a few words plainly."

Charles White, an eighteenth-century obstetrician and fellow of the Royal Society, celebrated the preeminence of whites as if there could be few things more delightful than his own pale image in the mirror: "Ascending the line of gradation, we come at last to the white European," he wrote, the "most removed from the brute creation . . . the most beautiful of the human race. No one will doubt his superiority in intellectual powers . . . Where shall we find, unless in the European, that nobly arched head, containing such a quantity of brain . . . ? Where the perpendicular face, the prominent nose, and round projecting chin? . . . Where, except on the bosom of the European woman, two such plump and snowy white hemispheres, tipt with vermillion?" It reads like a Gilbert & Sullivan parody, sung by a dancing chorus of periwigged gentlemen in knee breeches, and delivered with almost enough conviction for the intended audience to ignore the muffled cries of the slaves chained belowdecks.

At times, however, the consequences of such thinking came all too visibly to the surface. Robert Schomburgk was a German naturalist best known for discovering *Victoria regia*, a waterlily with great round leaves like serving platters. He happened to be at Anegada, the northernmost of the Virgin Islands, in 1831, when a passing Spanish slaver, the *Restauradora*, hit a reef and sank in shallow water. When he passed the spot soon after, "the clear and calm sea" revealed "numerous sharks, rockfish and barracuta . . . diving in the hold where the human carcases were still partly chained, to tear their share from the bodies of the unfortunate Africans."

CRANIOLOGICAL LONGINGS

Rogala, my little Ishogo slave, is sick, and will
die to-night: I know it. You have often asked for
an Ishogo skull, and now you shall have one.

           –A TRAFFICKER IN SKULLS,
        TO EXPLORER PAUL DU CHAILLU

I N 1822, RICHARD OWEN was a young medical apprentice in the gloomy,
Gothic castle that had become the jail for the city of Lancaster in
northwestern England. His funereal appearance—high brow, hollow
cheeks, a cleft chin—fit the setting. In his best-known photograph,
he cants his head forward, taking his facial angle well past vertical,
and looks up knowingly from deep-set eyes. The castle's busy gallows
served up plenty of fresh material for anatomical studies then. The jail
also provided the grist for "the Negro's head story," which Owen often
told, and almost certainly fictionalized when he published it under the
pen name Silas Seer in 1845.

By then, Owen ranked among the world's most celebrated scientists,
intellectual heir to both Georges Cuvier and the pioneering anatomist
John Hunter. He was in fact the Hunterian Professor at the Royal College of Surgeons, where Hunter's anatomical collection was enshrined.
Like Hunter, he routinely dissected all cadavers from the London Zoo.
He also had Hunter's breadth of expertise, writing brilliant descriptions of species from the pearly nautilus to the giant ground sloth. He

Richard Owen made a reputation for brilliant anatomical descriptions—and for bending the truth to his relentless ambition.

had coined the term "dinosaur" and would soon help create the first life-size re-creations of some of these giants. Like Cuvier with his "correlation of parts," he had also performed anatomical magic, studying a single bone fragment and declaring publicly that a bird the size of an ostrich had existed, or might still exist, in New Zealand. A few years later the extinct moa turned up, a 12-foot-tall bird that had weighed 500 pounds.

But at the height of his power, Owen somehow also found time to publish a garish story from his medical training, in *Hood's Magazine and Comic Miscellany*. It told of the family of a recently deceased sea captain living in a small house just down the steep hill from the castle gates. The sea captain had worked in the slave trade, and one winter evening his widow was trying to justify his livelihood to her daughter as they sat before the fire. Suddenly they heard a clattering of footsteps. A loud blow burst open the front door. The mother let out a shriek at the sight of a shadowy figure and "the phantom of a negro slave lying on the floor, which turned its ghastly head and glared upon her for a moment, with white protruding eyeballs." In the days after, the family showed their friends the bloodstain left behind on the floor as physical proof of this haunting.

But, as Owen told it, only he knew the true story of the ghostly apparition. After overcoming the horror of his first weeks of apprenticeship, he had begun to enjoy anatomy. It happened "that on the day when a negro patient in the gaol hospital had died, a treatise on the 'Varieties of the Human Race' fell into my hands, and greatly increased my craniological longings." He bribed the turnkey to allow him to return to the morgue after normal working hours, and went home to bide his time. Then, equipped with "a strong brown-paper bag, I sallied forth on a fine frosty evening in January to secure my specimen of the Ethiopian race."

As Owen climbed the spiral stairs of Hadrian's Tower back to the castle morgue, he speculated about "facial angles" and "prognathic jaws," key concepts in the bogus science of race. Owen wanted the evidence. Working by lantern light, alone in "the stone chamber of the dead," he cut off the deceased prisoner's head, placed it in the bag, and hid it under his cloak. He closed up the coffin and stopped on the way out to instruct the turnkey to bury it, reducing the danger of discovery.

But on the icy slope outside, "my foot slipped, and, being encumbered with my cloak, I lost my balance and fell forward with a shock which jerked the negro's head out of the bag, and sent it bounding down the slippery surface of the steep descent. As soon as I recovered my legs I raced desperately after it, but it was too late to arrest its progress. I saw it bounce against the door of a cottage facing the descent, which flew open and received me at the same time, as I was unable to stop my downward career. I heard shrieks, and saw the whisk of the garment of a female, who had rushed through the inner door; the room was empty; the ghastly head at my feet. I seized it and retreated, wrapping it in my cloak. I suppose I must have closed the door after me, but I never stopped till I reached the surgery."

For Owen and other collectors, the skulls of human beings, mainly from outside Europe, were just another class of natural history specimen, stripped of personal and cultural connections. "My Friend W^am MacGillivray wishes to posses some Sculls of Negroes, *Africans if possible*," John James Audubon wrote from London to John Bachman, a South Carolina naturalist and Lutheran minister, in 1834. "But at all events send him half a dozen or whatever you Can—he wishes also to have the *heads* of a few alligators of different Sizes—Turkey Buzards and Carrion Crow heads indeed Sculls or heads of any description."

That same note of shopping wholesale turned up in a letter John Cassin of the Academy of Natural Sciences of Philadelphia sent in 1843 to Thomas S. Savage, an Episcopal missionary and naturalist in Liberia: "If you could send a barrel or less of the crania of the natives with the names of the tribes when you know them, you would confer a very great favor and promote the study of the science of our race."(We don't know how Savage responded to this chilling request. But a few years later, he would bring back two of the most important skulls in the history of primatology, the subject of chapter 15.)

At times the skulls could be alarmingly fresh. The explorer Paul Du Chaillu was paying $3 per human skull in the 1860s in Gabon and sending them home only to dispute "the doctors . . . in my country who believed that negroes were apes almost the same as the gorilla." Then an eager seller promised him the skull of a gravely ill, but still living, slave named Rogala, and it apparently dawned on him that he might inadvertently be subsidizing murder. In New Zealand, Joseph Banks had been the first to collect the head of a chieftain with the intricate Maori face tattooing called *moko*. "It is impossible to avoid admiring the immense Elegance and Justness of the figures," he wrote, with the concentric spirals "resembling something of old Chasing upon gold or silver; all these finished with a masterly taste and execution." Subsequent European demand was so great that chieftains sometimes had slaves tattooed with moko, normally reserved for the nobility, then slaughtered them and smoked their faces, for sale to collectors. Grave robbing was also common. In North America, a traveler plundered a fresh Indian grave in 1837 and carried the skulls in his pack for two weeks, "in a highly unsavory condition, and when discovery would have involved danger, and probably death."

The American Golgotha

The latter skulls were destined for Samuel G. Morton, a Philadelphia Quaker, physician, and naturalist. Tall, pale, with blue-gray eyes and a modest demeanor, Morton was a devoted family man, "of no arrogant pretensions, and of a forgiving temper; charitable and respectful of others." He began studying skulls in 1830 and eventually accumulated 1000 of them, a collection his friends proudly described as "the American Golgotha." (The name of the hill on which Christ was crucified meant literally "place of the skull.") The collection was housed at the Academy of Natural Sciences of Philadelphia.

Morton believed he could characterize races by the capacity of their skulls, with a bigger skull meaning a bigger brain, a bigger brain meaning greater intellectual ability, and the biggest brains naturally belonging to Europeans. He made his measurements by filling the inverted skulls with white mustard seed, at first, and later, for greater uniformity, with No. 8 shot, a methodology considered remarkably objective and scientific for its time. At Morton's death in 1851, a New York newspaper remarked that "probably no scientific man in America enjoyed a higher reputation among scholars throughout the world." In the course of a eulogy, a friend remarked: "I trust I may further say I am quite sure the idea of disturbing any man's religious faith was foreign to his good heart."

But religious faith dictated accepting other races as fellow descendants of Adam and Eve. Morton preferred the heresy of keeping them separate and at a safe distance. He didn't openly declare that different human races were separate species until a few years before his death. But he argued from the start that races were as immutable as spe-

Samuel G. Morton was a Quaker whose work with human skulls made him a key figure in the bogus science of race.

cies. In his 1839 book *Crania Americana*, he doubted that the Biblical flood could have covered the entire Earth, or that the different races could have developed and dispersed in the short time since then. Other thinkers had concluded too hastily that racial differences were solely due to "vicissitudes of climate, locality, habits of life, and various collateral circumstances." He argued for "the reasonable conclusion that each race was adapted, from the beginning, to its peculiar local destination." In his 1844 book *Crania Aegyptiaca* he extended his argument to assert that the monuments of Egyptian civilization—a topic of heightened interest and influence then—had been the work of whites, not blacks.

Morton's work was deeply flawed. When paleontologist Stephen Jay Gould reanalyzed the data in 1978, he pointed out that measurements made with mustard seed can vary as much as 4 cubic inches for the same skull, with a variation of up to 1 cubic inch for lead shot, either method leaving "a wide berth for the influence of unconscious bias." Morton also neglected to distinguish between male and female skulls, and the normal size difference between genders skewed his results, particularly where he relied on only a few individual specimens to characterize whole groups. While Gould could find "no indication of fraud or conscious manipulation," he called Morton's work "a patchwork of assumption and finagling, controlled, probably unconsciously," by the urge to put "his folks on top, slaves on the bottom." (A more recent study has disputed the charge of finagling. Anthropologist John S. Michael argues that Morton went wrong not so much in his methods as in his acceptance of the bogus idea of racial classification.)

The Theory of Repugnance

Scholars have found nothing in Morton's background to explain his animosity toward different races. And yet he elevated racial repugnance to the level of a theory. For other writers, expressions of distaste were a way to assert their credentials as whites and to disguise the sexual attraction that so plainly troubled them. Thus Thomas Jefferson denigrated "the other race" for "that eternal monotony . . . the immoveable veil of black which covers all the emotions." He also invoked the standard bestial imagery with which people often vented their sexual and primatological confusion, writing that blacks themselves found whites more attractive, much as "the Oran-ootan" preferred "black women over those of his own species."

For John James Audubon, asserting his disgust for blacks may have been a way to deal not just with sexual attraction, but also with his own uncertain birth. He was said to be the illegitimate child of Jeanne Rabin, a 27-year-old French chambermaid on his father's sugar plantation in St. Dominique, now Haiti. But the elder Audubon also fathered a daughter Rose by a mistress of mixed race on the plantation. Rabin was dead by the time the family returned to France in the early 1790s, and Audubon *père* listed her as mother of Rose so he could register his daughter as white. It is possible that he did the same thing for his son.

Audubon was insecure enough on the point to describe his mother in later life as "a lady of Spanish extraction . . . as beautiful as she was wealthy." And as a young man strolling amid the racially mixed crowd on the levee in New Orleans, he made it a particular point that "the Citron hue of almost all is very disgusting to one who likes the rosy Yankee or English cheeks."

Samuel G. Morton seems to have had no such personal motive behind his "theory of repugnance," and his cautious, unpolemical pursuit of what seemed like science gave his ideas widespread acceptance. Among humans and animals alike, he wrote, "the separation of distinct species is sufficiently provided for by the *natural repugnance* between individuals of different kinds." This "moral as well as physical" repugnance was "proverbial among all nations of the European stock among whom negroes have been introduced." Ever the objective observer, he added that this repugnance was "almost equally natural to the Africans in their own country, towards such Europeans as have been thrown among them." The almost mutual horror had partially broken down only through "the moral degradation consequent to the state of slavery."

There was at least one awkward impediment to Morton's developing theory that human races constituted separate species: The streets of Philadelphia itself were alive with evidence that people of different races had been breeding together for centuries, with progeny even Morton acknowledged were "more or less fertile." In conventional taxonomy, that almost certainly made them members of the same species. So in 1847, Morton set out to refute the widely accepted reliance on interbreeding as a key part of the definition of a species, by demonstrating that separate species commonly produced fertile hybrids in the animal world, too.

His paper, "Hybridity in Animals, Considered in Reference to the Question of the Unity of the Human Species," cited crosses not just between zebras and horses, and between wolves and dogs, but also between a deer and a pig, producing India's spotted hog-deer. (In fact, *Cervus axis*, also known as the chital, is all-deer.) He accepted a published Russian account of the fertile offspring from the mating of a house cat and a marten. (This was overcoming repugnance of a high order, given that martens don't even much like one another. They are, moreover, mustelids, or weasels, belonging not just to a different species and genus from cats, but to a different taxonomic family.) Morton concluded that separate species could produce fertile hybrids "in proportion to their aptitude for domesticity," and that this was especially true for humans, as the most domesticated species.

The historian William Stanton has credited Morton, much too generously, with being at work on "essentially the same problem" as Charles Darwin—that is, how to explain "the vast number of species and the curious appearances of variations among them." In his 1960 book *The Leopard's Spots: Scientific Attitudes Toward Race in America 1815–1859,* Stanton added that Morton's thinking on hybrids "has since been largely vindicated," in that we now know hybridization between separate species can be a factor in the emergence of new species. But Morton's work was more accurately a prime instance of how the obsession with race dominated and distorted American thinking about "the species problem."

His close friend, the botanist Charles Pickering, took a broader interest in species but with equally perverse results. As a naturalist with the United States Exploring Expedition, he spent four years, from 1838 to 1842, traveling in the same waters Darwin had visited just a few years earlier, and considering some of the same questions about the adaptation of species to narrow habitats. The U.S. Ex Ex, as it became known, was the first great attempt to project American scientific expertise on the world stage.

Pickering's travels enabled him to write confidently about "millions of species, of plants & animals, scattered over this immense Globe"— a number that must have seemed absurd to his contemporaries. Just 80-odd years before, in his 1758 edition of *Systema Naturae,* Linnaeus had recognized only about 4400 animal species. But in plants alone, the U.S. Ex Ex brought back 50,000 specimens, belonging to 10,000 species, many of them previously unknown to science. James Dwight

Dana, traveling as a geologist, described 500 new species just in the lobsters, shrimps, crabs, and barnacles.

Pickering was impressed both with the abundance of life, according to Stanton, and "with the precision of adaptation." He noticed that different kinds of animals and plants were restricted not just to particular countries, but "also to much smaller areas—a fact which so impressed Darwin when he studied the birds of the Galapagos Islands."

But having helped gather this astonishing wealth of material, ripe with possible discoveries about evolution and biogeography, Pickering chose instead to follow the American obsession with race straight to the junk heap of biological history. On his travels, he counted not 5 but 11 separate human races, including the Negro, the Negrillo, the Hottentot, and the Ethiopian. With humans, as with plants and other animals, he believed that nothing had been "modified or moulded by Climate, but always adapted by nature precisely to that climate in which it is naturally found."

Pickering was eager to say that different human races constitute separate species, with separate origins. But because he required the approval of a federal government committee overseeing publications from the U.S. Ex Ex, he had to obscure this controversial conclusion in his 1848 book *The Races of Man and Their Geographical Distribution.* The result, as the physician and poet Oliver Wendell Holmes Sr. put it, was "amorphous as a fog, unstratified as a dumpling and heterogeneous as a low priced sausage."

Blowing Up the Bible

Some of Morton's other followers managed to say far more plainly what so-called American School ethnology was really all about. Josiah Clark Nott, a physician in Mobile, Alabama, was an imposing figure, with deep-set, visionary eyes and a thick billy-goat beard down over his starched white collar and black bow tie. He was endowed with a polemical frame of mind and a lynch-mob way with words, and he distilled Morton's research into lectures on what he called "niggerology." A slave-owner himself, he made polygenism—the theory that white and blacks originated as separate species—the basis for an ostensibly scientific defense of slavery. Nott piously declared that he loathed slavery in the abstract. But it was a kind of public service: The institution enabled a lesser human species to attain "their highest civilization." He added that "the Negroes of the South are now . . . the most contented population of the earth." Morton could easily have distanced himself from such twisted reasoning. Instead, he wrote of his "great pleasure and instruction" as Nott advanced his ideas where he himself had held back.

In the face of religious attacks on polygenism, Nott talked eagerly about "parson skinning" and worked to see the Biblical account of Creation "blown up." With scriptural nonsense cleared away, he wrote, "the whole field is open to us & if the argument is properly managed the world is ready for it." He wanted "to cut loose the natural history of mankind from the Bible," always with the aim of establishing the separate and inferior status of blacks. When Nott and another Morton acolyte, George R. Gliddon, published their *Types of Mankind; or, Ethnological Researches* in 1854, they grandly depicted themselves as waging the "last great battle between science and dogmatism."

The clergyman who provided the most rigorous argument against the Morton camp, defending scientific truth and (for once) religion, too, was John Bachman of Charleston, South Carolina—the same Bachman to whom Audubon had written requesting skulls of Africans and turkey buzzards. Bachman was a thoroughly trained naturalist and knew his way around the species question. When he wasn't tending his flock at St. John's Lutheran, he published studies in botany, ornithology, and mammalogy. He and Audubon also collaborated on a book, *Viviparous Quadrupeds of North America*, and published detailed descriptions of dozens of mammal species, making taxonomic distinctions of a high standard for the day. The families also collaborated in their personal lives, with two Audubon sons marrying two Bachman daughters.

Bachman carefully avoided religious arguments in making the case that all human races belong to a single species. Instead, he relied on the same scientific methods used to separate one species from another in the animal world. He pointed out, for instance, that the American swan and the trumpeter swan resemble each other so closely that they were long regarded as one species. Then dissection revealed significant differences in bone count and other internal structures, causing naturalists to split them apart as two separate species. "Let us now apply this rigid rule of investigation to the anatomy of the bones and the physiology of the various organs in the different races of men," he suggested. Then he did the numbers: The human breast bone consists of eight pieces in infancy, three pieces in youth, and a single piece in old age—regardless of race. The human skull contains eight bones, including four bones in the ear—again regardless of race. The number of milk teeth and their adult replacements is identical in all races.

Nightingales and mockingbirds differ enough in the structure of the larynx alone to distinguish them as separate species, Bachman wrote. But in humans the structure of the larynx is identical, and "the same power of speech, the same power of song, the same love of music, exists in races the most widely separated from each other in colour." Given the long list of shared characteristics found in all human races, he asked, "what grounds have we, on any principles of science, to deny their common origin?" He compared the differences among human races with those found in domestic cattle, which display "every conformation of skull" some with "immense horns . . . as long and as thick as the tusks of the elephant," others with no horns at all, some as big as small elephants, others the size of dogs, some with humps, some with long pendulous ears, some black, brown, white, or spotted—but all universally regarded as one species. "If they will make five, ten, or an hundred species of men," Bachman wondered, "why do they not carry out their principles and make five, ten, or an hundred species of common cows."

The argument grew louder and more rancorous through the 1850s. His Hippocratic oath did not keep Josiah Nott from voicing his wish to "kill of[f] Bachman," to "skin Bachman," to see him "cut up into sausage meat." After what he deemed a particularly effective riposte, he wrote of Bachman, "I really feel as if a viper had been killed in the fair garden of science, and I hope his death will be a warning to all such blasphemies against God's laws"—the laws, that is, that made blacks a separate, inferior species, and keeping them as slaves the work of righteousness.

"Scientific Moonshine"

In the face of such lynch-mob language, a newcomer entered the debate against Morton and his followers in 1854. Then about 25 years old, Frederick Douglass had grown up as a slave on a plantation on the Eastern Shore of Maryland, and in Baltimore, before making his escape to New York. He had been separated from his mother early on, and bore scars on his back from having been whipped. But his physical presence hardly suggested innate inferiority, nor any hint that he might belong to some other species. He was more than 6 feet tall, with olive brown skin, a shock of hair slanting across a notably broad forehead, and flashing, wounded eyes. The feminist Elizabeth Cady Stanton vividly recalled the first time she heard Douglass speak, at an antislavery meeting in Boston, where he swept his listeners from laughter to tears on the tide of his voice. "All the other speakers seemed tame after Frederick Douglass," Stanton wrote. "He stood there like an African prince, majestic in his wrath."

When the literary societies at Western Reserve College in Ohio took the bold step of inviting a black man to speak during commencement week, Douglass delivered an address on "The Claims of the Negro Ethnologically Considered." It was a devastating critique of the science of the day: "The evils most fostered by slavery and oppression, are precisely those which slaveholders and oppressors would transfer from their system to the inherent character of their victims." Douglass laid into Samuel G. Morton's supposed objectivity: "The European face is drawn in harmony with the highest ideas of beauty, dignity and intellect . . . The Negro, on the other hand, appears with features distorted, lips exaggerated, forehead depressed—and the whole expression of the countenance made to harmonize with the popular idea of Negro imbecility and degradation."

Douglass ridiculed Morton's assertion that the Egyptians had been white and claimed the glory of Egyptian civilization as the product of blacks and mulattos. By his presence as well as his words, he also demolished the attempt to deny the full measure of humanity to the Negro: "His speech, his reason, his power to acquire and to retain knowledge, his heaven-erected face, his habitudes, his hopes, his fears, his aspirations, his prophecies, plant between him and the brute creation, a distinction as eternal as it is palpable," he declared. "Away, therefore, with all the scientific moonshine that would connect men with monkeys" and particularly with the "sort of sliding scale, making one extreme brother to the ourang-ou-tang, and the other to angels."

In Worcester, Massachusetts, the newspaper *Spy*, an abolitionist mouthpiece, took note of the speech, though with faint praise. Douglass had shown "he was familiar with the general and natural history of man. His language was chaste, and his reasoning strong, able, and logical." Even abolitionists regarded him, Douglass recognized, as, at best, a lesser member of the species. Not a brother, but a chattel, a thing.

"*It* could speak," he remarked mordantly.

Chapter Thirteen

"A FOOL TO NATURE"

The enthusiasm of naturalists is very apt to surprise ordinary people.

—WILLIAM SWAINSON

THOMAS EDWARD, a shoemaker in Banff, on the northeast coast of
Scotland, had been afflicted with the love of "beasts" from earliest
childhood and on that account managed the singular achievement of
getting expelled from three schools before he finally abandoned formal
education at the advanced age of six. His infractions included applying
a horseleech to the leg of a fellow student and smuggling a jackdaw into
class, hidden in his trousers. (A jackdaw is a very large bird for a small
boy to hide in his trousers, but Edward was not easily discouraged.)
This predilection continued into his working life, when he angered
a boss by bringing moles to work and alarmed a fellow worker with
runaway caterpillars.

As an adult, Edward was a fierce-looking character, with deep-set
eyes, a high brow, hair that rolled back in a windswept wave, and a
ragged beard. His work in a shoe factory consumed the hours from 6
a.m. to 8 or 9 p.m. six days a week. Sundays, like his devout Scottish
neighbors, he spent in prayer and at rest. But at night he wandered
the north coast countryside collecting almost anything that lived. He
carried boxes for butterflies and beetles, a book for pressing botanical

The passion for natural history swept up people of all classes, among them the shoemaker Thomas Edward.

specimens, and a decrepit shotgun, its barrel lashed to the stock with heavy twine.

When he got tired, he often bedded down in the shelter of rocks, graveyards, or abandoned buildings. Sometimes he tucked himself feet first into a fox hole or badger hole in a sandy bank. Once, a badger came home and, finding him there, bared its teeth. With regret, Edward shot it (he did not like wasting ammunition). Another night in a rainstorm, he slept under a flat gravestone supported by four pillars. At the height of the storm, a weird moaning awakened him—not ghosts, but as he discovered when they raced across his legs, just a couple of cats.

Edward's zeal for the natural world ran even deeper than this might suggest: Whatever he killed, he wrapped in gun wadding and slept with it stored atop his head, under his hat. One morning he woke with "something cold pressing in betwixt my forehead and the edge of my hat." It was a nose, with a live weasel attached, hoping to get at some birds he had stored there. Twice, Edward grabbed the weasel and tossed him aside, and then went back to sleep. The third time, he got up and walked to a different field 100 yards away. The weasel followed him and on the fifth attempt, Edward wrote, "I suffered him to go on with his operations until I found my hat about to roll off. I then throttled, and eventually strangled, the audacious little creature, though my hand was again bitten severely."

Then he went back to sleep, with this new specimen no doubt safely stashed in the overhead compartment. The same thing happened on other occasions with rats and even a polecat, with which he wrestled for two hours. "I never wasted my powder and shot upon anything that I could take with my hands," he declared.

This makes Edward sound like some kind of backwoods sociopath, and he may have seemed that way to unfortunate strangers who happened to see him lurch up from among the gravestones just before dawn. But in fact he was a learned naturalist, elected by the Linnean Society to its small circle of associates. He was also a family man. His biographer, the self-help author Samuel Smiles, dryly noted that "he knew nothing about 'Malthus on Population'" but "merely followed his natural instincts." His wife Sophia "was bright and cheerful, and was always ready to welcome him from his wanderings." Thus they managed to produce eleven children together and spent much of their lives on the brink of starvation.

Other naturalists who beseeched him for specimens mostly seem to have been oblivious to his circumstances, though one benefactor supplied him with a microscope for his work. They often promised to send the books he requested, but seldom delivered. At one desperate point, after an exhibition of the specimens he had collected around Banff failed to earn the money he needed to feed his family, Edward attempted suicide. But as he waded into the sea to drown himself, the sight of a rare bird distracted him. He spent the next half hour chasing it and must have realized that he was too attached to life in all its splendid permutations to give up so easily. Back home, he went on supplying crustaceans to Mr. Bate, echinoderms to Mr. Norman, sponges to Mr. Bowerbank, sea squirts and mollusks to Mr. Alder of Newcastle, and so on. What Edward mainly got from them were the scientific names of the specimens he supplied and the chance to correspond with like-minded scientists. He also enjoyed the honor of having an isopod named for him, *Praniza edwardii* and a fish, *Couchia edwardii*. But since both proved to be synonyms for previously described species, the names later vanished. One book, *A History of the British Sessile-Eyed Crustacea*, praised him as "that indefatigable lover of nature" and cited him as the source for 20 new species in that group alone. Otherwise, Edward, like many collectors of that era, worked purely for love of the natural world. As he put it, "I have been a fool to nature all my life."

"*That Vampire*, Ennui"

The diagnosis was remarkably common then. By the middle of the nineteenth century, interest in nature and new species had grown from a pastime of explorers and the educated elite into a popular mania. In Britain particularly, faddish waves of enthusiasm raged through the populace: "One year they centred on mosses, the next madrepores," a kind of coral, Lynn Barber writes, in *The Heyday of Natural History*. "In the decade from 1845 to 1855 they moved successively from seaweeds to ferns to sea-anemones. In the next decade they switched bewilderingly to sea-serpents, gorillas and infusoria. These were all national crazes."

It may be hard to imagine a national craze for infusoria, which are minute aquatic creatures including ciliates, protozoa, and single-celled algae. People from that era would probably feel equally bemused, and with better reason, by our crazes for Tickle Me Elmo, pet rocks, or Sudoku. But affordable microscopes had recently opened up new worlds to middle-class curiosity. Likewise, ornate, glass-domed "ferneries" encouraged "pteridomania," the madness for collecting and keeping ferns. Even quirkier enthusiasms erupted in the provinces, according to Barber's *Heyday*, with limpet fever striking Bangor in northern Ireland in the 1830s, and a passion for keeping baby alligators seizing the otherwise sensible women of Southport, near Liverpool, in the 1870s. Books with seemingly oxymoronic titles—*A Popular History of British Zoophytes, or Corallines*—sold handsomely, thanks in part to lavish illustrations made possible by new printing technologies.

The status of natural history as an avocation had risen to such a degree that novelist Charles Kingsley could depict a London merchant's head clerk, "your very best man of business, given to reading of Scotch political economy and gifted with peculiarly clear notions on the currency question," who nonetheless gets arrested "in the act of wandering in Epping Forest at dead of night, with a dark lantern, a jar of strange sweet compound," and pockets crammed with pillboxes. For a "blameless entomologist" like the head clerk, painting tree trunks with a sugary concoction, or "sugaring for moths," was a perfectly reasonable way to attract choice insect specimens. The wardens who arrested him were merely too dim to realize it. But Kingsley may have been overstating the status of natural history when he depicted "a Cromarty stonemason" as "the most important man in the City of Edinburgh by dint of a work on fossil fishes" and when he declared that "the successful investigator of the minutest animals" was now "fit company for dukes and princes." (Though he did not name him, Kingsley's Cromarty stonemason was Hugh Miller, author of several popular geological books.)

The motives inspiring these new fools to nature varied widely. It was a time of social upheaval, as factory work took hold and people flocked to jobs in the cities. The Industrial Revolution induced a natural surge of romanticism about the leafier existence rapidly being left behind. Outings to the countryside and the seashore looked like a good way to get healthy exercise and recreation, combined with spiritual and intellectual uplift. The social status of natural history also made it an excellent means of social climbing and career building.

Nature was also a remedy for the newfound idleness of the rising middle classes. "The boredom of the affluent Victorian family is truly frightful to contemplate," Lynn Barber writes. Field guides to the natural world offered an engaging form of diversion. "The naturalist knows nothing of that *taedium vitae*,—that vampire, *ennui*, which renders life a burden to thousands," the Rev. David Landsborough advised, in his 1849 book, *A Popular History of the British Sea-weeds* (a sort of prequel to his book about zoophytes). "To him every hour is precious." Specimens from the natural world were of course also ornamental, and the Victorians liked few things better than tricking out their homes with stuffed pheasants and pinned butterflies under glass.

To supply this demand and to serve the more serious needs of the scientific community, specimen dealers set up shop around the world—Samuel Stevens in London; Maison Verreaux in Paris; Ward's Natural Science Establishment in Rochester, New York; Dr. August Müller in Berlin; and eventually hundreds of others. The proprietors were typically collectors themselves, often undertaking risky expeditions in their early years, and sometimes publishing the first scientific descriptions of new species they had brought home. (Such a description could boost demand among collectors seeking a "complete set" of the genus of cone shells or ground beetles or hawk moths in which they specialized.)

By paying for naturalists to supply them with specimens on a piece rate, these dealers helped to sustain an extensive network of collectors in the field. Their workrooms also served as a training ground for young naturalists and future museum staffers. Many of the museums themselves only came into existence through the intervention of dealers. Jules Verreaux, for instance, helped found the South African Museum in Cape Town, in 1825. Henry A. Ward of Ward's Natural Science Establishment not only provided specimens for new museums, according to science historian Mark Barrow, "he often helped plant the idea, locate suitable patrons, initiate subscription drives, design exhibits, and otherwise work to bring his ambitious schemes to fruition." In one such case, he badgered the department store magnate Marshall Field into founding the Field Museum in Chicago. By the end of the nineteenth century, the United States alone would have more than 100 natural history museums, in no small part due to the efforts of dealers. Natural history societies also sprang up everywhere.

Not everyone approved of this broadening excitement about the natural world. For George Ord, a difficult and deeply snobbish ornithologist at the Academy of Natural Sciences of Philadelphia, it was an intrusion. In November 1842, he wrote to his equally cranky English friend Charles Waterton, to fulminate against "an English fool, named Smithson, or some such name" who had endowed a new national museum for the promotion of science in the United States.

The nascent Smithsonian Institution was already busying itself, Ord sneered, "in setting up a grand meeting of the wisdom of the nation, on the model of the British Association, in order that 'the beams of science may be diffused to the remotest parts of the land.' The preliminary concourse will take place in a few days. There will be an immense stir; a grand *speechifycation*, characterized by rant, fustian and nonsense; the baboons of literature and science will play their pranks for the amusement of the mob, and then the farce will end. Science and learning in Washington? I should as soon expect to see them flourish within the purlieu of Newgate" (the notorious London prison). In his next sentence, Ord revealed the true source of his anger, as an envious museum curator: "It is this same National Institution which has laid its grasping paws upon the precious collections" of the U.S. Exploring Expedition, just returned from four years visiting the remotest parts of the Pacific. The business of discovering new species was becoming increasingly competitive, pitting collectors, museums, cities, and nations against one another in the quest for the next great prize.

Our Demanding Correspondents

For naturalists traveling abroad, it became routine to receive letters begging for specimens. The growing ranks of natural history enthusiasts were not content to limit themselves to what they could collect in their own neighborhoods, or purchase from the nearest specimen dealer. "I only wish I could impart some of my Botanic enthusiasm to you," a

Connecticut man wrote in 1841 to a neighbor, the Rev. Thomas S. Savage, who had gone off to work as a missionary and physician in Liberia and was displaying a misguided penchant for zoology. Resorting to guilt, the writer mentioned an acquaintance then in Alabama and wrote, "*He finds time* to make large collections of plants . . . *Think* and *act* on this *my Dear Friend.*" And finally, "If you should be so fortunate as to meet with that famous tree, the *Adansonia digitata* [the baobab], may I entreat you to secure several fine specimens of the flower, and leaves, and a sample of the bark." .

Most correspondents managed to couch these requests in the diplomatic language of mutual aid. The more scholarly ones often enclosed a scientific paper or two, and they offered morsels of gossip, mixed with collegial griping and practical tips. Another neighbor wrote to Savage advising: "I find but few Entomologists understand and appreciate the *value of Turpentine*. In hot weather Boxes of Insects are very *apt to mold* when kept closed, but the turpentine *prevents this* and never injures beetles. For the Butterflies it requires more care. "

Sometimes the advice was not so practical. Augustus Gould, a physician and prominent naturalist in Boston, had lately taken up the study of shells. "As your main object in studying the works of Nature seems to be the preservation of your health," he wrote to Savage, "you may perhaps be pleased to search occasionally for shells . . . always to be found in all fresh water streams and rivulets in all parts of the world." Such habitats were of course an ideal place for contracting malaria from mosquitoes, particularly in West Africa. But nobody knew that in the mid-nineteenth century.

"Have you any alligators in your neighborhood," another correspondent wrote Savage in 1850, when he had returned to the United States to take up duties at a church in Mississippi. "I have been for a long time desirous of getting the skeleton of one say five or six feet in length . . . All I should ask in the way of trouble would be to have the flesh cut off in a rough manner."

"Fantastic Trilobite"

If they were hunting for the secrets of life itself, the naturalists of the day were also having fun, banging across land and sea in what amounted to a great scavenger hunt to find and name new species. In the summer of 1845, for instance, the brilliant naturalist Edward Forbes was cruising the Scottish coast on a friend's yacht, dropping a dredge to collect specimens along the way. He quickly sketched and described 20 new species of medusae, or jellyfish, as well as several new starfish and shellfish. "One of the most interesting features of the excursion," his colleagues George Wilson and Archibald Geikie wrote in their *Memoir of Edward Forbes*, "was the discovery of several living molluscs, which had up to that time only been known in a fossil state." Then Forbes headed down the Irish Sea to join paleontologist Henry De la Beche at a dig in Hook Point, on the southeast coast of Ireland.

"The excitement of detecting new fossils, and recognizing old ones, of eliciting the grand physical structure of the country, and disentangling the intricacy of its details, kept the party busily at work," according to Wilson and Geikie. "The freshness of out-door labour, by breathing new life and heartiness into their spirits, braced them for physical exertion. No company, indeed, was likely to lack life when De la Beche and Forbes were members of it. They threw a charm around their work, even in its dullest parts."

Evenings "passed merrily away by the fireside of some cabin or village inn," with Forbes "chanting some of his humorous scientific songs" or entertaining the party by sketching gnomes, fairies, and whimsical versions of the species they had spent the day uncovering. Forbes even included such sketches in his best-known scientific work, *A History of British Star-fishes, and Other Animals of the Class Echinodermata*. One chapter featured an illustration of Cupid in a sea-going chariot drawn by a pair of sea creatures with bodies like snakes and heads like sea urchins (they were Ophiuridae). Another chapter ended with Puck playing his pipe for a couple of dancing brittle stars, one of them rest-

Edward Forbes sometimes dressed up his discoveries of new species with whimsical sketches of imaginary ones.

ing the back of a "hand" against out-thrust "hip." Elsewhere, he drew a stingray smoking a pipe and winking.

There was a clear dividing line between science and whimsy. But some naturalists crossed it with gleeful abandon. De la Beche, for example, drew a deft cartoon mocking a theory put forward by geologist Charles Lyell that the Earth's history was an endlessly repeating cycle. Lyell imagined that the iguanodon might some day "reappear in the woods, and the ichthyosaur in the sea, while the pterodactyle might flit again through umbrageous groves of tree-ferns." De la Beche's cartoon depicted a professor with the head of an ichthyosaur lecturing on the "insignificant" teeth and "trifling" jaws of a skull belonging to some "lower order of animals"—*Homo sapiens*.

Light verse seems to have come as readily as cartooning to these gifted naturalists. Earlier in 1845, for instance, Forbes, then 30, published a Valentine love poem in the *Literary Gazette*, depicting his own heart as a fossil :

> *Like some fantastic Trilobite*
> *That perished in Silurian sea,*
> *And long lay hid from mortal sight,*
> *So was the heart I yield to thee.*
> *Now from its stony matrix free,*
> *Thy paleontologic skill*
> *Once more hath call'd it forth to be*
> *The servant of thy will.*

The continual discovery of strange new forms of life didn't just foster a fanciful mood among the naturalists themselves. It also inspired the

A geologist's cartoon lampooned Charles Lyell's idea of Earth history as an endlessly repeating cycle.

flowering of a genre—literary nonsense. Before he became celebrated for *The Owl and the Pussycat* or *The Jumblies*, for instance, Edward Lear was a talented bird illustrator. His first book, *Illustrations of the Family of Psittacidae, or Parrots*, published in 1832 when he was 19, drew favorable comparisons with Audubon. When Lear turned to nonsense, he often dispatched his characters, like naturalists, on wild explorations to the back of beyond ("They sailed to the Western Sea, they did, / To a land all covered with trees"). He also had them devote considerable energy to collecting the oddities of the country:

> *And they bought a Pig, and some green Jack-daws,*
> *And a lovely Monkey with lollipop paws,*
> *And forty bottles of Ring-Bo-Ree,*
> *And no end of Stilton Cheese.*

Nonsense was almost a byproduct of natural history. The twin themes of exploration and taxonomy were "present in the genre as a whole, even in Lewis Carroll, who had no special interest in the subject," Jean-Jacques Lecercle writes in *Philosophy of Nonsense*. "The reader of *Alice's Adventures in Wonderland* is in the position of an explorer: the landscape is strikingly new, new planets swim into his ken, and a new species is encountered at every turn, each more exotic than the one before. Nonsense is full of fabulous beasts, mock turtles and garrulous eggs."

In Lecercle's view, Charles Darwin himself could sound at times as whimsical as Lewis Carroll. When he pulled the tail of an irascible lizard in the Galapagos, for instance, Darwin wrote: "At this he was greatly astonished, and soon shuffled up to see what was the matter; and then stared me in the face, as much as to say, 'What made you pull my tail?'" Likewise, in Patagonia, Darwin communed with the camel-like guanacos: "That they are curious is certain; for if a person lies on the ground, and plays strange antics, such as throwing his feet up in the air, they will always approach by degrees to reconnoitre him." Darwin was only 23 at the time, not the gloomy eminence of later years, but

Edward Lear's literary nonsense drew on natural history lore—and on his own early work as a bird illustrator.

Lecercle takes comfort in the idea "that the famous scientist should behave like Lear's 'Old Man of Port Grigor,' who 'Stood on his head till his waistcoat turned red.'"

Alfred Russel Wallace, coauthor of the theory of evolution by natural selection, was so enamored of Lewis Carroll that, late in life, with his travels to the Amazon and the East Indies far behind him, he named his house Tulgey Wood, after the haunt of the jujub bird and the frumious bandersnatch in "Jabberwocky":

> *And as in uffish thought he stood,*
> *The Jabberwock, with eyes of flame,*
> *Came whiffling through the tulgey wood,*
> *And burbled as it came!*

Lecercle concludes that nonsense was "part of a craze for discovering and classifying new species," and the scientists sometimes envied the writers their freedom—"Oh frabjous day! Calloo! Callay!"—to invent those species out of crazy cloth.

Thy Works Shall Praise Thee

For other naturalists then, studying real species was an extension of their religious faith. The Rev. William Paley's 1802 book *Natural Theology* had popularized the idea that nature revealed the mind of God. Like an ingenious clockmaker, the Creator had designed each species to be perfectly adapted to its habitat. Discovering this perfection was a sacred discipline, and the more detailed and scientific the description of these adaptations, the closer the naturalist came to God. Apart from seeming to be true, natural theology provided not just an excuse but a religious mandate for getting outside and chasing beetles. Otherwise, collectors were liable to suffer "from the barbs of soulless utilitarians," historian David Allen writes, for being caught up in the "compulsive fascination" of a seemingly useless pursuit. Their best defense, and one that was "ordinarily unanswerable," was "to discover some moral content" in their pastime "and so proclaim its edifying character."

Proponents of natural theology genuinely delighted in the myriad ways God had formed the biological world for human comfort. "An all-wise and beneficent Providence" had assigned birds, E. P. Thomson wrote in his 1845 *Notebook of a Naturalist*, "to free us from the clouds of insects, which would otherwise infest our dwellings, and destroy the labours of the field." Natural theology also provided a kind of economic argument for discovering the perfection of the natural world. One soulless utilitarian proposed, for instance, that the "productions of nature, varied and innumerable as they are, may each, in some future day, become the basis of extensive manufactures, and give life, employment, and wealth, to millions of human beings."

Nor had God neglected the comfort of the animals themselves. Collecting in Jamaica, the naturalist Phillip Henry Gosse contemplated the nictitating membrane, a transparent lens protecting the eyes of a tree frog. Living among the "sharp serrated spines" of wild pines, this frog moves "to and fro by violent headlong leaps, in which it needs to be guided by the sharpest sight," he wrote. "How interesting, then, is it to see that its gracious Creator has furnished it with a glassy window, which it may in a moment draw before its eye, for shelter from danger, without in the least hindering the clearness of its vision! . . . 'All thy works shall praise thee, O Lord!' "

Author Margaret Gatty chose to praise God by making herself an expert on seaweed, discovering new species and having others named in her honor. Gatty also became a highly successful author of children's books about the natural world. Her 1855 *Parables from Nature* went through more than 100 editions and multiple translations. Nature abounded in "wonderful adumbrations of divine truths," she wrote, and the metamorphosis of a grub into a dragonfly—clambering up from its dark underwater world, sprouting wings, and lifting off into the heavenly light—provided a scientific basis for belief in the afterlife of the human soul. ("Dear Ones!" says the grub, as he is about to become a dragonfly and follow a brother he has mourned, "I go, as he did, upwards, upwards, upwards!") She didn't mean to push the natural analogies too far, Gatty wrote, but getting the details right was fundamental. So even when she made a seashell into a talking character, she included a note identifying it as a member of the limpet genus *Patella*.

Other naturalists, setting out to glorify God by describing His creation, came to see things they had not expected. Contemplating the vast span of geologic time, they wondered why so many species had come and gone even before humans arrived on the scene to appreciate them. The spiral form of fossil ammonites delighted Hugh Miller, the Cromarty stonemason, paleontologist, and evangelical Christian. "But why so much beauty when there was no eye of man to see and admire? Does it not seem strange that the bays of our coasts should have been speckled by fleets of beautiful little animals, with their tiny sails spread to the wind and their pearly colors glancing to the sun, when there was no intelligent eye to look abroad and delight in their loveliness?"

Many believers struggled to hold on to the security of the traditional view of the world—created by God for the comfort of Man. The idea that the Earth had had a history before us was deeply disturbing—like the moment when a small child dimly realizes that his parents had other lives before the one in which he has happily cast himself as the star. But the discrepancies between Scripture and the facts of nature became more obvious each year.

One reader later recalled how much satisfaction he had gotten from reading *Natural Theology* as a student at Cambridge. He declared that its logical arguments had given him "as much delight as did Euclid," which he apparently did not intend as faint praise. He put these two authors above all others "in the education of my mind." The student who thus learned the critical importance of studying minute variations in nature was named Charles Darwin.

Chapter Fourteen

THE WORLD TURNED UPSIDE DOWN

It strikes me, that all our knowledge about the
structure of our Earth is very much like what an
old hen wd know of the hundred-acre field in a
corner of which she is scratching.

–CHARLES DARWIN

IN THE SPRING OF 1837, two months before Darwin would begin his
first notebook on the origin of species, a Scottish journalist with no
scientific training was already making a stab in print at the outlines
of evolutionary theory. Animal and vegetable life had "taken forms
suitable to contemporary circumstances," Robert Chambers wrote in
Chambers's Edinburgh Journal, the popular weekly he and his brother
published. And when circumstances changed, he continued, "living
things of both kinds have been destroyed in entire species without
mercy, new ones, suitable to the new circumstances, readily spring up,
to be in their turn swept away, and leave no copy." Later that year,
he speculated about competition and extinction, and about how fossil
species might have been transformed over time into totally different
creatures.

Back in 1831, Patrick Matthew, a Scottish landowner, had also theorized in print about what would later become popularly known as the
survival of the fittest: "Those individuals who possess not the requisite
strength, swiftness, hardihood, or cunning, fall prematurely without

reproducing" generally due to predation, hunger, or disease, "their place being occupied by the more perfect of their own kind." He thought "this natural process of selection" could lead to the evolution of new species: "the progeny of the same parents, under great differences of circumstance, might, in several generations, even become distinct species, incapable of co-reproduction." Darwin later called it a "complete but not developed anticipation" of his own theory—and understandably excused himself for "not having discovered" the precedent, which had appeared only in the appendix of Matthew's opus *On Naval Timber and Arboriculture*. (Matthew nonetheless modified his calling card to anoint himself "Discoverer of the Principle of Natural Selection.")

Evolutionary ideas, first loosely formulated decades earlier by Erasmus Darwin in England and Jean-Baptiste Lamarck in France, among others, were now massing beneath the surface, building up pressure, distorting the surface crust of conventional discourse, flaring up in odd corners of the intellectual world. Then, in October 1844, with the appearance of an anonymous tract called *Vestiges of the Natural History of Creation*, they burst out onto the public streets, up church aisles, and into coffee shops and gentlemen's clubs. *Vestiges* was an almost miraculous work, deeply flawed, brimming over with "dangerous" ideas, and yet wildly popular. It was a *Victorian Sensation*, as historian James A. Secord has titled his definitive history of the book.

The anonymous "Mr. Vestiges," as he became known, deftly wove evolution into a sweeping history of the cosmos, beginning in some primordial "fire mist." He did not get around to mentioning God until page 152 of the British first edition, and then mainly to dismiss as "ridiculous" the notion of the Almighty having "to interfere personally and specially on every occasion when a new shell-fish or reptile was to be ushered into existence." Instead, he proposed that "the Great Architect" had set the world in motion and then allowed God-given natural laws to take their course. Thus different species shared the same basic physiological structures and modified them to their own needs, with ribs, for instance, serving for locomotion in snakes, the proboscis becoming a prehensile instrument in elephants, and the same bony structure that formed the human hand serving to spread the wings in bats. At the very moment when Darwin thought that admitting to belief in the mutability of species was like "confessing a murder," the anonymous author of *Vestiges* plainly declared that humans had arisen from monkeys and apes.

"In many ways," Secord writes, "Darwin had been scooped." Though *Vestiges* is largely forgotten today, it would alter the course of public opinion, and set both Darwin and an unknown surveyor named Alfred Russel Wallace onto career paths that would converge, years later, in the triumph of evolutionary thinking.

"Mud-Worms and Monkeys"

But the real miracle of *Vestiges* was the immediate popular reception it received. "Mr. Vestiges" had taken scientific ideas associated with political radicals, atheists, and, even worse, the French (particularly the much ridiculed Lamarck), and somehow spun them into a confection Prince Albert could happily read aloud to Queen Victoria at Buckingham Palace. The author appealed to readers of all classes with the uplifting message that evolution was about progress and improvement, with *Homo sapiens* standing at the top of the heap (and still climbing). He could write, for instance, that the human brain is subject to natural law, but in the next breath add: "And how wondrous must the constitution of this apparatus be, which gives us consciousness of thought and of affection, which makes us familiar with the numberless things of earth, and enables us to rise in conception and communion to the councils of God himself!"

Oddly, the idea of low origins made some people almost giddy. Jocular types greeted each other on the street with phrases like, "Well, son of a cabbage, whither art thou progressing?" Augustus De Morgan, a puckish mathematician, worried about a kind of literal backsliding— whether "by constantly thinking of ourselves as descended from primaeval monkeys" we might "actually *get our tails again?*" If a colleague commented on *Vestiges* too knowingly, De Morgan thought it was great fun to peer around facetiously for signs of a posterior appendage.

Others took evolution more seriously. Rev. Baden Powell, the prominent Oxford mathematician and theologian, praised *Vestiges* in print for boldly applying natural laws to the history of life. On a museum visit, Florence Nightingale noticed that small flightless birds of the modern genus *Apteryx* had vestigial wings like those of the giant moa, an extinct bird that had recently been discovered. One species ran into another, she remarked, much "as *Vestiges* would have it."

Many clergymen railed from the pulpit against such evolutionary thinking. But scientists hated *Vestiges*, too, for its loose speculation and careless use of facts. Given the pervasive influence of Paley's *Natural Theology*, the religious and scientific critics of *Vestiges* were often one and the same. The Rev. Adam Sedgwick, who had been Darwin's geology professor at Cambridge, let loose a cry of anguish: "The world cannot bear to be turned upside down." He thought that the faulty logic of *Vestiges* revealed the hand of a female author, and this mistaken belief colored his rhetoric. He set out to stamp "with an iron heel upon the head of the filthy abortion, and put an end to its crawlings." In a blistering 85-page attack in the *Edinburgh Review* he effectively called *Vestiges* satanic: "and the serpent coils a false philosophy," inviting readers "to stretch out their hands and pluck the forbidden fruit."

The Rev. George B. Cheever also rolled out the heavy guns in the first American edition, blasting *Vestiges* for putting more faith "in the footprints of death than in the Word of Life." It was surely the only book ever commended to gentle readers with an introduction calling on Scripture for "a thunder-stroke of annihilation to this writer's speculations upon nature."

At a meeting of the British Association for the Advancement of Science in June 1845, the eminent astronomer John Herschel criticized the book's failure to explain *how* evolution might have occurred; *Vestiges* was about as miraculous as the Biblical account of Creation. (The anonymous "Mr. Vestiges " happened to be sitting in the front row during this attack, probably trying not to squirm.) The naturalist Edward Forbes remarked that an author "who succeeds in half-persuading the majority of his readers . . . that they and all mankind are the lineal descendants of mud-worms and monkeys . . . has a power within him which might be turned to better purposes." All it needed was the addition of actual knowledge. Darwin also disliked what he called "that strange unphilosophical, but capitally-written book." Some people even guessed that he was the author, "at which I ought to be much flattered & unflattered." He confided to his good friend the botanist Joseph D. Hooker that the author's "geology strikes me as bad, & his zoology far worse." Darwin and others in the scientific community soon suspected that the real author was an outsider, the Edinburgh journalist and publisher Robert Chambers.

Science for Everyman

The two Chambers brothers, sons of a failed manufacturer, had launched their *Edinburgh Journal* in 1832 with the aim of providing "a meal of healthful, useful and agreeable mental instruction" that even "the poorest labourer in the country" could afford and understand. Though he had no special training, Robert Chambers was an enthusiastic reporter of scientific news, with a strong interest in geology, astronomy, paleontology, and the popular pseudoscience of phrenology. As a naturalist, Robert's "obvious lack of taxonomic knowledge was balanced by the sense of wonder he skillfully conveyed to readers," writes science historian Joel Schwartz.

Both brothers suffered from a genetic defect, having been born with six fingers on each hand and six toes on each foot. Corrective surgery had left Robert lame, encouraging him to turn to books when other children were out playing. His condition may have made him particularly alert to the ways biological forces could transform a species. In *Vestiges*, he relied on the work of Charles Babbage, the mathematician who had devised a mechanical system for calculating mathematical data (he also gets credit for inventing the concept of a programmable computer). Chambers used Babbage's theories to explain how, after generations of parents' producing offspring much like themselves, a species might suddenly become different.

Babbage believed that, rather than jumping in with a miracle of cre-

Wild speculations hurt his argument, but the journalist Robert Chambers saw better than most scientists where biology was headed.

ation every time a new species was required, the Almighty had merely promulgated laws to introduce the right species at the right time. He asked readers to imagine a calculating machine that produces a series of numbers increasing by one digit each time until it reaches 100,000,001, then suddenly jumps by 10,001 digits and continues to follow a new, more complicated pattern thereafter, until that, too, changes, at some point further down the line.

To the modern ear, it sounds momentarily as if he was arguing that a species could evolve by mutations in its programming. The development of genetics much later in the century would reveal this to be precisely the case. But the idea of preordained shifts in natural law missed the randomness of genetic mutation, and neither Babbage nor Chambers proposed anything like natural selection to explain why some mutations might survive and others quickly vanish.

Chambers thought such shifts in natural law acted mainly on the development of the embryo. He cited the example of bees manipulating their larvae to turn some juveniles into workers and others into drones. He also talked—without mentioning his own hexadactylism—about how the wrong circumstances during development could sometimes lead to monstrous offspring in humans and other species. But he thought developmental forces could sometimes produce positive changes, too. He knew about the idea put forward by some scientists that embryos pass through the forms of more primitive animal groups—reptiles, fish—in the course of their development. And here he went off the taxonomic deep end: The right circumstances during development, he wrote, could cause a fish to develop a reptilian heart, or a goose to give its offspring the body of a rat, and thus produce a platypus. Small wonder the scientific community let out a howl of protest. But they also howled because Chambers was forcing them to confront the connections tying together the species they delighted in discovering.

Chambers saw himself as having a broader view of nature than professional scientists in their narrow specialties: "From year to year, and from age to age," he wrote, "we see them at work, adding no doubt much to the known, and advancing many important interests, but, at the same time, doing little for the establishment of comprehensive views of nature. Experiments in however narrow a walk, facts of whatever minuteness, make reputations in scientific societies; all beyond is regarded with suspicion and distrust."

He wasn't entirely wrong about his own breadth of vision. In *Vestiges*, "a dilettante perceived more clearly and sooner than most of the professional scientists the new direction that biology was taking at the time," the twentieth century paleontologist George Gaylord Simpson wrote. He may have "mingled legitimate evidence with false data, naïve arguments, wild speculations, and impossible theories," but the result was to take at least some of the sting out of the evolutionary theories

Darwin would present years later. Thus Simpson credited *Vestiges* with "a contribution to the history of opinion, as distinct from that of ideas." *Vestiges* also deserves credit, or blame, for having aggravated Darwin's characteristic caution.

"A Complete Scribbler"

Darwin did not need evolution to make his name. Following the return of the *Beagle* in October 1836, after five years of exploration, he scored an immediate triumph with his South American fossils, including giant ground sloths and extinct rodents the size of hippos. (Richard Owen, the former medical student with the craniological longings, was now established as an anatomist at the Royal College of Surgeons and handled the descriptions of the fossils for Darwin.) Early on, Darwin also delivered a talk to the Geological Society offering evidence that the coast of Chile had been formed from uplifted seafloor, and it earned him such scholarly acclaim that he felt, he said, "like a peacock admiring his own tail." He gained admission to the prestigious Athenaeum Club and confessed that the first time "I sat in that great drawing room, all on a sofa by myself, I felt just like a duke." It was a long way from dining alone with Capt. Robert FitzRoy or slinging his hammock in the 10-by-12-foot cabin he had shared with two junior officers aboard the *Beagle*.

Darwin had taken a bachelor flat around the corner from his brother Erasmus, at 36 Great Marlborough Street, where getting his findings into print turned him into "a complete scribbler" (to which he added "God help the Public"). Like any young man approaching 30, he also found time to contemplate the possibility of marriage, though from a species seeker's peculiar perspective: "As to a wife, that most interesting specimen of the whole series of vertebrate animals," he wrote to an old friend, "Providence only know[s] whether I shall ever capture one or be able to feed her if caught." He admitted that the vision of a cottage and "some white object like a petticoat" often drove granite, trap, and other geological topics "out of my head in the most unphilosophical manner."

Perhaps with the idea of keeping both geology and the opposite sex in view simultaneously, Darwin became a regular visitor at the Blooms-bury home of geologist Leonard Horner and his five highly educated daughters. (One of the Horner girls, Mary, was already married to the geologist Charles Lyell, who would become a key Darwin ally.) But Darwin's father steered him instead toward Emma Wedgwood, his 30-year-old first cousin, good-natured and with a handsome dowry. The two married early in 1839 and were soon looking for their own first home in the same Bloomsbury neighborhood, though Emma prudently advised against living too close to "the Horneritas."

Darwin's reputation increased with the publication in May 1839 of his *Beagle* journal, an engaging mix of geology, natural history, and travel. It was originally no more than a third volume in FitzRoy's *Narrative of the Surveying Voyages of His Majesty's Ships Adventure and Beagle*. But Darwin's account was popular enough for a pirated edition to appear separately later that year, and it soon left FitzRoy behind, taking on a life of its own under the title *Journal of Researches* (the modern title *The Voyage of the Beagle* became popular only in the twentieth century).

Stability

In private Darwin had also begun to develop his own ideas about how species could change over time. It had started as the *Beagle* beat its way up the west coast of Africa toward home. Reviewing his ornithological notes, Darwin paused to remark how curious it was that each of three mockingbird species from the Galapagos seemed to occur only on its own island. And though he had failed to observe it himself, he recalled that the locals had reported a similar pattern in tortoises. That kind of branching, with a species in one island seeming to give rise to a slightly different form on the next island over, suggested that the zoology of archipelagoes deserved a closer look, "for such facts undermine the stability of species." It was his first concession to "transmutation"—the idea that species could change with time and habitat. It was also his first hint of natural selection. Characteristically, he had second thoughts, and went back and revised it to say "would" undermine the stability of species.

The species question continued to trouble him back home, and in July 1837, he opened the first of many notebooks devoted to how and why species vary. He was hardly alone in pursuing these questions. That same year, John Herschel, the most famous scientist in England, had publicly endorsed the idea that there might be a natural process of creation, "in contradistinction to a miraculous" one, to explain "that mystery of mysteries," the origin of species. But Darwin saw what that natural process might be. His breakthrough came on September 28, 1838, at Great Marlborough Street, while reading the demographer T. R. Malthus on factors limiting human population growth. Among animals, Darwin suddenly realized, hunger, predation, and other "checks" on population could provide "a force like a hundred thousand wedges," thrusting out weaker individuals and creating gaps where better-adapted individuals could thrive. But refining this insight, and testing it against the evidence, would take years of work.

Before they married, Darwin confessed his transmutationist leanings to Emma, a devout liberal Christian. She worried that this heresy would separate them in eternity (presumably by consigning him to hellfire) but married him anyway. Darwin otherwise kept his ideas on the origin of species to himself. He and Emma moved to a rented brick row house on Upper Gower Street, which they nicknamed "Macaw Cottage" for its gaudy decor. They lived there in "extreme quietness," and Darwin added that "if one is quiet in London, there is nothing like its quietness—there is a grandeur about its smoky fogs, and the dull distant sounds of cabs and coaches." From the back garden, Darwin could see the main building of University College London, where his onetime mentor Robert Grant had become a professor of zoology. Grant had taught him basic field biology at the University of Edinburgh. But Darwin managed to avoid him for the three years he lived on Gower Street; apparently he didn't want his career tainted by Grant's radical beliefs—including an early brand of evolutionary thinking.

"More Kicks Than Half-pennies"

In 1842, Charles and Emma moved with their three children to an old parsonage, called Down House, on a hill above the rustic village of Downe, outside London. There, by 1844, he had expanded his own evolutionary ideas into a manuscript of 231 pages, working out the details of how the process he called natural selection could give rise in time to new species without divine intervention. He was already disinclined to publish such controversial ideas. But just then the publication of *Vestiges* gave him a foretaste of the outrage he was likely to provoke. Evolution was a threat not just to Scripture, but to the social order, particularly in the hands of radicals who regarded it as a way to undermine the idea of a divinely ordained social hierarchy. Darwin himself sat comfortably in the upper ranks of that hierarchy. He was a product of inherited wealth and his closest colleagues were other gentlemen naturalists, including the clergy.

"Anglican dons believed that God actively sustained the natural and social hierarchies from on high," Adrian Desmond and James Moore write in their biography *Darwin*. "Destroy this overruling Providence, deny this supernatural sanction for the status quo, introduce a leveling evolution, and civilization would collapse. And, more to the point, Church privileges." The Rev. Adam Sedgwick "apocalyptically predicted 'ruin and confusion in such a creed.'" Undermining the stability of species, warned Sedgwick, would "undermine the whole moral and social fabric," bringing "discord and deadly mischief in its train."

Fear of the reaction from the religious and political establishment caused Darwin to abandon his essay on evolution in 1845, putting it away in the game closet under the stairs at Down House. It would remain there with his children's tennis racquets and croquet mallets for more than a decade, ticking like a bomb.

Beyond the reluctance to disrupt his entire social class (including his old professor), Darwin also hesitated because the loose, speculative reasoning in *Vestiges* drove home the need for detailed evidence. Joseph Hooker, the one colleague with whom he had shared his evolutionary ideas, remarked to him that no one has "a right to examine the question of species who has not minutely described many." The implied criticism did not stop Darwin from pushing forward. "Though I shall get more kicks than half-pennies," he wrote, "I will, life serving, attempt my work." But he took the comment personally and must have recognized how defensive his reply to Hooker sounded: "My only comfort is . . . that I have dabbled in several branches of Nat. Hist: & seen good specific men [that is, taxonomists] work out my species & know something of geology."

Describing himself as a dabbler wasn't just a matter of self-deprecation. We tend now to think that Darwin's discovery of the different finch species inhabiting the islands of the Galapagos was one of the great turning points in human knowledge about the world. But it certainly had not seemed that way to Darwin himself during his visit in September and October 1835. He categorized the various small brown birds that he collected as wrens, grosbeaks, blackbirds, and finches and confessed that they were difficult to tell apart, leaving him in a state of "inexplicable confusion." He did not even bother to note the exact island where he collected each of them.

Back in London in January 1837, he turned his bird collection over to the ornithologist John Gould. At a Zoological Society meeting just eight days later, Gould reported that the birds were in fact all finches, of a dozen new species, differentiated mainly by the shape of their bills. Depending on the Goulds of the world, with their specialist knowledge of particular animal groups, was of course essential. But to avoid future scholars categorizing "Mr. D" with "Mr. Vestiges," Darwin saw that he needed to undertake some serious taxonomy of his own.

Meeting Mr. Arthrobalanus

On a beach in Chile 10 years earlier, he had picked up a conch with tiny holes bored in its shell. Examining one of these holes through a microscope, he found a small creature glued in headfirst, with its legs waving in the air. It looked like a barnacle. But no one had ever described a barnacle before without the familiar, hard exterior. Trying

The sensation set off by "Mr. Vestiges" caused Charles Darwin to slow down his theorizing, but it also opened the public mind to evolutionary ideas.

to sort out the idiosyncracies of the creature he named "Mr. Arthro-balanus" caused Darwin to back away from theorizing and settle down to describing the extremely minute differences within one invertebrate group, the barnacles, or Cirripedia.

Darwin was also settling down in other ways. At Down House, he set up his microscope in the window of his study, and persuaded museums and collectors from around the world to send him their barnacle specimens. In her book *Darwin and the Barnacle*, Rebecca Stott argues that barnacles would prove crucial to Darwin in working out his theory and also in securing his credibility within the scientific community.

He was, Stott suggests, becoming a bit of a barnacle himself. "My life goes on like Clockwork," he wrote to his old traveling companion Fitz-Roy, on the tenth anniversary of their return home in the *Beagle*, "and I am fixed on the spot where I shall end it." His barnacles occupied his attention to such an extent that his young son George thought it was what all adult work must be about. Visiting another child in the neighborhood, he inquired, "Where does [your father] do his barnacles?"

Invertebrates were useful for teasing out biological questions because they display such extreme variety, often differing dramatically from one tiny ecological niche to the next. Barnacles were particularly interesting just then because their hard shell-like exterior had caused Linnaeus, Cuvier, and others to misclassify them as mollusks. But in 1830 a naturalist named John Vaughan Thompson had collected live larval crustaceans near Cork, Ireland, kept them in captivity, and watched in astonishment as they went through metamorphosis, became barnacles, and glued themselves to the bottom of the jar. Barnacles were not mollusks after all, but house-bound crustaceans. Thus Darwin's interest in the group was timely, and what had started with Mr. Arthrobalanus soon turned into a revision of the entire class Cirripedia, numbering hundreds of species.

The careful work of dissecting and describing species often gratified him. "I have been getting on well with my beloved cirripedia," he wrote Hooker in 1848. The minute transitions he was finding from species to species moved him to add, "I don't care what you say, my species theory is all gospel." But barnacles also drove him mad. Sorting through their endless diversity in 1850, he muttered darkly about "this confounded variation." Two years later, still tangled up in taxonomic minutiae, he exclaimed, "I hate a Barnacle as no man ever did before." Another year after that, he added, "I have gnashed my teeth, cursed species, & asked what sin I had committed to be so punished."

He and Emma meanwhile filled the place with children—10 altogether, 7 of whom survived to adulthood. Despite the prevailing image of Darwin as a perennial invalid, or as a gloomy éminence grise, the mood of the house in those years seems to have been playful. While Darwin struggled with his barnacles, his offspring happily raided his supplies and sometimes lit out with his microscope seat, a burled-wood stool on brass wheels. They used it, with a cane for an oar, to go punting around the first floor of the house.

Darwin also seems to have engaged in a form of punting. In photographs of the study, his chair, high-backed and narrow, looks a little severe. But it stood on long, bird-like iron legs borrowed from a bedstead, with wheels for scooting from bookshelf to table, or for checking out visitors from the window. Darwin also regularly went roaming from his study on foot, to dip snuff from a jug in the hall, or to check the mail, delivered several times a day then. (The Darwin Correspondence Project counts 14,500 letters sent or received during his life.) He sometimes played billiards with his butler, Joseph Parslow, and he wrote, "I find it does me a deal of good, & drives the horrid species out of my head."

But even if the manuscript was tucked away in the game closet under the stairs, and even if his barnacles were partly a delaying tactic, to forestall facing the kind of reception "Mr. Vestiges" had endured, still Darwin's horrid species kept softly, insistently, calling him back.

"An Ingenious Hypothesis"

In the Welsh village of Neath in 1845, a railway surveyor was also reading *Vestiges*. Alfred Russel Wallace was just 22, from a downwardly mobile family, with a somewhat indiscriminate appetite for progressive political causes and "scientific" topics including botany, mesmerism, and phrenology. During a brief stint as a schoolmaster in the East Midlands city of Leicester, he had befriended the young entomologist Henry Walter Bates, who introduced him to beetle collecting and to the startling idea that perhaps 1000 beetle species existed within 10 miles of the city. Wallace soon obtained a collecting bottle and specimen pins of his own and joined Bates in his entomological outings.

The two continued to correspond after Wallace moved to Neath to take up the surveying business of a brother who had died suddenly. The speculative "railway mania" was at its height in the mid-1840s, and Wallace was earning good money investigating potential new routes.

(Charles and Emma Darwin were among those investing in this new technology.) But he still found time for natural history. Reading *Vestiges* got Wallace thinking about how all those beetle species could have come into existence. When Bates disparaged the book, Wallace replied, "I do not consider it a hasty generalization but rather as an ingenious hypothesis strongly supported by some striking facts and analogies, but which remains to be proved."

Testing the idea that natural laws could drive evolutionary change provided a motive beyond mere curiosity for hunting species. "It furnishes a subject for every observer of nature to attend to," Wallace wrote. "Every fact he observes will make either for or against it, and it thus serves both as an incitement to the collection of facts, and an object to which they can be applied when collecting."

The next year, Wallace wrote to Bates about two books that were stirring up his eagerness to visit the tropics, Darwin's *Voyage of the Beagle* and Alexander von Humboldt's *Personal Narrative*. (Humboldt's account of his 1799–1804 travels in Central and South America was the inspiration for many naturalists, including Darwin himself, who later wrote, "Nothing ever stimulated my zeal so much.") After another year gathering beetles and other specimens in the British countryside, and a flying visit to the natural history museums in Paris and London, Wallace wrote to Bates: "I begin to feel rather dissatisfied with a mere local collection . . . I should like to take some one family to study thoroughly, principally with a view to the theory of the origin of species."

They agreed to undertake an expedition. *A Voyage up the River Amazon*, published in 1847 by the American W. H. Edwards, steered their attention to that region. Lacking the family fortune and social connections that had served Darwin so well, Wallace and Bates made arrangements to finance their travels by collecting and selling specimens. Sir William Hooker, director of the Royal Botanic Gardens at Kew (and father of Darwin's friend Joseph), gave them a letter of introduction and agreed to pay for properly prepared specimens. Edward Doubleday, the keeper of Lepidoptera at the British Museum, likewise patronized them. (They probably got patronized in both senses of the word; they were after all raw beginners, fit only to become foot soldiers in the scientific enterprise, and there was the small matter of social class.) By good fortune, Bates found a capable specimen dealer named Samuel Stevens to dispose of their duplicates on a commission basis. Stevens, whose shop was on Bedford Street around the corner from the British Museum, expected to sell a typical insect specimen for four pence, with three pence going back to the collectors in the field. Undeterred by the promise of such dazzling riches, Wallace and Bates sailed for the Amazon in April 1848.

From then on Wallace, slogging through distant jungles, and Darwin, fixed at the microscope in his study in Downe, would be racing to answer the same fundamental questions, in pursuit of the same great prize—the key to understanding the origin of species.

Chapter Fifteen

A PRIMATE NAMED SAVAGE

How much doth the hideous monkey resemble us!
—QUINTUS ENNIUS, C. 200 B.C.

O N JULY 16, 1847, a missionary newly arrived in New York City from
West Africa packed a collection of bones in a box and shipped them
off to a colleague in Massachusetts. In a letter, the Rev. Thomas S. Savage
admitted to being "quite unwell," probably meaning "utterly wretched."
He had already endured tropical diseases in Liberia off and on for more
than a decade, and he'd seen his first two wives languish and die there,
most likely of malaria. He wasn't the sort to complain lightly.

In any case, his weakened state was evident. He had at first mis-
placed the list of contents for the box. And despite his original plan "to
describe the bones of the animal myself," he had to ask his colleague
to handle that chore. Describing the creature's habits "will be about all
that I shall be able to do." A few weeks later, as he recuperated at his
family home in Middletown, Connecticut, he wrote again: "Will you
inform me whether you received the two canine teeth of the best male
cranium? I remember that one came out, but cannot tell whether I
replaced it. I have not seen it since I sent the bones."

What he sent was sensational enough. In mid-August at the Boston
Society of Natural History, Savage and his coauthor, Harvard anatomist Jeffries Wyman, together presented one of the most startling and
important discoveries in the science of life on Earth, a disturbingly

227

Thomas S. Savage and Jeffries Wyman were deeply modest men, but their discovery shocked contemporary sensibilities.

familiar creature that would soon enter popular lore and also play a key part in the coming debate over Darwinian evolution.

They called their new species "gorilla."

A casual reader browsing through the biological literature today could easily get the impression that the discovery of the gorilla was the work of Richard Owen, the ambitious British anatomist best known for coining the name "dinosaur" and for disputing Darwin. Credit also sometimes goes to the French-American explorer Paul Du Chaillu, who in the 1850s became the first Westerner to give the outside world an eyewitness account of gorillas in the wild. One source even seems to credit the discovery to an otherwise unknown British sea captain named George Wagstaff, who acquired several gorilla skulls while trading in 1847 on the coast of Gabon, at the same time as Savage: "Before these items were collected, gorillas were thought mythical . . . Captain Wagstaff died soon after arriving in Bristol and the full story of these first skulls went with him to the grave."

In fact, the scientific discovery of the gorilla was entirely a collaboration between Savage and Wyman, one from Yale, the other from Harvard, both trained as physicians, both from old and somewhat starchy New England stock. At a time when many scientists competed fiercely to discover new species, even stealing credit from one another, the two American naturalists were meticulously fair and almost comically modest. Thus when Wyman, the professional anatomist, entered the new species into the annals of natural history, he sidestepped scientific immortality and named Savage as the describer: *Troglodytes gorilla* Savage. Savage responded with courtly dismay: "I . . . very much regret that you have done it; for it cannot be an honest act on my part to desire to appropriate to myself that to which I have no claim. I look upon you as the describer." For their gentlemanly behavior, history has largely forgotten them.

Savage was the key to the discovery, in the right place and with the right background to recognize from a single skull the existence of the largest primate on Earth, a creature Richard Owen would later describe as "the most portentous & diabolical caricature of humanity that an atrabilious poet ever conceived or a naturalist ever realized." Except for vague rumors, this behemoth had somehow escaped the attention of Europeans traveling the coast of West Africa for the previous 400 years.

Hunting Souls and Species

By the time he made his discovery, Savage was an old Africa hand. He'd arrived on the continent for the first time at the age of 32, on Christmas Day, 1836. Though he was just three years out of Yale Medical School, he'd taken up his station at Cape Palmas, Liberia, primarily as an Episcopal missionary (he had studied at the Theological Seminary of Virginia), and only secondarily as a doctor. He served in Liberia with what seems now like an odd mix of Christian colonial fervor and considerable scientific curiosity.

As a missionary and doctor, he came to know "the African's mind & tongue," which in Cape Palmas meant Grebo. As a naturalist, he also brought a keen eye and unusual depth of knowledge to the discovery of

African species. He regularly shipped insect specimens back to a friend in England. Later, he would publish the first detailed description of the behavior of driver ants, not a study for the faint of heart: When he presented his pinky to the decapitated head of one soldier ant, it bit so powerfully "that the point of the mandibles met beneath the cuticle." Then it withdrew and started digging in with alternating knife strokes, "wounding and cutting wider and deeper."

He was no less intrepid about his religion. In 1839, four months into a new tour of duty, malaria killed his 28-year-old first wife, Susan. No record survives of his own sense of loss, nor do we know what he told her parents, who had already lost seven other children in infancy and now had just one daughter remaining. (She would die a year later.) But an elegy written by a friend, a fellow clergyman, suggests the strength of the religious feelings that had motivated them both. "If the blood of the martyrs be the seed of the church," then from Susan's grave, the friend declared, "there yet may spring a noiseless band of heavenly soldiery who will carry the war into Africa and plant the ensigns of the gospel high on the pagan hills." Savage remarried in 1842 and continued to serve at Cape Palmas even after seeing his second wife succumb to tropical disease.

While Savage proselytized in Africa, Jeffries Wyman devoted himself to anatomical studies. He spent a year studying in Paris and then, under the tutelage of Richard Owen, in London. For five years he taught anatomy and physiology at a medical school in Richmond, Virginia, where he marked time, hoping his Boston friends would secure him a position on the faculty at Harvard.

Savage and Wyman probably met in August 1843 when the missionary returned by way of Boston for 10 months of recovery at home. They collaborated almost immediately on a paper about chimpanzees. "The breasts were flabby and slightly protuberant," Savage reported, sounding more like a doctor than a missionary. He went on to comment learnedly on the tendency of other biologists to confuse the placement of the big toe in chimpanzees with that of the orangutan. (Though both primates were known to European scientists, they remained poorly understood.) He also challenged the great French anatomist Georges Cuvier's interpretation of the chimpanzee's prominent brow ridge. That self-assurance, and the detailed knowledge about chimpanzee anatomy, would soon prove critical to the discovery of the gorilla.

"An Animal of an Extraordinary Character"

His moment came, oddly, after he had already sailed for home early in 1847, at the end of his years in Africa. The ship he took from Liberia headed eastward at first, under the belly of West Africa, to the Gaboon (now Gabon) River. There it was "unexpectedly detained" for more than a month. Savage stayed at the house of a friend and fellow missionary, Rev. John L. Wilson, in a village just south of what is now the capital city of Libreville.

Wilson and his wife Jane kept a deer, a porcupine, and other species as pets, and also apparently decorated their home with African specimens and curiosities. One skull immediately caught Savage's attention. It was too large for a chimpanzee, with huge, glowering eye sockets, a high bony sagittal ridge running back like a Mohawk across the top of the skull, and a nuchal crest like a broad shelf across the back—anchor points for huge jaw and neck muscles.

Savage questioned the local hunters, who told him about "a monkey-like animal, remarkable for its size, ferocity and habits." The shape of the skull, combined with "information derived from several intelligent natives," convinced him that he was looking at "a new species of Orang." (The Malay word *orang*, meaning "man," made famous by the Southeast Asian orangutan, or "man of the forest," was still broadly applied to all great apes.)

Savage had corresponded in the past with Richard Owen at the Royal College of Surgeons and Samuel Stutchbury, curator of the Bristol Institution for the Advancement of Science. Now he wrote to both. "I have found the existence of an animal of an extraordinary character in this locality," he told Owen, and sent along detailed drawings of the skull, asking him to compare it with an orangutan skull in the Royal College collection.

Savage never actually laid eyes on a living gorilla. He was waiting for his ship to head home, and there was no gorilla population in the immediate vicinity. But he was a good interviewer. Instead of the "marvellous accounts given by the natives . . . to credulous traders," local hunters gave him a remarkably accurate description of the gorilla's appearance and behavior. Among other things, he learned that a single adult male dominated each band of animals, and that the hair tended to go gray with age.

"The gait," Savage wrote, "is shuffling, the motion of the body, which is never upright as in man, but bent forward, is somewhat rolling, or from side to side." The gorilla "has the power of moving the scalp freely forward and back, and when enraged, is said to contract it strongly over the brow . . . so as to present an indescribably ferocious aspect." But he added, "The silly stories about their carrying off women from the native towns, and vanquishing the elephants . . . are unhesitatingly denied."

Savage's pursuit of this information, and the feelers he put out for specimens, attracted the attention of other traders in Gabon, probably including Capt. Wagstaff. When a local hunter eventually presented Savage with skulls and an assortment of other bones of the new species, a bidding war ensued. Wagstaff, commanding a brig called the *John Cabot*, was assembling a cargo consisting mainly of elephant tusks, ebony, and nuts. But sea captains also customarily eked out their wages with "private trade" in anything they calculated would sell back home, including natural curiosities.

Writing to Wyman soon after his return to the United States, Savage complained: "I had three competitors for them, two sea captains & one missionary, who raised by their eagerness the expectations of the natives to a high pitch. Nor should I have succeeded in getting them at all, had it not been for" his host and fellow missionary, J. L. Wilson, "who exerted his influence with the chief & master of the slave who killed the animals. Others stood ready to pay almost any sum for them, & I believe they constitute the only set that has ever been taken out of the Gaboon." The bones had cost him $25, a substantial sum, and

Savage reported that the chief had also agreed to free the hunter from slavery.

But for Savage, it was "more than I am able to lose." Though he wanted to donate the bones to the cabinet of the Boston Society of Natural History, and was "sorry to be under the necessity of requiring anything for them," he was obliged to offer them to anyone who could reimburse his costs. Wyman, who had just obtained the faculty position at Harvard for which he had been patiently waiting, put payment in the mail by the next post.

Savage had hoped the species would become commonly known by its Gabon name, anglicized *engé-ena*. But he left the choice of the scientific name to Wyman, who noted that the ancient Carthaginian explorer Hanno had used the word "gorilla" for "wild men" living on the coast of West Africa. Thus the gorilla may have ended up with an African name in any case, *gorel* being a word used by the Fulani of West Africa for "little men," or Pygmies.

Unseemly Ambition

In their letters back and forth, Wyman suggested that Savage write to the *Annals and Magazine of Natural History* in London to establish priority for his discovery. "I will do so, but I do not view it of any consequence," Savage replied, a little naively. "Should any one be fortunate enough to have discovered it & forestal me, I must say that I shall have no regrets, inasmuch as it will be secured to science."

He would feel differently later. Indeed, their correspondence suggests that, beneath their courtly manners, both Savage and Wyman cared deeply about being first. In February 1848, Richard Owen published what he apparently took to be the first scientific description of the new species—based on skulls supplied to him by Stutchbury. Wyman later admitted to being "not a little surprised" by this maneuver, as Owen "must have known that we either had or intended to describe the crania & bones brought back by yourself from Africa." In truth, he could hardly have missed it: The placeholder notice suggested by

Wyman had appeared the previous October in an issue of *Annals and Magazine of Natural History* dominated by Owen's own article on the plesiosaurus. It was the first printed reference to the gorilla, under the headline "new Orang-Outang," and announced that a description would appear shortly in the *Journal of the Boston Society of Natural History*.

When Savage saw Owen's article, he reacted with uncharacteristic rancor, directed not at Owen but at Stutchbury. Only a few years earlier, Savage had arranged to have the body of a pregnant chimp packed in a cask and shipped to Stutchbury. It was no small favor, and Stutchbury had professed deep gratitude: "I scarcely know how and certainly never can return you proper thanks for the very kind manner you responded to the desire of a perfect stranger." But when the moment to return the favor eventually came, in the form of Savage's letter from Gabon seeking information about a new "orang," Stutchbury had never even bothered to reply. Instead, he had commissioned Wagstaff and other captains to bring him specimens. It looked to Savage as if Stutchbury had leapt at the possibility of a huge new primate and done his best to steal credit for the discovery. In a note to Wyman, Savage wrote, "Had I not taken the precaution . . . at your suggestion to announce . . . the discovery" in *Annals and Magazine of Natural History*, then "Mr. Stutchbury's efforts" would have given the discovery "to himself or Capt. Wagstaff."

From Owen's perspective, there was nothing improper about it. His article freely recounted how Stutchbury "had requested some of the captains of vessels trading from Bristol to the Gaboon river to make inquiries respecting the species and endeavour to obtain specimens of it." Wagstaff had in fact eventually succeeded in getting additional skulls, after Savage's departure for the United States. He delivered them to Stutchbury on his return to Bristol in November 1847, then promptly died "of African fever." Stutchbury in turn handed the skulls to Richard Owen, as the most qualified anatomist for a proper scientific description.

Owen was by then already earning an ugly reputation among his fellow naturalists. Just the year before, he had presided over a meeting of the Royal Society at which his own paper about a squidlike fossil had been nominated for the Royal Medal for the advancement of natural knowledge. After Owen won this honor, it turned out that the fossil described and named in his paper had actually been described four years earlier by an amateur paleontologist at a meeting Owen attended.

But when it came to the gorilla, Savage thought the British anatomist had acted "in the whole matter like a gentleman." Having received the skulls from Stutchbury, Owen quickly published his description in the *Transactions of the Zoological Society of London*. But if Stutchbury had been intending to steal the glory of the discovery, as Savage suspected, then Owen's proposed name for the animal was a clear rebuke. He called it *Troglodytes savagei*, "after Dr. Thos S. Savage, by whom it had been discovered and its existence made known to Professor Owen."

The evidence suggests that Savage was being too harsh on Stutchbury, whose own notes about the discovery also give proper credit to Savage. But Savage may also have been too kind to Owen. Wyman seemed to think so, despite his own close relationship to Owen. In 1866, in a lined notebook, he prepared a handwritten account of the discovery. He never apparently intended it for publication; that wasn't his style. But for the record, he carefully underscored the relevant dates: "A joint memoir was presented by us to the Boston Society of Natural History <u>August 18th 1847</u>. . . . An account . . . was presented by Prof. O. to the Zoological Society of London <u>Feb. 22 1848</u> six months after our memoir had been read in Boston."

On learning of the previous publication, Owen had no choice but to acknowledge that the American naturalists had described the species first. That meant the species name *gorilla* would prevail, with *savagei* becoming a mere synonym. (This was no doubt a good thing. In coining the name *dinosaurs*, meaning "terrible lizards," Owen had already demonstrated a penchant for what one historian has called "flesh-creeping suggestiveness." But gorillas suffered right from the start under an exaggerated reputation for ferocity, and naming them "savages" surely would not have helped.) A few months after his own article, Owen added

Wyman's description of the bones into the *Transactions of the Royal Zoological Society*. But in his notebook, Wyman pointedly remarked, "It does not appear however either in the Proceedings or the Transactions at what time our memoir was published nor that we had anticipated him in our description." (In fact, Owen went on publicly giving credit to Stutchbury and himself for bringing the gorilla's existence "clearly to light," the adverb perhaps indicating that the prior American discovery had been lost in mid-Atlantic fog.)

Wyman added, "The credit of the discovery clearly belongs to Mr. Wilson & Dr. Savage chiefly to the latter, who first became convinced of the fact that the species was new, & who first brought it to the notice of naturalists. The species therefore stands recorded *Troglodytes gorilla*, Savage." A few years later, French zoologist Geoffroy Saint-Hilaire separated the new ape from the chimpanzees (then also in the genus *Troglodytes*) and elevated it to its own genus, *Gorilla*. Eager naturalists proceeded to describe a plethora of new species for this large ape over the rest of the nineteenth century. In 1929, a zoological revision by a Harvard primatologist winnowed down this excess of names to the two forms most commonly recognized today, the eastern gorilla, *Gorilla beringei*, and the western gorilla, *Gorilla gorilla*, to which, overcoming undue modesty, he added Wyman's name as co-describer.

"Gorilla Quadrille"

While the scientists squabbled about credit for the discovery, laymen grappled with the deeply disturbing question of how much humans looked and acted like other primates. Seeing an orangutan sip tea at the London Zoo in 1842, Queen Victoria pronounced the beast "frightful and painfully and disagreeably human." But the gorilla, above all primates, would force these uneasy feelings about our own origins into furious public debate, often driven by powerful undercurrents of class and race.

The racism was evident from the start. In his section of the 1847 paper announcing the gorilla to the outside world, Wyman wrote that any anatomist "who takes the trouble to compare the skeletons of the

Negro and Orang, cannot fail to be struck at sight with the wide gap which separates them." But then he added, "Negro and Orang do afford the points where man and the brute . . . most nearly approach each other." Like Edward Tyson, the seventeenth-century anatomist who first dissected a chimpanzee, Wyman was perhaps trying to minimize the similarity between gorillas and humans by deflecting it onto "men of the lowest rank." But other, more polemical, writers used that sort of thinking to justify keeping blacks as slaves. Likewise, the humor magazine *Punch* turned Irish nationalists into "Mr. O'Rangoutang" and "Mr. G. O'Rilla." Having dehumanized one such nationalist in this fashion, a *Punch* cartoon bluntly inquired, "Shouldn't he be extinguished at once?"

The gorilla would loom increasingly large in public debate over the next 15 years, and, for Richard Owen, it would become the chief weapon in a bizarre war against evolutionary thinking. Despite his own lower-middle-class background, Owen had risen to the top of London society, with a powerful position at the British Museum, a residence provided by the Queen, and the privilege of lecturing to the royal children on zoology. The establishment turned to him, as the nation's leading anatomist, to defend the status quo from the rising forces of evolution.

Owen dutifully argued for a divine "archetypal light" guiding the development of species. He maintained that the prominent brow ridge of the gorilla skull clearly distinguished it from the human skull, and along with other differences "must stand in contravention of the hypothesis of transmutation." Later, as the evolutionary debate heated up, he would use the gorilla's brain as evidence for the distinct, God-given nature of human beings: we differed from the gorillas and other apes, he announced, on the basis of three brain structures found only in humans. In fact, no such structures existed. But Owen continued to argue the case even after he got caught distorting and fabricating his evidence.

The gorilla would become a creature not just of serious scientific discussion but of sideshows and tabloid newspapers, especially after the

French-American explorer Paul Du Chaillu arrived on the scene with complete gorilla specimens. A "gorilla ballet" went thumping across the London stage in the early 1860s, and in parlors around Europe and the United States, amateur pianists performed a "Gorilla Quadrille" (in which the gorilla was a "darkie" with all the grotesque "doo-dah" stereotypes).

The two men who had started it all stayed apart from the fray, possibly out of dismay. Savage's African writings, with their frequent and respectful reliance on local knowledge, suggest that he would have had no part in the racial misuse of the gorilla. Nor did his discovery seem to cause him any religious doubt. For others, the gorilla would come to threaten the central place of humanity in a divinely ordained universe almost as profoundly as had the realization that the sun did not revolve around the Earth. But not, apparently, for Savage, who returned to his calling as a clergyman. He helped raise four children by his third wife and served as a pastor in Mississippi, Maryland, and finally Rhinecliff, New York, where he died at the age of 76. His tombstone there describes him as a "pioneer missionary" and makes no reference to his medical or scientific work. Wyman meanwhile became a quiet advocate of Darwinian theory, but continued with his anatomical work and left others to make the case in public. At his death, the poet James Russell Lowell eulogized Wyman in a sonnet that might have applied to either man: "simple, modest, manly, true, / Safe from the Many, honored by the Few."

No one knows what happened to two of the four gorilla skulls Savage brought home in 1847. In his letter to Wyman, he had asked that they be set aside for J. L. Wilson, his host in Gabon. So perhaps they are gathering dust as curios in some family member's home. The other two skulls, male and female, are stored now on a bed of white polyethylene foam, in a metal drawer in a climate-controlled room at Harvard's Museum of Comparative Zoology. They are the type specimens, the models by which the species is defined for scientists everywhere. The female skull has been cut in half, from front to back, as if Wyman at some point went looking to see even the vaguest cranial evidence for the brain differences Owen kept going on about.

The male skull is largely intact, glowering and a little forlorn, with the collection number neatly inked by Wyman on the zygomatic arch. The surface of the skull is mottled with black flecks, and polished with handling, as if the hunter has just pulled it out of his kit bag after a long journey. The canine tooth on the upper left side is as thick as a finger. But just as Savage had feared in that feverish summer of 1847, the right canine tooth is still missing.

Chapter Sixteen

"SPECIES MEN"

Something hidden. Go and find it. Go and
  look behind the Ranges.
Something lost behind the Ranges. Lost and
  waiting for you. Go!
   —RUDYARD KIPLING, "THE EXPLORER"

THE AMAZON is astonishingly broad and powerful, with the black water of tributaries jostling side by side with the muddy main river, until they finally swirl together miles downstream. Violent storms rise up suddenly, making the river wild as an ocean. The surging water frequently undercuts river banks, toppling trees into the water. Canoemen, the field naturalist Henry Walter Bates wrote, "live in constant dread of the '*terras cahidas*,' or landslips, which occasionally take place along the steep earthy banks, especially when the waters are rising." He was inclined to dismiss the stories that "these avalanches of earth and trees" could swamp even larger vessels. But one morning before dawn "an unusual sound resembling the roar of artillery" startled him out of his sleep. It felt at first like an earthquake, "for, though the night was breathlessly calm, the broad river was much agitated and the vessel rolled heavily."

The "thundering peal" of explosions echoed back and forth along the river, with "a long, continued dull rumbling" in the intervals. When day broke, Bates looked to the opposite riverbank, 3 miles off, and saw that "Large masses of forest, including trees of colossal size, probably 200 feet in height, were rocking to and fro, and falling headlong one after the other into the water." The impact sent out a sort of Amazonian tsunami that undermined other parts of the bank, extending the landslip over a mile or two of coast. "And thus the crashes continued, swaying to and fro, with little prospect of a termination" as their boat went out of sight up river two hours later.

The Best Years of Their Lives

Bates had a knack for sailing through great perils like that, often badly shaken, but never sunk. He and his traveling companion Alfred Russel Wallace shared an unflappable quality, an ability to be agreeable in the face of any hardship, and they endured an abundance of them. Delight always trumped misery in the travel memoirs they both produced for eager readers back home.

Bates, like Wallace, had come from a progressive family background, Unitarian, committed to religious and civil liberties. Put to work as an apprentice at age 14, he started the day sweeping out a warehouse at 7 a.m. and remained on the job most days till 8 p.m. But he was intent on self-improvement, and regularly woke at 4 a.m. to teach himself Homer in the original Greek. He also attended night classes at the Leicester branch of the Mechanics Institute, which local clergy denounced for encouraging "infidel, republican, and leveling principles." (That is, it advocated educating the poor.) He thus worked himself into a state of exhaustion, and his family only assented to the Amazonian expedition, incredibly, on the advice of a family physician that tropical weather might improve Henry's tattered health.

Bates characteristically continued to work hard in the Amazon and was constitutionally inclined to minimize the suffering. When biting flies were a maddening nuisance, for instance, he could make it sound almost like a blessing that "all the exposed parts of my body" were "so closely covered with black punctures that the little bloodsuckers could not very easily find an unoccupied place to operate upon." Life was good, even when it was wretched.

And that attitude helped make Bates and Wallace the great endurance species seekers of the century. Beginning with the two years they spent traveling together, Bates would pass what he called "eleven of the best years of my life" in the Amazon (1848–1859). Wallace would spend four years there (1848–1852), and another eight in the East Indies (1854–1863). They seemed to thrive in climates that frequently killed other European visitors in weeks or months. (One such victim was Wallace's brother Herbert, who came out to join him as a collector in 1850 and died the following year, age 22, of yellow fever.)

Bates's parents, belatedly anxious about the likelihood of death and disease, made the trip to London to visit Samuel Stevens, the specimen dealer who represented their son. They were helicopter parents

In the Amazon, Henry Walter Bates was once mobbed by curl-crested toucans.

before there were helicopters: Bates père, who ran the family hosiery business, grilled Stevens about the financial rewards of species hunting and suggested that such an endeavor might be good enough for "proved failures" like Alfred Russel Wallace. *His* son could do better, and more safely, manufacturing socks.

Wallace and Bates decided to separate in March 1850. Neither mentioned the cause of the split, and they remained friends. Money may have been an issue, since the difficulties and delays of transatlantic commerce meant they had received no payments for the specimens they had shipped home during their entire first year in the field. Wallace biographer Peter Raby suggests that temperamental differences may also have emerged over their months together in the field: "Bates was comparatively at ease in company, more tolerant, readier to absorb the atmosphere and accumulate knowledge gradually; Wallace more driven, impatient, competitive." But the split may simply have been a practical choice, dictated in part by a split in the river itself. They would be able to cover more territory, and avoid the inevitable duplication of effort from traveling together, if Bates continued west on the upper Amazon toward Peru, and Wallace split off to explore the Rio Negro, the other great tributary of the Amazon, heading northwest toward the Colombia-Venezuela border.

"Stranger on a Strange Errand"

Bates regretted the loss of an intellectual equal for kicking around the meaning of that day's discoveries. But he was otherwise content with the methodical, unremunerative business of finding new species. (In one stretch of 20 months, he ran a total profit of £27.) He headed out into the forest each morning dressed in boots, trousers, an old hat, and a colored shirt with a pin cushion on the front for keeping six different sizes of insect pins at the ready. He carried a shotgun over his left shoulder, one barrel loaded with No. 10 shot, the other with No. 4, for anything from a small bird to an animal the size of a goose. In his right hand, he carried his butterfly net. A leather bag at his left side held ammunition and a box for insect specimens. A game bag on his right held further supplies, with leather thongs for hanging lizards, snakes, frogs, large birds, and other specimens.

He must have been an odd, otherworldly sight for people living along the river. But he also had a knack for fitting in. A village youngster once borrowed his hat and shirt to put on a performance as a bespectacled entomologist flailing about the forest with his net and collecting bag. Bates laughed along with everybody else, entirely aware of his image as "a solitary stranger on a strange errand."

He was a tolerant, curious, uncomplaining traveler, with an appreciative eye for local customs. "There were, of course, many drawbacks to the amenities of the place as a residence for a European," he admitted. The pleasure of a leisurely bath in the river could, for instance, be compromised by a lurking alligator. Bates was obliged, when cleaning between his toes, to keep an "eye fixed on that of the monster, which stares with a disgusting leer along the surface of the water." He wasn't drumming up false drama for readers. Later, an alligator seized and ate a drunken man from the riverbank in a village where Bates was staying.

He also acknowledged that "it was rather alarming, in entomologizing about the trunks of trees, to suddenly encounter, on turning round . . . a pair of glittering eyes and a forked tongue within a few inches of one's head." He once stepped on the tail of a highly venomous jaracara: "It turned around and bit my trousers; and a young Indian lad, who was behind me, dexterously cut it through with his knife before it had time to free itself." But he wrote that "there was scarcely any danger from wild animals" and "it seems almost ridiculous to refute the idea of danger from the natives in a country where even incivility to an unoffending stranger is a rarity."

River travel, though, could be frightening. On one trip, Bates's *cuberta*, a big canoelike boat, had been tied up for the night when "a horrible uproar" awakened the crew, as a sudden gale swept across the river and slammed the *cuberta* into the river bank. One of the crew leapt ashore and used a pole to fend off the boat and push it around

Mosquitoes and other nighttime visitors made for restless sleep.

a point, "swinging himself dexterously aboard by the bowsprit as it passed." The boat moved out into the relative safety of midstream as trees came crashing down along the shoreline.

Bates sometimes seemed more anxious for his specimens than for his life. Once, trying to make port under full sail just ahead of a storm, "the old boat lurched alarmingly, the rigging gave way, and down fell boom and sail with a crash, encumbering us with the wreck. We were then obliged to have recourse to oars; and as soon as we were near the land, fearing that the crazy vessel would sink before reaching port, I begged Senor Machado to send me ashore in the boat with the more precious portion of my collections."

Seeing Beyond Appearances

What he was protecting amounted in the end to 14,712 specimens, mostly insects, representing 8000 new species. He sent them off in periodic batches to Stevens, his dealer in London, who shrewdly promoted them by publishing excerpts from Bates's letters in key journals. "The collections were unrivalled," a friend and fellow entomologist, W. L. Distant, later recalled, "and one can still hear echoes . . . of the intense interest with which Bates' consignments were anticipated. The banks of the great river were at last telling the tale of their inhabitants to the zoologists of Europe, for the collections were widely circulated."

Writing at century's end, Distant recalled how different the scientific world had been then. The comprehensive museum collections that people would soon take for granted did not yet exist. There were no complete series of species for constructing careful monographs of a given genus. People had only the vaguest notion of species distribution, and, unlike Bates, most field collectors lacked the patience or training to gather specimens and observe behaviors at the same time. "It was the age of the iconographer of remarkable forms," Distant wrote, and the term was not nearly as flattering as it sounds.

Stevens's buyers were typically "closet" naturalists—that is, relatively well-to-do sorts who never left home and never saw the insects they studied, except on a pin. They left the field collectors—sometimes dismissed as "species men"—to take the risks and do the work, while they sat at home and got the glory of describing the new species in scientific journals. This suited Bates just fine. He wasn't interested in merely cataloguing nature. Many such naturalists seemed to him to be mere "species grubbers" like "collectors of postage stamps & crockery."

For Bates, the iconography of remarkable forms was only a beginning. He aimed to understand how nature really works, and his travels gave him the advantage of knowing—at times with the hairs standing up on the back of his neck—how different species lived in their native habitats. Once, searching among the branches of a tree, for instance, he felt his heart leap on suddenly encountering a huge caterpillar stretching itself out and doing a brilliant imitation of a small but deadly snake. ("I carried off the Caterpillar and alarmed every one in the village where I was living," he wrote with boyish glee.) Other times, he swung his net at what he took to be one type of butterfly, only to examine it in his hand and realize with "an exclamation of surprise" that it was actually a different species altogether.

And this was precisely the sort of mystery he liked to puzzle over. Other naturalists had noted that certain long-winged butterflies had

Lacking the right social connections cost Bates a job as a museum naturalist, but he went on to help other explorers as an administrator at the Royal Geographical Society.

strikingly similar color patterns, and also seemed to fly alike, with a slow, fluttering motion. So these species had all gotten tossed together in the family *Heliconidae*. (Edward Doubleday at the British Museum, who had encouraged Wallace and Bates when they were raw beginners, was among those who did the tossing.) But when Bates looked more closely at features other than wing color, the classification made no sense. The Heliconids, for instance, all stand on two pairs of legs, with their puny forelegs tucked up in front. But some of their imitators bestrode a leaf on six normal-size legs, like most other butterflies. Others differed in the way the veins ran through the wings. So Bates began sorting these imitators into different taxonomic groups, some of them quite distant from the Heliconids. He also speculated about how such different species could come to look so strikingly similar.

Having been fooled himself, he noticed how wing coloration seemed to affect the survival of butterflies in the forest. "I never saw the flocks of slow-flying *Heliconidae* in the woods persecuted by birds or Dragonflies . . . nor when at rest on leaves, did they appear to be molested by Lizards or the predacious [robberflies] . . . which were very often seen pouncing on Butterflies of other families." He surmised that something made them unpalatable to predators ("They have all a peculiar smell," he wrote) and that wing coloration was a way of announcing it as a warning to potential predators.

When there were enough such butterflies in a neighborhood brightly advertising their nastiness, predators got the message and steered clear. So a smaller number of unrelated butterflies could also benefit by adapting over time to borrow the alarm signal and tag along "disguised in their dress, and thus share their immunity." This kind of protective disguise is now known as "Batesian mimicry." To understand the idea, Bates suggested that his readers imagine "a Pigeon to exist with the general figure and plumage of a Hawk."

But it was more complicated than that. His travels had demonstrated that the unpalatable butterflies could wear a different color pattern, or even belong to a different species, from one riverbend to the next. So the unrelated butterflies had to mimic only very local models. Bates described these neighborhood-scale adaptations as "one of the most beautiful phenomena in Nature." He failed, however, to deduce how such localized adaptations could have occurred. Alone in the Amazon, without libraries, specimen drawers, or intellectual companionship, he explained it inadequately as resulting from "the direct action of the local conditions." In fact, the direct action consisted of predators gradually weeding out the butterflies that lacked the protective wing coloring, causing the mimics to flourish in their place. Bates was witnessing a prime instance of what would soon come to be known as evolution by natural selection.

"A Little Career of Looseness"

He was also increasingly experiencing long difficult bouts of loneliness, when the lack of social and intellectual stimulation defeated his natural optimism. He depended on a parcel of letters and reading material from England arriving by steamer every two to four months, and rationed his reading "lest it should be finished before the next arrival, and leave me utterly destitute. I went over the periodicals, the *Athenaeum*, for instance, with great deliberation, going through every number three times; the first time devouring the more interesting articles; the second, the whole of the remainder; and the third, reading all the advertisements from beginning to end."

When his reading ran out, the endless forest could seem as depressing as it was otherwise delightful. The sound of river dolphins "rolling, blowing, and snorting, especially at night," haunted him with a sense of "sea-wide vastness and desolation." Bird calls intensified the "feeling of solitude" by their "pensive or mysterious character." Sometimes a sudden scream would pierce the stillness from some creature seized by a "tiger-cat or stealthy boa constrictor." And at dawn, the howler monkeys let loose their "fearful and harrowing noise, under which it is difficult to keep up one's buoyancy of spirit."

He got along well with both settlers and Indians along the river, drinking with them and sharing their other amusements. There was nothing priggish about him. "Coyness is not always a sign of innocence in these people, for most of the half-caste women of the Upper Amazons lead a little career of looseness before they marry and settle down for life," he wrote, with no hint of disapproval. "The women do not lose reputation unless they become utterly depraved." He was himself a young man in his twenties and often commented appreciatively on the women of the Amazon with their "dark expressive eyes, and remarkably rich heads of hair," their "mingled squalor, luxuriance and beauty."

But whether he also enjoyed a "little career of looseness" among them, he was too reticent to say.

At times, he managed to share the excitement of his work with the locals. Traveling up the Tapajos, a tributary of the Amazon, he fetched from his canoe his copy of the two-volume *Knight's Pictorial Museum of Animated Nature* and showed the engravings to a chieftain, his wives, and a rapidly growing crowd of women and children. Elsewhere he had characterized the Amazonian Indians as suffering from "coldness of desire and deadness of feeling, want of curiosity and slowness of intellect." But the spectacle of strange new species clearly excited them.

"It was no light task to go through the whole of the illustrations, but they would not allow me to miss a page, making me turn back when I tried to skip. The pictures of the elephant, camels, orangutangs, and tigers, seemed most to astonish them; but they were interested in almost everything, down even to the shells and insects . . . Their way of expressing surprise was a clicking sound made with the teeth, similar to the one we ourselves use, or a subdued exclamation, Hm! hm! Before I finished, from fifty to sixty had assembled."

But the moments of connection were too infrequent and too superficial to sustain him. Bates was "obliged at last to come to the conclusion that the contemplation of Nature, alone is not sufficient to fill the human heart and mind." Apart from loneliness and disease, long periods of severe hunger had also sapped his mood. At one point, he'd been driven by "the hardest necessity" to eat smoked spider monkey, which turned out to be both the best meat he had ever tasted and also disturbingly near to cannibalism: "My monkeys lasted me about a fortnight, the last joint being an arm with the clenched fist, which I used with great economy, hanging it in the intervals, between my frugal meals, on a nail in the cabin."

He booked his passage home in June 1859. His travels had transformed him, he warned his parents, from a fresh-faced 23-year-old to an "oldish, yellow-faced man with big whiskers," in ruined health. He kept to himself his thoughts on the prospect of industrial England with its "long gray twilights, murky atmosphere. . . . Factory chimneys and crowds of grimy operatives, rung to work in early morning by factory bells . . . confined rooms, artificial cares, and slavish conventionalities. To live again amidst these dull scenes, I was quitting a country of perpetual summer."

From the civility of the Amazonian tribes, he was also returning to the occasionally ferocious in-fighting of the British natural history community. John E. Gray, keeper of zoology at the British Museum, who seldom had a kind word for any field biologist, would later complain about Bates's supposed moral, intellectual, and physical failings, saying that he had "spent all his time in idleness & licentiousness amongst the natives on the Amazons," hadn't collected 8000 new species, and in any case had been supported largely by the British Museum. (In fact, the museum had merely been a customer for specimens sent by Bates to Stevens.)

His parents at least rejoiced to have him safely home in Leicester. His naturalist friends would eventually find him an administrative post at the Royal Geographical Society in London, after a poet with the right social connections, but no scientific experience, beat him out of a zoological job at the British Museum.

For now, though, Bates would work in the family hosiery business.

Chapter Seventeen

"LABOURER IN THE FIELD"

I'd be an Indian here, and live content
To fish, and hunt, and paddle my canoe,
And see my children grow, like young wild fawns,
In health of body and in peace of mind,
Rich without wealth, and happy without gold!

—A. W. WALLACE, ON THE RIO NEGRO

"I'M AFRAID THE ship's on fire. Come and see what you think of it," the captain said. It was just after breakfast, August 6, 1852, and the writer recounting this awful moment was Alfred Russel Wallace. He was the only passenger on the 235-ton brig *Helen*, a fine piece of kindling and canvas bearing a cargo of balsam resin, aflame, in the middle of the Atlantic Ocean. The lifeboats were in such disrepair that the cook had to provide corks for filling the holes. As the crew scrambled for thole pins and rudders, Wallace wandered numbly down to his cabin, through the suffocating smoke and heat, to retrieve a single tin box with a few notebooks and drawings from his travels. He left behind three years of journals and a large folio of drawings and notes. In the hold of the ship were boxes and boxes of species never before seen outside the Amazon, plus a 50-foot-long palm leaf carefully packed for display at the British Museum. He had gathered it all by means of long, difficult travel, complicated by malaria, yellow fever, dysentery, and other hardships. He was in fact still recovering from a bout of fever as the ship burned, and he felt "a kind of apathy about saving anything."

When the time came, Wallace went over the stern on a rope, tearing up his hands as he slid down into a boat that was "rising and falling and swaying about with the swell of the ocean." Next morning, having lingered to see if the flames of their ship would attract help, "we set up our little masts, and rigged our sails, and, bidding adieu to the still burning wreck of our ship, went gaily bounding along before a light east wind," with Wallace alternately bailing the boat (salt water in raw wounds) and turning his naturalist's eye to the passing jellyfish and boobies, for all the world like the Jumblies gone to sea in a sieve.

The extent of his loss did not dawn on him, Wallace wrote, until they were finally rescued, seven days later, by a ship bound for London from Cuba. Perhaps "rescued" is too strong a word, since their savior was a plodding old hulk, with chunks of rotten wood tearing from the hull like wet bread. They were in such imminent danger of foundering in one storm that the captain of Wallace's former ship felt obliged to sleep with an ax by his side, "to cut away the masts in case we capsize in the night."

Even so, Wallace now felt secure enough to reflect on his loss: "How many times, when almost overcome by the ague [malaria], had I crawled into the forest and been rewarded by some unknown and beautiful species! How many places, which no European foot but my own had trodden, would have been recalled to my memory by the rare birds and insects they had furnished to my collection! How many weary days and weeks had I passed, upheld only by the fond hope of bringing home many new and beautiful forms from those wild regions . . . And now everything was gone." It was as if Darwin's *Beagle* had sunk with all his Galapagos treasures still unmined for scientific insights. But "I tried to think as little as possible about what might have been," Wallace later wrote, "and to occupy myself with the state of things which actually existed."

Arriving back in England that October, after almost three months at sea, including 10 days in an open lifeboat, Alfred Russel Wallace was tattered, unwashed, emaciated—and also jubilant. "Oh glorious day!" he exclaimed, going ashore at Deal in Kent. "Oh beef-steaks and damson-tart, a paradise for hungry sinners." In London, his agent Samuel Stevens got him a new suit of clothes and had his mother feed him back to health at the family home. Stevens had taken the precaution of insuring all shipments from his collectors. So Wallace at least had the £200 insurance payout, small compensation for his loss, but enough to live on for now.

Aboard his rescue ship, Wallace had made two sensible resolutions— to avoid ever again trusting his life to the sea and "to make up for lost time by enjoying myself as much as possible for a while." He kept neither. Within days of his return, he was already contemplating his next expedition. Over the next year, he also busied himself writing four scientific papers and two books, one a technical treatise on Amazonian palms, the other his *Narrative of Travels on the Amazon and Rio Negro*, cobbled together from letters home and from memory.

He was, as he later put it, "the young man in a hurry," and it showed. At a meeting of the Zoological Society of London in December 1852, just two months after his return, Wallace gave his fellow naturalists a cordial earful. He had found to his dismay that the labels in museums and in natural history books seldom recorded more than the vaguest hint of where a specimen came from—"Brazil," "Peru," even "S. America."

"If we have 'River Amazon' or 'Quito' attached to a specimen," Wallace said, "we may think ourselves fortunate to get anything so definite . . . though we have nothing to tell us whether the one came from the north or south of the Amazon, or the other from the east or the west of the Andes." Such geographic barriers, as he knew from his own travels, often demarcated "distinct zoological districts."

"Native hunters" were "perfectly acquainted" with the need to seek out the correct habitat when they wanted a particular species. But the scientific world remained indifferent to this essential fact. "There is scarcely an animal," said Wallace, "whose exact geographical limits we

Alfred Russel Wallace became the greatest field biologist of the nineteenth century, as well as a brilliant theoretician.

can mark out on the map." He was introducing his listeners to what we now know as biogeography—the study of how species are related to one another both in space (by mapping their precise geographic range) and also in time (with the help of geology, and later genetics). He was, in truth, introducing them to what it means to be a species.

Though Wallace was invariably polite, it would have been hard to miss his point that just about everybody else in the naturalist community was doing it wrong. And for the temerity of this criticism, he got a quick comeuppance. Among those in the audience was John E. Gray, keeper of zoology at the British Museum, closet naturalist, and, not at all coincidentally, the world's first postage stamp collector. (He gets the title for having purchased as a keepsake the British "penny black" when it became the first postage stamp ever sold on May 1, 1840.) Gray's idea of intrepid adventure often involved finding ways to undercut his counterparts in the field. He was the same Gray who had gone at William Burchell's boxes of African bird specimens with hammer and chisel. To the Zoological Society audience, Gray now casually remarked: "Why, we have specimens collected by Mr. Wallace himself marked 'Rio Negro' only."

Wallace was too flustered to make much of a reply. But he later wrote to Henry Bates that those vaguely labeled specimens came from early in his time at Barra (now Manaus) before he had realized that the opposite banks of a river could be home to different species. Local hunters soon taught him that certain monkey species lived north of the Amazon and east of the Rio Negro, but never crossed to the opposite banks—where other monkey species, often closely related, took their places. Likewise, "little groups of two or three closely allied species" of hummingbirds and toucans often occurred "in the same or closely adjoining districts." Palm tree and insect species displayed similar distribution patterns, seeming to branch out from one district to the next.

Conventional naturalists still mostly treated new species as the result of separate and seemingly random acts of creation by God. But Wallace was seeing connections and asking what they signified. Why did clusters of similar species all occur within a single small area? Why did species often vary only slightly from one island to the next? "Are very closely allied species ever separated by a wide interval of country?" he asked the Zoological Society audience. "What physical features determine the boundaries of species and of genera? Do the isothermal lines ever accurately bound the range of species, or are they altogether independent of them?" (Isothermal lines indicate temperature zones on a map. So he was wondering if a change in the mean temperature from one area to the next could form a boundary between species.) "What are the circumstances which render certain rivers and certain mountain ranges the limits of numerous species, while others are not? None of these questions can be satisfactorily answered till we have the range of numerous species accurately determined."

This idea that they'd been going about their business wrong irritated other naturalists, not least because it came from a field collector who earned his wages like a shoemaker on a piece-rate basis. "The professional experts in the museums of London, and the connoisseurs of the rectories and country houses" did not even want to allow the likes of Bates and Wallace into their learned societies, according to Wallace biographer Peter Raby. Edward Newman, the president of the Entomological Society, had to admonish members in 1854 for their snobberies. But even Newman's passionate defense managed to patronize the field collector as "the real labourer in the field," capable of "observations as well as manual industry," and deserving of respect "in whatever station of life." The very idea that crass field collectors required payment for their butterflies and beetles apparently made connoisseurs shudder with horror. But "their motives are no more to be called in question," Newman admonished, "than those of the artist or the author, who receives the just reward for his well-directed labours."

East into the Sun

Wallace was in fact thinking far more deeply about species than the experts and connoisseurs who bought his specimens. Though he was careful not to say so out loud, he was still focused on testing the idea, put forward by Mr. Vestiges, that natural laws could drive evolutionary change. Such a law was already forming in his mind, and he might well have come to it, and to the idea of natural selection, far more rapidly except for the loss of so much valuable evidence: "Of the smaller perching-birds and insects, which doubtless would have afforded many interesting facts corroborative of those already mentioned, I have nothing to say," he wrote in his *Narrative of Travels on the Amazon and Rio Negro*, "as my extensive collection of specimens . . . all ticketed for my own use, have been lost; and of course in such a question as this, the exact determination of species is everything."

He cast about for ways to renew his attack on what he later called "the most difficult and . . . interesting problem in the natural history of the earth"—the origin of species. The Malay Archipelago, sweeping from Malaysia to Papua New Guinea, seemed to offer "the very finest field for the exploring and collecting naturalist" on account of its "wonderful richness" and relatively unexplored state. It bridged the gap between the very different fauna of Asia and Australia, and its 17,500 islands offered an almost infinite variety of habitats, of all sizes, and all degrees of isolation. Wallace needed, as he later explained to his bewildered family, to "visit & explore the largest number of islands possible & collect animals from the greatest number of localities in order to arrive at any definite results" about the geography of species. So much for never again trusting his life to the sea.

Early in 1854, after little more than a year back in Europe, Wallace boarded a Peninsular and Oriental steamer, bound by way of Egypt for Singapore. The Royal Geographic Society had called in some favors to get him a first-class ticket, a rare taste of luxury. But Wallace would once again need to support himself upon his arrival as a scientific "labourer in the field," earning his pennies by sending back rare and beautiful specimens to the John E. Grays of the natural history world. (The specimens would at least be labeled with the precise locality.) By the start of 1855, Wallace was holed up during the monsoon in a small house at the mouth of the Sarawak River, just opposite the blue mass of Santubong Mountain, on the north coast of Borneo.

His books had arrived belatedly by the long route around Africa, and now he took time to consult them and brood over his findings about the puzzling distribution of hummingbirds, toucans, monkeys, and other species in the Amazon. The resulting article in that September's *Annals and Magazine of Natural History* proposed a simple law: "Every species has come into existence coincident both in space and time with a pre-existing closely allied species." They hadn't just dropped down from heaven.

Wallace titled his article "On the Law Which Has Regulated the Introduction of New Species." A knack for compelling titles clearly eluded him. But the text struck an unmistakable note of urgency: "Hitherto no attempt has been made to explain these singular phænomena, or to show how they have arisen. Why are the genera of Palms and of Orchids in almost every case confined to one hemisphere? Why are the closely allied species of brown-backed Trogons all found in the East, and the green-backed in the West? Why are the Macaws and the Cockatoos similarly restricted? Insects furnish a countless number of analogous examples . . . and in all, the most closely allied species [are] found in geographical proximity. The question forces itself upon every thinking mind,—why are these things so?"

Wallace avoided the language of evolution. Instead of saying new species had "evolved" he said "created" and instead of connecting them to an "ancestral species" or "common ancestor" he used the unfamiliar word "antitype." This vocabulary obscured his logical conclusion—that allied species occur close together because one species has evolved from another. Wallace also neglected to propose a mechanism for how this kind of evolution could occur. So even Darwin missed the point when he read the article, scribbling "nothing very new" and "it seems all creation with him." For his trouble, Wallace received a note from Stevens gently relaying country house and rectory complaints that he was "theorizing" when he ought to be out getting beautiful new butterflies.

Notes from the Field

Wallace recorded both his developing theories and his field notes in a journal now kept at the Linnean Society in London. The cardboard covers, quarter-bound with faded old leather, are falling apart, and many of the pages are loose and with brittle, broken edges. But the greatest field collector of the nineteenth century still lives between the lines. Capturing an *Ornithoptera*, "the largest, the most perfect, and the most beautiful of butterflies" in the Aru Islands near New Guinea, for instance, Wallace recorded the moment of discovery as we like to imagine it: "I trembled with excitement as I saw it come majestically toward me & could hardly believe I had really obtained it till I had taken it out of my net & gazed upon its gorgeous wings of velvety black & brilliant green, its golden body & crimson breast. It was six and a half inches across its expanded wings & I have certainly never seen a more gorgeous insect." Wallace thought he had discovered a new species and named it *Ornithoptera poseidon.* "I had almost by heart the characters of all the known species," he wrote, "& I thought I could not be deceived in pronouncing this to be a new one."

Even when he was describing something not particularly new or spectacular, Wallace's journal entries made his delight in nature obvious. One day he brought home a blade of grass covered with the plant-sucking insects called aphids. With their antennae folded back, "the whole group looked like a lot of long eared white rabbits nibbling at some very short grass," he wrote. An ant herded the aphids and harvested the sticky sweet liquid they excreted, racing about with a great "rusling & tapping & expectant gaping quite ludicrous to look at till at last an aphis gently raised up his tail & out came a drop of transparent viscous liquid." The ant instantly "seized the luscious drop which hung for some seconds between his extended mandibles while he slowly sucked it in. He then brushed his face & cleaned his jaws with his fore feet & again commenced watching his flock."

But the painful and tedious realities of the collecting life also found their way into Wallace's notes. "Wherever I go dogs bark, children scream, women run & men stare with astonishment as though I were some strange & terrible cannibal monster," he lamented, while visiting the island of Celebes (now Sulawesi), just east of Borneo. The spectacle of his pale skin was so unsettling that he learned to hide behind a tree when pack horses or buffalo approached. Otherwise, they were likely to "stick out their necks" at the sight of him, break loose, and trample anything in their path. The "excessive terror" he inspired, day after day, was dispiriting "to a person who likes not to be disliked, & who has never been accustomed to consider himself an ogre or any other monster."

At the village called Wanambai on the Aru Islands, "a funny old man" became indignant even at the name of Wallace's home country. He tried out pronunciations: "Ung-lung," "ang-lang," and "angor-lang," but finally threw up his hands. "That *can't* be the name of your country, you are deceiving us . . . My country is Wanambai, any body can say Wanambai . . . but Ngling!" Then the old man started to question Wallace about what he intended to do with the birds, insects, and other animals he was so carefully preserving. When Wallace gave him an honest explanation, the old man replied, "*They all come to life again, don't they?*" Thus, Wallace wrote, "I was set down as a conjuror and was unable to repel the charge." When the chorus again went up, "Ung lung! It can't be," Wallace despaired and dragged himself off to sleep.

Though he was often busy late into the night preparing specimens, Wallace nonetheless found time to comment in his journal on an astonishing variety of books and articles, from Tullom's *Marvels of Science* ("Absurd book full of error") to Dufaur's *History of Prostitution* (an intriguing choice when spending years on the road, but Wallace noted only a reference to early humans having a tuft of hair at the base of the spine in place of their lost monkey tails). The journal also served as a forum for testing out grand schemes: "Note for determining species population of Globe," "Formation of a complete library of Nat. Hist," "Plan to stop the further increase of Synonyms."

The last item, the proliferation of duplicate names for the same species, particularly vexed him as "a blot upon our science." The rules of scientific description might not require a naturalist to say exactly where a species lived. Those rules did, however, require listing "all the various errors that have been made respecting it, & quoting every work in which the object it distinguishes has been mentioned or described."

Since species often wound up with duplicate names because the original description had appeared in an obscure journal or even a newspaper, Wallace wanted to limit the publication of species descriptions to just three periodicals for each major country. But even Wallace inevitably committed synonyms, despite his considerable efforts to avoid doing so. His *Ornithoptera poseidon*, for instance, turned out to be a subspecies already named by another entomologist; it's now known as *Ornithoptera priamus arruana*.

Meanwhile Wallace also contended with some of the more alarming distractions of travel, like the custom, when a man in the town of Macassar fell into desperate circumstances, of "running amok," that is, lashing out at strangers with a kris, or sword, on the theory that it was "best and easiest to die fighting and revenge himself against society by killing as many as he can first." It was common for 20 innocent bystanders to die in an amok. One of the bird-stuffers Wallace had hired soon shipped out for Singapore "from the idea that his life was not worth many months purchase among such a bloodthirsty & uncivilized people." Wallace of course stayed on and went about his collecting.

Chapter Eighteen

THE SLOW POWER OF NATURAL FORCES

> The life of wild animals is a struggle for exis-
> tence ... in which the weakest and least per-
> fectly organized must always succumb.
>
> —ALFRED RUSSEL WALLACE,
> FEBRUARY 1858

T HOUGH WALLACE thought it had sunk without notice, his 1855 paper on "the Introduction of New Species" had in fact stirred up interest in important circles. Charles Lyell, Darwin's friend and mentor, took Wallace seriously enough to open his own series of notebooks on the species question.

Lyell had long espoused the Creationist dogma that all species were adapted from the start to the places of their origin and did not change significantly thereafter. But his antievolutionary convictions were beginning to waver. His first notebook entry, two days after reading the Wallace article, disputed the idea that limb rudiments in a snake-like reptile were evidence for its evolution from a quadruped ancestor. "Arguments against such variability of species are too powerful," he wrote—and seemed almost to add, "Aren't they?" Wallace meanwhile was jotting notes to himself about how just such limb rudiments in whales revealed their descent from quadruped mammals, not fish. For the next few months, at a distance of 8000 miles, Lyell and Wallace harrowed each other's thoughts.

The two men inhabited distinctly different worlds, and not just geo-graphically. Lyell had "a Lord Chancellor Manner," one acquaintance wrote. He was also "clubbable and cultured; a friend to peers and Prime Ministers," according to Darwin biographer Adrian Desmond. "He was a lawyer by training and a gentleman by status: he lived on his capital and made geology his vocation." Wallace meanwhile was a friend to funny old men in remote villages, and of course still a glorified manual laborer who lived on catching butterflies. His temperament was also far more impetuous, more taken with new ideas.

But Lyell's 1200-page *Principles of Geology* was a continuing influ-ence. With lawyerly precision, it made the case that natural rather than miraculous forces had caused the raising of seabeds, delving of can-yons, and upthrusting of mountains. Lyell thought geologic changes had occurred gradually, from forces still operating in the modern world. He debunked Georges Cuvier's romantic vision of an Earth alternat-ing between epochs of catastrophic upheaval (when waves of extinc-tion swept across the planet) and periods of relative calm (when new species sprang up). In contrast to this Catastrophist worldview, Lyell's Uniformitarians saw a steadier, slower process of change, with the past not all that different from the present, give or take a few extinctions. But Catastrophists and Uniformitarians alike believed that new species, and particularly human beings, were the result of "special creations" by God, and also mostly permanent in character.

Lyell had devoted the second volume of his *Principles* to refuting the evolutionary thinking of Jean-Baptiste Lamarck. But reading and rereading *Principles* in the field, Wallace thought that the slow power of natural forces could produce major changes not just in geological phe-nomena but also in living plants and animals, even leading to the origin of new species. (Darwin had also taken this idea from Lyell, insisting in his species manuscript on gradual changes produced by natural causes.) It bothered Wallace that Lyell did not also see it. The reliance on "spe-cial creations" set Wallace off on a transmutationist tear in his journal: "In a small group of islands not very distant from the main land, like the Galapagos, we find animals & plants different from those of any other country but resembling those of the nearest land. If they are special creations why should they resemble those of the nearest land? Does not that fact point to an origin from that land." It was just a quick note to himself, jotted down too fast for proper punctuation. But it wasn't really a question, anyway.

From one island to the next, Wallace's thoughts repeatedly came back to Lyell, often in a spirit of contention. The geologist's talk of the "balance of species" pushed Wallace to the brink. "This phrase is utterly without meaning," he began. "Some species are very rare & others very abundant. Where is the balance? Some species exclude all others in particular tracts. Where is the balance. When the locust devastates vast regions, & causes the death of animals & man what is the meaning of saying the balance is preserved." And then the key phrase (italics added): "To human apprehension this is no balance but *a struggle in which one often exterminates another.*" In his state of high critical dudgeon, Wallace seemed to miss, for the moment, the full import of his own words. What he was describing was natural selection.

Pouters and Spitalfield Weavers

Lyell meanwhile went for a visit to Down House early in April of 1856, and Darwin showed him around his extensive collection of fancy pigeons. This was a new passion, and almost as intense as his former love-hate affair with barnacles. His motives were the same: Barnacles had provided him with evidence of the unlimited variation produced by natural selection. Now fancy pigeons offered "the most wonderful case of variation" achieved by 5000 years of artificial breeding. Pigeon-keeping was a lower-class pastime. But the normally reclusive Darwin immersed himself in this world, frequenting the taverns and other low abodes where fanciers met and becoming "hand & glove with . . . Spital-field weavers & all sorts of odd specimens of the Human species, who fancy Pigeons."

Darwin himself was surely the oddest of them. He had quickly come to regard his pigeons, he told Lyell, as "the greatest treat, in my opinion, which can be offered to human being." He and his daughter

Darwin saw in pigeons "the most wonderful case of variation."

Etty delighted, like any pigeon fanciers, in "watching them outside," for their feathers and flight. But Darwin's interests as a naturalist also obliged him to "skeletonise them & watch their insides." He wanted to know how pouters, fantails, almond tumblers, and other breeds varied in bone structure, blood cells, and other traits. It was the "most dreadful work," with the horror of snuffing out beloved birds compounded by putrid flesh and bad smells. But he considered it necessary to help develop his theory.

As he showed Lyell around the birdhouse, Darwin explained that the many fancy pigeon breeds had all descended from a single species, the rock pigeon (*Columba livia*). But they now differed enough from their ancestral species and one another, he argued, to constitute three distinct genera and 15 separate species. Lyell, who was a geologist, not a naturalist, may have overstated what Darwin was telling him; Darwin was mainly excited to have such vivid evidence that enormous variety could develop from a single species. In his *Origin of Species*, he later repeated that ornithologists would designate the different breeds as separate species, if they encountered such variety in the wild. But he still wasn't any kind of ornithologist himself, and he never attempted to propose species names for these breeds. Bird taxonomists today regard fancy pigeons not as separate species, but merely as breeds of C. *livia*, much as Chihuahuas and Irish wolfhounds alike are merely breeds of dog. In any case, the conversation was a turning point in his friendship with Lyell.

For almost 20 years, Darwin had managed to conceal the full extent of his evolutionary thinking from Lyell, the leading antievolutionary voice of his generation. But now the shocking truth was out: Darwin believed in the almost infinite possibility of species to vary and evolve, by natural or artificial selection. Lyell was alarmed. This wasn't some cockeyed journalist like Mr. Vestiges talking, or some dubious continental like Lamarck. Darwin was a cautious and highly regarded naturalist, and, no small thing, a member of his own social class. Lyell also knew that this kind of talk wasn't just about fancy pigeons evolving from C. *livia*; it was about humans evolving from "an ourang." He grew more alarmed later that month, when Darwin convened a gathering at Down House where he subtly lobbied for the evolutionary cause with his guests, the biologist and writer T. H. Huxley, the botanist Joseph Hooker, and the entomologist T. Vernon Wollaston, who had just published a book on variation in beetle species.

Lyell heard about this meeting almost immediately when he ran into Huxley at a Philosophical Club meeting in London. There he also heard enough to suspect that other young naturalists were losing their faith in the old doctrine—his doctrine—that species had a fixed, unchangeable nature. In a note to Darwin, Lyell worried that he and his guests "grew more & more unorthodox." And to a friend, he wrote, "I cannot easily see how they can go so far and not embrace the whole Lamarckian doctrine." (The friend wrote back that even Darwin must acknowledge some limit on variation; he would hardly "maintain that a Moss may be modified into a Magnolia, or an oyster into an alderman.")

But Lyell also could not help seeing that Darwin's "species-making" mechanism—natural selection—might actually make sense. So despite his own lingering antievolutionary beliefs, he did the right thing as a scholar and friend, urging Darwin to publish at least "some small fragment of your data . . . & so out with the theory & let it take date—& be cited—& understood." If Darwin didn't strike now, somebody else would. "Whether Darwin persuades you and me to renounce our faith in species," Lyell wrote to Hooker that summer, "I foresee that many will go over to the indefinite modifiability doctrine."

By now, Wallace and Darwin were also corresponding. Through his agent, Wallace sent Darwin poultry specimens from Bali and Lombok. Darwin replied in May 1857 with encouragement, cautiously praising Wallace's paper on the introduction of species: "I can plainly see that we have thought much alike & to a certain extent have come to similar conclusions . . . I agree to the truth of almost every word of your paper." But he also gently warned Wallace off: "This summer will make the 20th year (!) since I opened my first note-book" on the species question, he wrote, adding that it might take him another two years to go to press. Darwin wrote again later that year to encourage Wallace's work on species distribution and to empathize with him on the lack of response within the naturalist community to his theoretical writing: "So very few naturalists care for anything beyond the mere description of species."

That Tumultuous Year: 1857

Events threatened to pass them both by. Debate about the special status of the human species, simmering since *Vestiges*, was now boiling over. Richard Owen, who had dissected more apes than anyone else, had previously reported that the anatomical evidence separating humans from apes matched what God had "revealed to us as to our own origin and zoological relations." Now, in a couple of lectures at the Linnean Society, he went even further. He may have intended this new foray as a preemptive strike against Darwin's developing manuscript, and against the sort of evolutionary chatter Lyell was hearing. In any case, Owen now proposed that humans belonged to a separate subclass from all other mammals, the Archencephala, or "ruling brains." He would repeat his basic argument—that our brains are structurally different from theirs—often over the next few years, with fervor inversely proportional to his evidence. Unfortunately for Owen, the naturalist T. H. Huxley was sitting in the Linnean Society audience, fighting to hold down his disbelief. Huxley was 32 years old, intense, and impatient, "handsome as an Apollo," according to one acquaintance, with a mind "like Saladin's sword which cut through the cushion." Owen's scheme of classification looked to him "like a Corinthian portico in cow-dung," and he quickly set out to bring it down. It was the beginning of a confrontation that would unfold in a devastating and highly public fashion over the next few years.

That same year, German researchers announced that they had discovered the fossil remains of a brutal-looking human in the Neander (or "New Man") Valley near Dusseldorf. It had a brain capacity comparable to modern humans, but a prominent brow ridge like that of the gorilla—another little nightmare to trouble Richard Owen's sleep. Some scientists argued that it was simply a deformed variety of human, an idiot perhaps, or a "Mongolian Cossack" suffering from rickets. But an Anglo-Irish geologist soon entered it into the biological record as a new human species, named for its place of discovery, *Homo neanderthalensis*. The variation Darwin had discovered in barnacles, and Wallace in butterflies and beetles, now also threatened to turn up in human beings.

Richard Owen wasn't the only member of the naturalist community trying to hold back the evolutionary deluge. That tumultuous year before the Darwinian storm broke saw one particularly poignant and quixotic attempt to reconcile Scripture and science. Philip Henry Gosse was among the most obsessive naturalists of the day. When his only child was born in 1849, he noted it this way in his journal, "E. delivered of a son. Received green swallow from Jamaica." Gosse was also fearless in the pursuit of species. Other collectors might content themselves picking up shells on the beach; Gosse plunged into the crashing surf. In a typical passage from one of his many popular books, he wrote about prying shells from the rocks with the help of "an old rounded table-knife" and "a smart blow" with a mallet: "Yet the operation was not without difficulty from the very sharp projections of the rock, and the force of the surf, which frequently dashed violently over my head, and once or twice knocked me down. I had to catch the momentary intervals of the waves, to dislodge my booty; and sometimes a sea coming at the moment I had done it, washed it from my grasp."

Years of such passionate collecting, engagingly recounted in print, had made Gosse one of the best-selling authors of the day, and also given him broad expertise in both marine biology and ornithology. In a note sent that summer of 1857 asking for help with a point of evidence, Darwin addressed him as "you who have so watched all sea-creatures." Gosse had seen ample evidence of the "slow modification of forms" produced over eons "in all departments of organic nature," his son Edmund later wrote in a classic memoir *Father and Son: A Study of Two Temperaments*. So it was natural for both Darwin and Hooker to seek out his support for the evolutionary cause. As a scientist, "every instinct in his intelligence went out to greet the new light" of natural selection, according to Edmund. But Gosse was devoutly, even fanatically, religious. He also believed that he was about to cut through the evolutionary chatter and cast the bright light of divine truth on the species question with an entirely different theory of life on Earth.

Derision

Gosse's new book was called *Omphalos* (Greek for "navel"), and "never was a book cast upon the waters with greater anticipations of success," Edmund, who was only eight at the time, later wrote. "My Father and my Father alone possessed . . . the key which could smoothly open the lock of geological mystery." And it was astonishingly simple: Puzzling over the question of how fossil animals had gotten buried in the Earth, to a depth that seemed to translate to millions of years, Gosse concluded, in effect, that God had hidden them there like Easter eggs, to be discovered by his happy children. At the moment of Creation, Gosse wrote, "the Creator had before his mind a projection of the whole life-history of the globe." Then he called that globe into being not at its beginning, but at some random moment in its history. Thus it appeared at that moment as if "all the preceding eras of its history had been real."

This was somewhat more logical than it sounds. The trees in the Garden of Eden could hardly have stood upright without their concentric annual rings, Gosse reasoned. So God must have created them fully-formed, with rings representing years through which they had not actually lived. Likewise, he had created Adam and Eve, like anatomically correct dolls, with navels tying them to the mothers who had not borne them. Shells created out of nothing nonetheless had growth lines from their nonexistent past. The Earth itself came with a ready-made prehistory "of animals that never really existed" buried in stratigraphic layers within its crust. So perfect was this divine representation of a fictitious past that it even included coprolites, the fossilized excrement of animals that had never lived.

Not surprisingly (except to Gosse himself), the world, both religious and atheist, Catastrophist and Uniformitarian, drunk and sober, responded with derision. Even Gosse's friend, Charles Kingsley, a clergyman, wrote that he could not give up his geology in exchange for the notion "that God has written on the rocks one enormous and superfluous lie." Gosse had always been "the spoiled darling of the public, the constant favourite of the press," his son Edmund wrote. Now he was sea-tossed in the worst possible way: Though he had been "lifted . . . on the great biological wave," Gosse "never dreamed of letting go his clutch of the ancient tradition, but hung there, strained and buffeted" in the surf.

Gosse retreated to a new home, barely furnished, on the Devon coast. He and the young son who was his greatest admirer were alone in the world, wife and mother Emily Bowes Gosse having died of breast cancer earlier that year. And in "that grim season," Gosse became a holy monster, "angry with God," and tormented by the "dislocation of his intellectual system." Being caught between science and religion was a common disorder then. But Gosse took out his pain by renouncing all pleasure except prayer, often at his young son's bedside. ("I cannot help thinking that he liked to hear himself speak to God in the presence of an admiring listener," Edmund recalled.) When they talked together over the fire, murder was their usual topic. ("I wonder whether little boys of eight, soon to go up-stairs alone at night, often discuss violent crime with a widower-papa?")

Christmas celebrations were forbidden. ("The very word is Popish. Christ's Mass!" Gosse fulminated.) But that Christmas Day 1857, the servants, "secretly rebellious, made a small plum-pudding for themselves" and kindly slipped a piece to young Edmund. He wolfed it down. But his beleaguered conscience soon forced him to confess to having "eaten of flesh offered to idols." Furious, Gosse burst into the kitchen, seized what was left of the dessert, and dragging Edmund by one hand, carried it out and flung it into the dust heap, and "raked it deep down."

For father and son, the only consolation, the only permitted pleasure besides prayer, was the return to species seeking, in tidal pools along the coastline: "Then the rocks between tide and tide were submarine gardens of a beauty that seemed often to be fabulous," wrote Edmund, who managed to please his father, as a nine-year-old, by discovering a new genus there. When they drew aside the curtain of weeds from the surface of a pool, they would "for a moment see its sides and floor paven with living blossoms, ivory-white, rosy-red, orange and amethyst, yet all that panoply would melt away, furled into the hollow rock, if we so much as dropped a pebble in to disturb the magic dream."

The Fittest Would Survive

Early in the New Year, at the other end of the Earth, A. R. Wallace's travels took him to a ramshackle hut with a leaky roof on the coast of a mountainous island he called Gilolo, now known as Halmahera, just west of Papua New Guinea. It was one of the original Spice Islands, once the world's only source of cloves, nutmeg, and other spices, and the object of long squabbling among Middle Eastern and European colonial powers. But it was still largely unexplored, and Wallace arranged to spend a month there. He seems to have passed much of the trip prostrate, wrapped in blankets against the alternating hot and cold fits of malaria. Sickness made him think, if only by forcing a pause in his restless collecting.

As he lay there he mulled over the species question and, one day, the same book that had inspired Darwin came to mind—T. R. Malthus's *Essay on the Principle of Population.* "It occurred to me to ask the question, Why do some die and some live," he later recalled. Thinking about how the healthiest individuals survive disease, and the strongest or swiftest escape from predators, "it suddenly flashed upon me . . . in every generation the inferior would inevitably be killed off and the superior would remain–that is, *the fittest would survive.*" Over the next three days, literally in a fever, he wrote out the idea. On March 9, 1858, back on the volcanic island of Ternate, the commercial center for the region, he posted it off to Darwin.

It was arguably the greatest career miscalculation in the history of science. Wallace was clearly flattered to be treated as a colleague by the eminent Charles Darwin. He could perhaps contemplate no greater success than contributing to the manuscript Darwin had been working at (and dawdling over) for 20 years. One of Darwin's recent letters had also mentioned Lyell's favorable impression of Wallace's work, and in his cover letter Wallace asked Darwin to show the new manuscript to Lyell, if he deemed it worthy. But had he simply followed his practice with his previous articles, sending the manuscript via Stevens to the editors of *Annals and Magazine of Natural History*, credit for the discovery of natural selection would have been entirely his, and the name Wallace might now be as famous as Darwin.

Wallace may simply have been too distracted by his species seeking to think about the manuscript more strategically. Upon his return to Ternate, he immediately became caught up in plans for "my four month campaign" in New Guinea, where he would be assisted by "four servants, two hunters, a cook, and a wood cutter." His shopping list was long and difficult to fill in the "miserable poverty of Ternate." Among many other items, he needed 15 pounds of gunpowder, 4000 percussion caps and a bag of bullets, 10 pounds of arsenic and other preservatives, two narrow cases for bird skins, 10 boxes for storing insects,

6000 insect pins, and such ordinary provisions as 8 pounds of coffee, 40 pounds of sugar, two bottles of vinegar, and one of soy. He planned to stock up on sago cakes, a cheap starchy food made from palm pith, en route. (Always dreaming up grand schemes, he wondered, "Would it pay to be delivered to England" at a penny a pound, as a "cheap food for pigs & cattle"?) By the end of March Wallace was off exploring "those dark forests" of New Guinea that had given birth "to the most extraordinary & the most beautiful of the feathered inhabitants of the earth, the birds of paradise."

"All My Originality"

One morning a few months later, in mid-June 1858, Charles Darwin wandered out of his study to leaf through the mail on the hall table. A fat envelope awaited his attention, containing Wallace's 20-page handwritten manuscript, "On the Tendency of Varieties to Depart Indefinitely from the Original Type." Darwin read it with dawning recognition—and horror.

"The life of wild animals is a struggle for existence," Wallace wrote, and "the weakest and least perfectly organized must always succumb." He described some of the variations that occur normally within a species, and theorized about how different forms could determine whether animals lived or died: An antelope with shorter or weaker legs would be easier prey for big cats. A passenger pigeon with less powerful wings would have a harder time finding enough food, "and in both cases the result must necessarily be a diminution of the population of the modified species." On the other hand, a change in circumstances—a drought, a plague of locusts, or the appearance of some new predator—could make the parent form of a species extinct and enable some modified offshoot to "rapidly increase in numbers and occupy the place of the extinct species and variety."

Wallace devoted a lengthy section of his essay to showing how his theory differed from Lamarckian evolution: It wasn't about giraffes getting longer necks because they "desired" to reach higher vegetation. On the contrary, individual giraffes with somewhat longer necks simply got favored over time because they could secure "a fresh range of pasture over the same ground as their shorter-necked companions, and on the first scarcity of food were thereby enabled to outlive them." It was, in a nutshell, natural selection.

Darwin had long recognized that someone might beat him to the natural selection jackpot and "fancied that I had grand enough soul not to care," he later wrote. But now he saw how mistaken he had been. "All my originality, whatever it may amount to, will be smashed," he lamented in a note to Lyell. Darwin ventured that he would be "extremely glad now" to publish a brief account of his own lengthy manuscript, but that "I would far rather burn my whole book than that [Wallace] or any man sh^d think that I had behaved in a paltry spirit." The threat to his life's work could hardly have come at a worse moment. Darwin's daughter Etty, 15, was frighteningly ill with diphtheria. His 18-month-old son, Charles, would soon lie dead of scarlet fever. Lyell and the botanist Joseph Hooker, Darwin's closest friend, took the matter in hand.

"Eminent Men"

What happened next was, for some modern critics, a case of gentlemen friends using their social class and professional status to protect one of their own. Lyell and Hooker conspired, they argue, to deny a little-known outsider's rightful claim to priority. But Wallace himself did not feel that way when he found out about it three months later. (He wrote to his mother, somewhat too obsequiously, that the collaboration with Darwin arranged by Lyell and Hooker "insures me the acquaintance of these eminent men on my return home.") He was, he later wrote, honored that his "sudden intuition" had received credit "on the same level with the prolonged labours of Darwin." His essay had been "hastily written and immediately sent off." But Darwin had worked for years "to present the theory to the world with such a body of systematised facts and arguments as would almost compel conviction."

Instead of publishing his essay outright, Wallace had merely sent a letter from one naturalist to another. And on those grounds, Darwin had

Charles Lyell and Joseph Hooker discreetly secured Darwin's priority over Wallace.

fair claim to priority. On September 5, 1857, six months before Wallace sent his letter, Darwin had written to the American botanist Asa Gray, outlining his theory. His wording foreshadowed Wallace's almost point for point: "I cannot doubt that during millions of generations individuals of a species will be born with some slight variation profitable to some part of its economy . . . this variation . . . will be slowly increased by the accumulative action of Natural selection; and the variety thus formed will either coexist with, or more commonly will exterminate its parent form." In addition to this letter, Darwin had also showed his 1844 manuscript on natural selection to Hooker.

The compromise cobbled together by Lyell and Hooker called for a joint presentation before a meeting of the Linnean Society a few days later, on July 1, 1858. Lyell and Hooker wrote an introductory letter in which the "two indefatigable naturalists," Darwin and Wallace, started out on equal footing, having "independently and unknown to one another, conceived the same very ingenious theory." But they also made it clear that Darwin had come first, emphasizing the 1844 manuscript, "the contents of which we had both of us been privy to for many years"—a stretch in Lyell's case, as he had only heard about Darwin's theory for the first time in April 1856. The letter ended by emphasizing Darwin's "years of reflection," incidentally reducing Wallace to "his able correspondent."

The reading took place in a narrow, stuffy ballroom at Burlington House, just off Piccadilly Circus. An audience of about 30 men heard an excerpt from Darwin's 1844 manuscript, then an abstract from his letter to Asa Gray, and finally Wallace's essay. Neither author was present. Darwin had buried his son that day. Wallace was hunting for bird of paradise specimens in New Guinea ("Terrible wet weather . . . nothing to eat & all of us ill"). Darwin himself later expressed surprise at the order of the presentation, having thought that the Lyell-Hooker letter and his own letter to Asa Gray "were to be only an appendix to Wallace's paper." He had half-written a letter conceding all priority to Wallace "& sh^d. certainly not have changed it" except for the maneuvering by Lyell and Hooker. "I assure you I feel it & shall not forget it," he wrote to Hooker, with good reason.

At Burlington House, the Linnean Society meeting ended, after a long series of other scientific papers, with no discussion of natural selection. The society's president went home muttering about the lack of any "striking discoveries" that year. And so began the greatest revolution in the history of science.

God and Beetle Wings

Sixteen months later, on November 24, 1859, Darwin finally published his great work *On the Origin of Species by Means of Natural Selection*, and the unthinkable became common knowledge. He didn't just supply the mechanism, the *how*, of evolution, which he and Wallace had both discovered; his painstaking work on barnacles, pigeons, and a vast array of other species collected by naturalists over the previous century, combined with his reputation from the *Beagle* voyage, made the idea credible.

Even before publication, the clergyman, naturalist, and novelist Charles Kingsley already saw that evolutionary thinking and religious faith were separate and capable of coexisting: "If you be right, I must give up much that I have believed & written," Kingsley wrote, in a letter thanking Darwin for an advance copy of the book. "In that I care little . . . Let us know what *is* . . . I have gradually learnt to see that it is just as noble a conception of Deity, to believe that he created primal forms capable of self development into all forms needful . . . as to believe that he required a fresh act of intervention" to fill every gap caused by the natural processes "he himself had made. I question whether the former be not the loftier thought." It might be better, that is, to believe in a God who promulgated laws and let them take their natural course, than to believe in a God obliged, as Buffon had put it, to busy himself about "the way a beetle's wing should fold."

But evolutionary thinking inevitably struck those of weaker faith as an assault on religion, much as it does today. They read into it the loss of the special relationship with God. One day in February 1859, Richard Owen got a potent reminder what that loss could mean. That morning, at a levee, or formal reception, in Buckingham Palace, Owen was presented to Queen Victoria and Prince Albert: "Her M. looked very well and in good humour—the youngest-looking grandmother I ever saw," he wrote to his sister Catherine. "I had a gracious smile, & the Prince shook hands . . . The Queen's train was of crimson velvet: her lace flounces over white satin: a beautiful tiara of opals & diamonds on her brow . . . It was a bright sunshiny day, & the old Palace looked very gay." Being presented to the Queen was a bit like ascending to heaven and coming face-to-face with a welcoming God; it was a sort of dress rehearsal for the afterlife.

But that afternoon Owen and the naturalist Frank Buckland had an appointment at St. Martin-in-the-Fields Church, where the crypt was being cleared out on the grounds that it had become a health hazard. They hoped to rescue the coffin of the great surgeon and naturalist John Hunter and move it to a place of suitable honor. "But such a scene! A score of Irish labourers hauling along the coffins higgledy piggledy from one dark recess to the other," Owen wrote. The sexton pointed to a coffin where a profiteering undertaker had failed to include the requested lead lining. "Putting his foot upon it he pressed down & drew back the top & one side exposing to view the black shriveled remains of the 'Hon. Lady—' " her facial features now echoed in a seething, larval mask "of the Dermestes or darkling beetles, an inch thick—faugh! I quitted the scene." Small wonder Owen resisted evolutionary thinking so fiercely: In place of the prospect of being one with God, the idea of being one with nature was appalling.

Our Tastes, Our Needs, Our Vainglory

Characteristically, Darwin gave credit in his book to Wallace, and also to Malthus, Lamarck, and even the anonymous "Mr. Vestiges." Reading the copy Darwin sent to him in New Guinea, Wallace was plainly thrilled: "Mr. Darwin has given the world *a new science*, and his name should, in my opinion, stand above that of every philosopher of ancient or modern times." He seems to have felt no twinge of envy or possessiveness about the idea that would bring Darwin such fame. (Nor for that matter, did fame bring the reclusive Darwin much joy.)

After his return to England in 1862, Wallace no doubt experienced moments, in the belittling world of Britain's highly stratified class system, when a lesser man might have become wounded and resentful. Henry Walter Bates described one such encounter with the prosperous and patrician Charles Lyell, on a Sunday visit to the London Zoological Gardens in 1863. "He was wriggling about in his usual way, with spyglass raised by fits and starts to the eye," Bates wrote. Lyell said, "Mr. Wallace, I believe—ah—"

"My name's Bates," the younger man replied.

"Oh, I beg pardon," said Lyell, "I always confound you two."

Bates thought this a little odd, "for we have often met, and I was once his guest at the Geological Club dinner." Moreover, Bates had made a sensation that year with *The Naturalist on the River Amazons*, described by Darwin in a note to Hooker as "the best Nat. Hist. Travels ever published in England." Wallace meanwhile had merely codiscovered the greatest scientific idea of the century—and also been to Lyell's home for lunch that summer. Maybe Lyell confounded two such luminaries because of their triple-barreled names, Alfred Russel Wallace and Henry Walter Bates, or their shared interests in butterflies, beetles, and the Amazon. But Lyell was, as Bates also noted, "proud of his aristocratic friends and acquaintances," and in that world, neither Wallace nor Bates could ever really count for much.

From a modern perspective, though, Wallace had class issues of his own, like almost all field naturalists. In their book *The Bird Collectors*, Barbara and Richard Mearns celebrate the unsung contributions to science by native collectors, and they single out the ornithologist Frederick Jackson for the appropriateness of his response when a species was named Jackson's Weaver (*Ploceus jacksoni*) in his honor: "Little credit is due to me for having brought this new species to light, as the specimen was brought to me by a little Taveita boy, tied by the legs along with several other of the common yellow species."

Wallace behaved much more typically, the Mearnses write, when he "claimed the glory by right of his superior ornithological knowledge, and as employer of his team of assistants" for having discovered Wallace's Standardwing (*Semioptera wallacii*), a new bird of paradise, and the only species in its genus. It happened in late 1858 or early 1859, just as the natural selection story was unfolding back home. In his book, *The Malay Archipelago*, Wallace credited his field assistant Ali for collecting the bird, but immediately added (the Mearnses' italics), "I now saw that *I had got* a great prize, no less than a completely new form of the Bird of Paradise." Later in the book, he described it simply as "discovered by myself."

But there is something anachronistic in suggesting that Ali would have resented Wallace for this, or that Wallace ought to have resented Darwin. Each had hitched his fortunes to a brighter star, and each got precisely what he had bargained for—in Ali's case, seven years of employment, along with the excitement (and danger) of travel in a good cause, and in Wallace's case, a chance to make himself heard on a larger stage. "My paper would never have convinced anybody or been noticed as more than an ingenious speculation," Wallace admitted frankly, in an 1864 letter to Darwin, "whereas your book has revolutionized the study of Natural History, & carried away captive the best men of the present age."

Great discoveries seldom occur in the romantic way we like to imagine—the bolt from the blue, the lone genius running through the streets crying, "Eureka!" Like evolution itself, science more often advances by small steps, and with different lines converging on the same solution. It is a social enterprise, and whether we like it or not, thoroughly hierarchical. Ideas pile up in the air, the cumulative product of illiterate native hunters, virtuous and vainglorious field naturalists, and inglorious taxonomists, almost all of them soon forgotten.

Then someone with the power of deep thought and the luxury to indulge it, like Darwin, or someone with extraordinary courage and insight, like Wallace, gathers up these ideas, synthesizes them, and makes them visible to the world. Giving due credit no doubt matters, or people would not be arguing about it 150 years after the fact. But both Darwin and Wallace depended on examples from the natural world discovered by countless other naturalists. And though those naturalists certainly vied for priority, even over the naming of beetles, they also ultimately needed, in Charles Kingsley's words, to accept "each fact and discovery, not as our own possession, but as the possession of its Creator"—as the possession of the Earth itself—"independent of us, our tastes, our needs, or our vain-glory."

One man's name came to stand for all the ideas encompassed in Darwinism. But those ideas represented a broad triumph of the human mind—and of the species seekers in particular.

Chapter Nineteen

THE GORILLA WAR

He is an American, and the suspicion of negro
sympathies hangs about him in many ways.

–LONDON CORRESPONDENT,
NEW YORK TIMES

O N THE EVENING of October 1, 1861, a popular preacher took
the podium of London's vast Metropolitan Tabernacle, before a
standing-room-only crowd of 6000 paying customers. Three other luminaries shared the platform—a politician, a controversial explorer back
from Africa, and a stuffed gorilla, which the London *Times* described
as "placed in the attitude of speaking, with its arms outstretched and
its hands grasping the front rail of the platform," roughly in the spot
normally occupied by the preacher himself.

It was one of the odder spectacles ever presented in a house of wor-
ship and the flirtation with sacrilege sent a mood of nervous hilarity
rippling up into the balconies. For most people in the gaslit tabernacle,
this was their first glimpse of a gorilla. Darwinian evolution was still
a new idea, and the almost simultaneous arrival of an ape so closely
resembling humans was startling. Introducing the Rev. Charles Spurgeon, the politician archly wondered if a future "Mr. Gorilla would
lecture on some Mr. Spurgeon, instead of Mr. Spurgeon lecturing on a
gorilla."

For most people, that night was also their first glimpse of the

Though he suffered a kind of scientific lynching, Paul Du Chaillu made important contributions to the natural history of West Africa.

explorer, Paul Du Chaillu, who had become almost as notorious as the gorilla. He was a slight, wiry figure of French origin and adopted-American nationality, about 30 years old, with a thick moustache, a prominent brow, bright, flashing eyes, and a charismatic presence both on stage and off. Du Chaillu had brought back Europe's first eyewitness account of gorilla behavior in the wild, along with skins and bones of more than 20 gorillas from a three-and-a-half-year expedition in the West African nation of Gabon. The influential anatomist Richard Owen, superintendent of the natural history division at the British Museum, had introduced the young explorer to the British public early in 1861. Du Chaillu had gone on to win an enthusiastic following on the lecture circuit with colorful tales of hunting ferocious animals and befriending cannibals. People were more excited about Du Chaillu's adventures, *The Times* reported, than they had been about the discovery of the ruins at Pompeii. In the United States, the politician Edwin M. Stanton was soon calling President Abraham Lincoln "the original gorilla," and joking that Du Chaillu was a fool to go to Africa for what he could easily have found in Springfield, Illinois.

But in England, Du Chaillu soon became the object of one of the most sustained and venomous attacks in the history of science. In place of the veiled insults that are standard in the scholarly world, critics slammed him as a braggart, a plagiarist, and a charlatan. His chief antagonist, John E. Gray, the keeper of zoology at the British Museum, argued that Du Chaillu hadn't discovered anything new, hadn't traveled the distances he claimed, probably hadn't killed a gorilla himself, and may never actually have visited Africa at all.

Naturalists had plenty of reasons to nurse their feelings of resentment: Envy at Du Chaillu's popular success, disdain for his adventure-story writing style, dismay at what were either embarrassing factual errors or outright whoppers, skepticism about his depiction of gorillas as ferocious monsters, and annoyance that an amateur with little or no scientific training would trespass on the territory of trained naturalists. Du Chaillu had also inadvertently gotten caught in the crossfire between two dueling scientists—Owen and Gray—at the British Museum. And he suffered collateral damage as his gorilla became a flashpoint for debate about Darwinism and the origin of the human species.

Many of the complaints against Du Chaillu were valid. He exaggerated his own accomplishments and gave too little credit to other naturalists and explorers, including some he plagiarized. His depiction of gorillas as ferocious monsters would also distort attitudes toward these animals for generations afterward—incidentally providing raw material for the "King Kong" Hollywood legend of the twentieth century. But something else was troubling Du Chaillu's antagonists, something largely suppressed in print but whispered about on both sides of the Atlantic. The ample literature of what quickly came to be called the "gorilla war" is full of references to personal attacks that needed to be omitted and facts that were "inexpedient" to publish. Much later, the explorer Mary Kingsley, who considered Du Chaillu "terribly under-rated," noted that she had a manuscript biography "written by an old enemy of his and sent to me for publication which would have blown the roof off any publisher's house in London—not that I have shown it to anyone."

Why all the whispering? What charge could have been so explosive? It was hardly an unspeakable crime in the mid-nineteenth century for an amateur to put on the appearance of being a scientist. But Du Chaillu's critics also suspected him of another kind of deception—one that, combined with his considerable appeal to women, seemed to threaten their own lofty status as white men.

"Some Hellish Dream Creature"

Du Chaillu had lived in Gabon from 1848 to 1852 when his father was a merchant and consular agent there for the French government. He attended the American Protestant Missionary School near Libreville, living as a teenager in the home of the same Rev. John L. and Jane Wilson who had previously helped Thomas S. Savage discover the gorilla. The Wilsons' interest in natural history apparently rubbed off on their student. Du Chaillu packed bird and mammal specimens in his luggage when the family eventually arranged for him to go to America as a French tutor.

There his African adventure stories soon made a hit, both in person and in the pages of *The New York Tribune*. The ornithologist John Cassin of the Academy of Natural Sciences of Philadelphia sought him out and published descriptions of some of Du Chaillu's West African bird specimens. That relationship encouraged Du Chaillu to plan an expedition to Gabon, for which he sailed in the fall of 1855. He was probably about 24 then, roughly the age at which Darwin, Wallace, Bates, and many other naturalists undertook their first expeditions. He lacked their disciplined focus as collectors—a youthful fascination with beetles had served to concentrate their minds—and, unlike them, he had little interest in theory. But Cassin reported in the Academy's *Proceed-*

ings that "Mr. Du Chaillu is amply provided with the necessary equipment . . . through the liberality of the gentlemen of this Academy."

Over the next three-and-a-half years, Du Chaillu wrote in his best-selling *Exploration and Adventures in Equatorial Africa*, "I traveled—always on foot, and unaccompanied by other white men—about 8,000 miles. I shot, stuffed, and brought home over 2,000 birds, of which more than 60 are new species, and I killed upwards of 1,000 quadrupeds, of which 200 were stuffed and brought home, with more than 60 hitherto unknown to science. I suffered fifty attacks of the African fever, taking, to cure myself, more than fourteen ounces of quinine. Of famine, long-continued exposures to the heavy tropical rains, and attacks of ferocious ants and venomous flies, it is not worth while to speak."

Du Chaillu claimed to have made his way into the wilderness on the strength of local knowledge and languages he had picked up during his previous stay on the coast of Gabon and through his father's connections as a merchant. Local chieftains in the interior escorted him as "their white man," showing him off much as Londoners showed off visiting Tahitians or Eskimos. His close rapport with different tribes enabled him, he said, to bring back tales of local customs no outsider had ever witnessed.

Among the Mpongwe, for instance, he was present at the election of a king, after which the people immediately gathered round to heap abuse on their new leader. "Some spit in his face; some beat him with their fists; some kicked him; others threw disgusting objects at him." They were getting in preemptive strikes, according to Du Chaillu, because they would soon have no choice but to do the king's bidding. Among the Fan (or Fang), he reported seeing the heaped up remains of cannibal feasts, and a woman who carried a human thigh "just as she would go to market and carry thence a roast or steak."

Du Chaillu devoted considerable energy to collecting and preserving scientific trophies, in the form of new species, and he described his finds evocatively. A new bee-eater, for instance, had a "breast of gorgeous roseate hue" and flew about "like a lump of fire." A manatee was

10 feet long, "of a dark lead color," its smooth skin covered "with single bristly hairs" and its flesh "something like pork but finer grained and of a sweeter flavor."

The average reader was more excited about perilous adventures, which Du Chaillu also delivered abundantly. Once, for instance, he fell with his foot caught in a root as a wild buffalo charged: "I had been nervous a moment before; but now turning to meet the enemy, felt at once my nerves firm as a rock . . . I waited a second more, till he was within five yards of me, and then fired at his head. He gave one loud, hoarse bellow, and . . . tumbled at my feet, almost touching me, a mass of dead flesh."

In the same spirit of bloody-minded adventure, Du Chaillu presented the gorilla as resembling "some hellish dream creature—a being of that hideous order, half-man, half-beast, which we find pictured by old artists in some representations of the infernal regions." Its roar was like "the roll of distant thunder." One old male "beat his fists upon his vast breast" as he advanced, his "features working with rage," a "gloomy, treacherous" cast to his "mischievous gray eyes." At six yards, Du Chaillu and his party opened fire, and the monster tumbled "almost at our feet, upon his face, dead."

On the other hand, Du Chaillu also dispelled familiar myths: "The gorilla does not lurk in trees by the roadside, and drag up unsuspicious passersby in its claws, and choke them to death in its vice-like paws; it does not attack the elephant, and beat him to death with sticks; it does not carry off women from the native villages . . . and the numerous stories of its attacking in great numbers have not a grain of truth in them." Despite its huge fangs, the gorilla was a "strict vegetarian," he reported, and his examination of stomach contents never turned up "aught but berries, pine-apple leaves, and other vegetable matter."

Malignancy

In England, a reviewer for *The Athenaeum* praised *Exploration and Adventures in Equatorial Africa* as "a very amusing book." But the attack began a week later, when John E. Gray wrote in the same journal that

Du Chaillu's "qualifications as a naturalist were of the lowest order," that his new species were already well known, and that his ferocious creatures were in fact "mild and inoffensive." Gray worried that "Natural History may be converted into a romance rather than a science" on account of tall tales and the grave taxonomic sin of perpetrating "useless synonyms."

In part, this attack was simply another instance of the closet naturalist Gray indulging his penchant for making field naturalists miserable, as he had previously done with Alfred Russel Wallace, Henry Walter Bates, and the African explorer William J. Burchell. Even Darwin, who hardly ever spoke ill of anyone, would later call Gray an "old malignant fool" for such behavior. The botanist Joseph D. Hooker likewise commented that "I never heard such a slanderer in my whole life," though he also noted that Gray had "all the attributes of malignancy except malignance." He merely suffered from "a loose-tongued habit" and a strange compulsion to voice his stern view of the truth about almost anything. Gray, the story went, once spotted an Irishwoman holding a fat lapdog on a London omnibus. "Madam, you feed that dog too much," he remarked, to which she replied, in the same bracing spirit, "Indeed, sir, then I only do for my dog what you do for yourself."

Regarding Du Chaillu, Gray seemed to be especially vexed at the "new traveller's" immediate success in the upper levels of society and science. Sir Roderick Murchison, the president of the Royal Geographic Society, had nominated Du Chaillu for a fellowship, praising "his clear and animated descriptions." Du Chaillu had also lectured at the Royal Institution, with Richard Owen delivering a lecture the next night employing Du Chaillu's gorillas to further his own arguments about the separation between apes and humans. (Characteristically for that era, Owen's lecture on "The Gorilla and The Negro" repeated his contention that apes lacked brain parts present even in the "lowest variety of Human Race.")

Almost any cause Owen supported was likely to irritate Gray, who had been the most prominent scientist at the British Museum until Owen's appointment as superintendent of the Natural History Divi-

John E. Gray made it his mission to ruin Du Chaillu's reputation.

sion there a few years earlier. T. H. Huxley had predicted at the time that "heart-burnings and jealousies . . . beyond all conception" would result. "Owen is both feared and hated." And everyone knew about Gray's knack for quarreling. If the two of them became "officers in the same institution, in a year or two," it would be like the Kilkenny cats who fought till there was nothing left but their tails. Du Chaillu gave Gray a scratching post to work on in the meantime.

Gray, clearly a man with a mission, extended his attack in a lengthy letter to *The Athenaeum* each week for the next month, as well as one to the *Times* a few days later. He pointed out correctly that several of the illustrations in Du Chaillu's book, including one presented as a new species of chimpanzee, were in fact copied from the works of other naturalists. This might not matter, he wrote, if the book were titled *Adventures of the Gorilla Slayer*. He objected only because it was "the work of a professedly scientific traveler and naturalist." He claimed not to bear "the slightest ill will to Mr. Du Chaillu," but also wrote that Du Chaillu "must stand convicted of falsification both of facts and dates." Du Chaillu replied that Gray's disdainful scrutiny of his many specimens smacked of "the naturalist who works at home, safely and luxuriously lodged in his museum." It also reminded him "of the ape that grins a malicious snarl at the hand that has just given it a dainty."

Early in July, in the thick of the attack, Du Chaillu gave a talk to the Ethnological Society, and an improbably angry debate arose about his description of a harp strung with tree roots. T. A. Malone, a scientist in the audience, doubted whether such an instrument could actually produce music. Malone's questioning quickly escalated into a broad assault on the truthfulness and plagiaristic borrowings of the author, or rather, "the literary compiler of the book." Infuriated by this "gross personal attack," the speaker came down from the stage. Malone wrote afterward that he was astonished to find himself confronted by a "little figure with dark threatening eyes and hands." Du Chaillu then spat in his face. It was "conduct most unbecoming," as Du Chaillu quickly admitted in a public apology in the *Times*, and it stoked the outrage of his enemies.

The "gorilla war" seethed on through summer and into winter, particularly after Richard Owen had the British Museum purchase many of Du Chaillu's specimens on behalf of John E. Gray's own department. This was a territorial intrusion of the most public sort. Owen and Gray were soon battling in print about whether Du Chaillu's gorillas had been shot in the chest, as his book claimed, or in the back, as Gray thought likely for an ungentlemanly, plagiarizing, falsifier of facts (characterizations he might just as happily have applied to Owen himself). Gray fumed that Owen had arranged Du Chaillu's entire reception in English society, including publication of his book. Owen had also paid generously—about £600—for specimens "which [Du Chaillu] had been unable to sell in America, and which had been refused, when offered to them, by the Berlin and other Continental museums."

Others soon weighed in against Du Chaillu. The most damning attack arrived from Gabon itself, where the trader, explorer and amateur naturalist R.B.N. Walker denounced the book as an attempt to "humbug the scientific world." He could testify that Du Chaillu was

PUNCH, OR THE LONDON CHARIVARI.—May 25, 1861.

THE LION OF THE SEASON.

Alarmed Flunkey. "MR. G G-G-O-O-O-RILLA!"

The "lion" of the 1861 season presented a disturbingly familiar face.

ignorant of tribal languages because he had himself done the translating when Du Chaillu interviewed Mpongwe emigrants on the coast. He had also traveled to F'an villages (Walker's spelling) without ever seeing human remains. Du Chaillu, he added, had repeatedly seen a captive gorilla at Walker's own trading station and knew that these creatures were not the monsters he had described, but capable of being "perfectly tame, docile, and tractable."

The Athenaeum chose to omit parts of Walker's letter "to avoid mixing up personal matters more than is necessary with the question of accuracy as to scientific facts." But it included without further explanation the cryptic assertion that Walker had known Du Chaillu for years and possessed "moreover, from reliable sources, information the most exact as to his antecedents."

Going in Blackface

Du Chaillu was always vague about his origins, saying variously that he was born in Paris, New York, or New Orleans, usually in 1835. But it's more likely, according to detailed research conducted in the 1970s by historian Henry H. Bucher Jr., that he was born in 1831 on the island of Bourbon (now Réunion) in the Indian Ocean. His father was a merchant and slaveholder there. Bucher concludes that Paul Du Chaillu was the merchant's child by a woman of mixed race, not his wife and very possibly a slave. Concealing this background was "an understandable choice during the heyday of scientific racism," when some white researchers treated other races as inferior species. Bucher concluded that Du Chaillu had "good reason to deny, or purposely obscure, information that would have classified him as a 'quadroon' in the nineteenth century." Other considerations may also have dictated deception. In France, to which his father took him for a time in the 1840s, illegitimate children could not inherit.

Du Chaillu's "apparent lack of formal education added to his insecurity," Bucher writes, resulting in "an intense need to attain measurable success in his own right." This may have made him more pliant if, as later reported "on reliable authority," his New York publisher twice obliged him to revise his manuscript before deeming it "exciting enough to be of general interest." His publisher may also have paired him with a ghost writer, Du Chaillu's critics suggested, because of what one former student fondly described as his "often exceedingly queer English." If so, that ghost writer would probably have been Samuel Kneeland, a professor of natural history at the Massachusetts Institute of Technology, who worked with Du Chaillu both on a scientific article in 1860 and on at least one later book.

Kneeland's slapdash style fits this theory. He tended to retail curious anecdotes about the animal world, often apocryphal. Kneeland was also secretary of the Boston Society of Natural History and thus familiar with the writing of Rev. Thomas S. Savage. That might explain why Du Chaillu's description of driver ants seems to have been lifted almost verbatim from Savage's 1849 article on the species. Du Chaillu might naturally have welcomed Kneeland as a ghost writer to compensate for insecurity about his own lack of education. But Kneeland could also have added the odd veneer of academic science that John E. Gray found so disturbing on top of Du Chaillu's otherwise tolerable adventure writing.

Du Chaillu's own hand, and his insecurities, show up in another of the book's running themes that is at least equally disturbing—his peculiar consciousness about race. He wrote, for instance, that as his party approached one remote village, "the women caught sight of me and ran screaming into the houses. It is curious that nothing excites so much terror in an interior African village as the appearance of a white man." But where Alfred Russel Wallace hid behind trees, Du Chaillu felt obliged to prepare for the hunt by blackening his "face and hands with powdered charcoal and oil." That plus his "blue drilling shirt and trousers and black shoes made me as dark" as any African.

Going in blackface may have helped Du Chaillu fit in. But it was also a way to assert that he was white, a word he used with almost compulsive repetitiveness. In one village that had somehow gotten past the fear of whites, for instance, he recounted how the chief proclaimed, "'You are the first white man that settled among us, and we love you.' To which all the people answered, 'Yes, we love him! He is our white man, and we have no other white man.'"

Du Chaillu naturally noted the similarity between humans and other primates. He could not eat monkey because "it was too much like cannibalism" or "too much like roast-baby" (at least "until you get very hungry"). But like the race scientists at the Academy of Natural Sciences of Philadelphia, he deflected this similarity mainly onto his African hosts—only to have them at one point return the compliment. Du Chaillu had just killed a female of what he took to be a new species of chimpanzee, and he marveled that her infant's face was "pure white—*very* white indeed—pallid, but as white as a white child's." Du

Chaillu stood "wonderingly staring at the white face of the creature," when two of his hunters joined him, laughing: "Look, Chelly! Look at your friend. Every time we kill gorilla, you tell us, 'Look at your black friend!' Now you see, look at your white friend . . . See white face of your cousin from the bush! He is nearer to you than gorilla is to us."

"A Negro in the Vat"

Behind the scenes, as the controversy over Du Chaillu's book unfolded, his critics were also obsessed with questions of race. The South American explorer Charles Waterton offered no evidence, but wrote: "I suspect strongly that the traveler has been nothing but a trader on the western coast of Africa, possibly engaged in kidnapping negroes." At the same time, others suggested that Du Chaillu was of African origin himself. In a letter to Waterton, the cranky Philadelphia naturalist George Ord blamed African blood for Du Chaillu's overactive imagination.

Ord was an officer of the Academy of Natural Sciences of Philadelphia, where Samuel G. Morton had recently conducted his work measuring skulls and producing ostensibly scientific evidence for those working to classify Africans as a separate species. Du Chaillu's career as an explorer had gotten its start there, and the Academy had at first gladly accepted his shipments of specimens from Africa and also passed duplicates on to other museums without his consent. But despite having publicly pledged its support in 1857, the Academy balked in 1859 when Du Chaillu presented a bill of $866.50 for his travel expenses. A special committee disclaimed any responsibility, with the cryptic explanation that it was "inexpedient to report the facts." Historian Henry H. Bucher theorizes that the "mysterious and rapid" end to Du Chaillu's close association with the Academy in 1860 resulted from "a committee member's discovery of his maternal ancestry."

The letter Ord sent to Waterton in 1861 supports this theory. Some of his learned colleagues had taken note when Du Chaillu was in Philadelphia of "the conformation of his head, and his features," Ord wrote, and they saw "evidence of a spurious origin." (Even Du Chaillu's friend, the banker and anthropologist Edward Clodd, would later crassly comment that the explorer's "diminutive stature, his Negroid face, and his swarthy complexion made him look somewhat akin to our simian relatives.") Ord added: "If it be a fact that he is a mongrel, or a *mustee*, as the mixed races are termed in the West Indies, then we may account for his wondrous narratives; for I have observed that it is a characteristic of the negro race, and their admixtures, to be affected to habits of romance."

Curiously, the same issues of *The Athenaeum* in which the attack on Du Chaillu was playing out also featured a running plagiarism argument about the source of a stage melodrama called "The Octoroon." It told the story of a dazzling New Orleans beauty "educated in every refinement and luxury." She was "almost a perfect white, her mother being a quadroon." In all three contesting versions of this tale, an "underhanded Yankee overseer" seeks to possess the heroine through purchase on the slave market. And in each case, a dashing sea captain foils the nefarious plot and carries the beauty off to freedom—and presumably to love. Audiences apparently felt comfortable taking the heroine's side because she was seven-eighths white. But what if the sexes had been reversed, with a white woman falling for a mixed-race man—a man like Du Chaillu, say?

Du Chaillu was an enthusiastic socializer, of undeniable charm, whose address books in later life were full of notes about "calls to make," "notes to send," and new acquaintances, both male ("lawyer good fellow") and female ("of medium height with dark chestnut hair . . . an exquisite figure . . . graceful"). Samuel Kneeland later passed the word "sub rosa" to an acquaintance: "He is, or was, rather too fond of women & the demi-monde: <u>entre nous deux vous savez et cave</u>"—that is, "Between you and me, keep it in mind and watch out." Womanizing, plus race, was of course an inflammatory combination.

Suspicions about Du Chaillu also simmered beneath the surface in England. The mathematician Augustus de Morgan found the running battle over Du Chaillu so entertaining that he sent a congratulatory note to his friend William Hepworth Dixon, editor of *The Athenaeum*: "This Gorilla matter is a godsend," just when the journal was seeming stodgy and predictable, he wrote. "Bear this one in your editorial mind—you cannot always have Du Chaillu—but you may have *a* Chaillu."

Then, without actually stating the racial gossip outright, he drove the point home with a ghastly joke: A customer comes to the brewer demanding a batch as good as the last one, just after the brewer has discovered a dead Negro at the bottom of the vat. "It's all very well to say as good as the last!" the brewer replies. "But where am I to get a negro every time I brew?" De Morgan summed up the editorial lesson: "That is the question—a negro in the vat every time!"

Return to Africa

Many new authors would have buckled under the sort of literary lynching Du Chaillu had experienced. He later admitted to having been "hurt to the quick" by "unfair and ungenerous criticisms." But he didn't self-destruct. Instead, he trained himself in the latest scientific methods; purchased the best photographic, astronomical, and meteorological equipment; and set out to redeem his reputation in the scientific world.

By October 1863, "after three years' recreation in the civilized countries of Europe and North America," he was back in Gabon, for a new expedition to the interior. Things began badly when the canoe carrying Du Chaillu ashore capsized, ruining vital supplies, including much of his scientific equipment. He was forced to bide his time for a year along the coast as he waited to be resupplied from England. But he busied himself hunting and gathering specimens, and soon shipped back to England several new gorillas, giant pangolins, manatees, a catlike carnivore called the genet, 4500 insects, plus a live African bush hog and two live fishing eagles. The gorillas went to the British Museum, and Du Chaillu wrote that he meant them "to repay the debt of gratitude which I owed to the large-hearted British nation who had so generously welcomed me when I arrived in England, an unknown traveler, from my former arduous journey." He was thinking of Roderick Murchison, who had given the Royal Geographical Society's endorsement to this second expedition, and of Richard Owen. He would later name a mountain after each of them.

Du Chaillu and his party headed back into the interior in October 1864, and this time the explorer made detailed astronomical and meteorological readings en route, and kept his journal in triplicate, with each copy entrusted to a different porter to avoid a total loss. It would turn out to be a sensible precaution. The party encountered difficulties almost from the start, including a standoff at gunpoint over the philandering of Du Chaillu's porters among local women, a series of robberies in the villages where they stayed, and drawn-out negotiations for permission to move from one community's territory to the next. His hosts were often suspicious and sometimes hostile. But instead of relying on force, Du Chaillu generally deployed his characteristic kindness and generosity, as well as a genuine affection for the local people, to charm his way inland.

A smallpox epidemic was afflicting West Africa that year, and many people blamed Du Chaillu, the *oguizi*, or white spirit, "bleached by the grave," for bringing the deadly plague among them. (They may have been right. A French ship had brought the disease to the Gabon River. Using porters and assistants from the coast could have helped spread it inland.) Finally in the ninth month of the expedition, warriors from a village on Du Chaillu's proposed route refused to allow him to proceed, and in the confrontation that ensued one of Du Chaillu's men killed two local men. During a brief attempt at peace making, Du Chaillu prepared to withdraw, loading his seven assistants "with my most valuable articles, my journals, photographs, natural history specimens, and a few of my lighter goods." His own burden included "five chronometers, a sextant, two revolvers, rifle, with another gun slung at my back, and a heavy load of ammunition."

Negotiations soon broke down. "A general shout arose of 'war!' and every man rushed for his spear or his bow. I gave the order for the retreat." Du Chaillu and his men fled through the forest, in a running exchange of arrows and bullets that went on for nine hours. Du Chaillu himself suffered two arrow wounds, and one of his men was also injured. As they ran for their lives, the party tossed away anything that might slow them down. But the explorer begged his men, even as he thought he might be dying, to preserve his journal for delivery "into the hands of the white men on the coast." Du Chaillu and his party killed or wounded an unknown number of their pursuers before finally driving them off in a "last stand" on a hilltop.

The flight back down to the sea took seven weeks, often through countryside emptied by smallpox. In one village, Du Chaillu drew back a mosquito net and found a skeleton lying under it. Human skulls and bones littered the forest trails, "gnawed and scattered by prowling hyenas and leopards." Miraculously, though, Du Chaillu and his men all survived, as did one copy of his journal. Six days after arriving at the coast, at the end of September 1865, Du Chaillu loaded his collections aboard a ship bound for England and fled Africa forever.

In his 1867 book *A Journey to Ashango-land, and Further Penetration into Equatorial Africa*, Du Chaillu soon renewed his quarrel with John E. Gray. He gloated particularly over the giant river otter specimens he'd found and shot on December 28, 1864, early in his drive to the interior. In his book, he recalled how on his previous trip he had been able to bring back only a tattered skin of the animal. He had described it in a scientific journal as part of the otter genus *Cynogale*, which Gray had established in 1837. But at the same time, Du Chaillu had also coined his own new provisional genus, *Potamogale*, in case more complete specimens bore out his impression that this species actually differed significantly from *Cynogale*. Gray promptly dismissed the idea that it was any kind of otter at all. He classified it as a rodent instead and renamed it *Mystomys* at first, and then, finding that this name was already taken, as *Mythomys*. It was, Du Chaillu recalled, a characteristically unsubtle "commemoration of my supposed fabulous statement."

This time Du Chaillu had saved the bones as well as the skins of both his specimens "and wished that I could at once have sent them to London to vindicate my statements." But other specimens had reached Edinburgh in the meantime and, even better, they had caused a professor there to publish an account asserting, after considerable deferential throat-clearing on Gray's behalf, the accuracy of Du Chaillu's classification: "Mr. Du Chaillu's name of *Potamogale* stands: it has thus precedence over Gray's name of *Mythomys*; and the laws of natural-history nomenclature compel us to accept it." It was a vindication for Du Chaillu, and he featured a drawing of the creature as the frontispiece of his book.

Du Chaillu also collected the first-known specimens of many other mammals, including a bushbaby (*Euoticus elegantulus*), a hammer-headed fruit bat (*Hypsignathus monstrosus*), and a pygmy squirrel (*Myosciurus pumilio*). He collected the type specimens for no fewer than 39 valid bird species. "In an overall Afrotropical context," says Robert Dowsett, a specialist in West African ornithology, "this rates as a very significant contribution indeed, bettered by few."

Du Chaillu's misrepresentation of gorillas as ferocious monsters remains a sore point for primatologists to this day. But later travelers eventually confirmed many of his other supposedly "wondrous narratives." Traveling in Gabon with Du Chaillu's first book in hand, "I jealously looked into his every statement," the irascible explorer Richard Burton reported in 1876, "and his numerous friends will be pleased to see how many of his assertions are confirmed by my experience." Likewise in the 1890s the intrepid British explorer Mary Kingsley, modestly attired in a heavy, ankle-length Victorian dress, followed in Du Chaillu's footsteps and "never came across anything while in this region that discredited Du Chaillu's narrative on the whole." Kingsley offered critics at home this tart advice: "Have you been there? No! Then go there or to whatever place you happen to believe in! and till then—shut up."

Du Chaillu himself continued to travel after *A Journey to Ashango-land*. In truth, "he never attempted to establish a real home," according to his biographer Michel Vaucaire. "He lived in hotels or stayed with friends, a most beloved and amusing guest," mainly in New York and Chicago. "The explorer became the ideal bachelor, fond of good food, good wine, and attractive women." He remained a popular lecturer "in spite of his peculiar accent," and also continued to produce books for both adults and children. But after the battering he had received in England, he largely abandoned natural history. Instead, he focused his energies as a travel writer and scholar mainly on blue-eyed and pale-skinned Scandinavia, about as far from Africa as he could get. He died of a stroke at St. Petersburg, Russia, in 1903.

Decades later, the Gabonese writer Abbé André Raponda Walker reported a curious discovery. (Coincidentally, Walker was the child of a Mpongwe woman by R.B.N. Walker, the same trader who had possessed "information the most exact" about Du Chaillu's antecedents.) Climbing a mountain in the country of the Eshira ethnic group, Catholic missionaries had found Paul Du Chaillu's name carved on a large rock at the summit, supporting evidence for the itinerary the explorer had so loosely outlined in his first book. The mountains Du Chaillu had dedicated to Owen and Murchison had long since reverted to their local names. But the Eshira, Walker noted, still affectionately called this place *Mukongu-Polu*—that is, "Paul's Mountain."

Chapter Twenty

BIG NOSES AND SHORT TEA-DRINKERS

> I am told about thefts and murders. . . . For
> safety only and to cool evil fancies I shall be
> careful to keep my gun much in evidence.
>
> —PÈRE ARMAND DAVID

IN MARCH 1869, on a brilliant spring morning promising release from the hard Sichuan winter, a French naturalist and missionary named Père Armand David and his Chinese assistant Ouang Thomé set out to climb a mountain. The path soon ended at the foot of a cascade fed by snow melt, where they paused to snack on a crust of bread soaked in ice water. Then they started to climb, looking for a safe route across the cascade. Soon they were pulling themselves up from rock to rock, sidestepping along the roots of trees and clinging to branches, with their guns slung over their shoulders.

The two of them were accustomed to enduring dangerous conditions in the pursuit of new species. But four hours in, they were both repenting the decision to attempt "this horribly uncomfortable ascent." Any surface not vertical was caked in frozen snow. Badly scratched, soaked to the skin, their strength utterly sapped, they turned back too late and began to retrace their route down. They slipped and fell repeatedly, at times sinking waist-deep in melting snow. Twice, a branch snapped and David managed to catch hold of Thomé with one hand as he was already slipping over the edge of the abyss. "If we live through this," Thomé remarked, "we'll never die."

Around three in the afternoon, the sun disappeared in a thick fog. Earlier, they'd been grateful that the trees and shrubs at least blocked their downward view of the gorge into which they seemed likely to plunge at any moment. Now they continued blindly down the 3000-foot wall, repeatedly forced to climb back up when they came to a foothold that was too narrow, or without shrubs for a handhold.

Back at the foot of the cascade, "soaked with sweat and water," and with a burning thirst despite the snow they'd been eating, they thanked God "for letting us escape from danger." But they still had another two hours of trekking back to shelter. This corner of Sichuan Province was home to leopards and tigers. But wild bulls and the ferocious Himalayan black bear were a more serious threat around the mountain David called Hong-chan-tin. He spotted footprints in the snow, but he was too depleted to pay much attention. In any case, their ammunition was wet, and snow filled the barrels of their guns. It was soon dark, as they floundered through streams fattened into rivers by the melting snow. They had just begun feeling for a cavity in the rock where they could huddle for the night, when they heard a voice. "We had not known there was a human habitation in these wild gorges," David wrote, but the "good people of this little wooden house" soon had them eating potatoes and maize cakes. Then they set them up before a "colossal fire" to warm themselves and fervently recite their prayers. For their trouble, the day had gained the explorers a total of two animals—a new species of the birds known as nutcrackers and a new squirrel.

Even for Père Armand David, a Catholic missionary of the Vincentian (or Lazarist) order, this was probably not quite enough. But he later made a note in his journal about how precious "the most minute objects, the most insignificant details" were to understanding the problem of life on our planet. "A dot, a comma, a small line are not important in themselves but have value in relation to the whole, and can change radically its final significance." The fine points of Christian dogma clearly mattered to him on the religious side of his life. But he saw that studying the distribution of species—not the Bible—was a better way to gain new insights into the past. "The study of spiders alone proved that Italy had touched the continent of Africa in ancient times," he wrote, adding that data about mammals, birds, and insects had enabled Alfred Russel Wallace to draw a line on the map separating the islands of the East Indies with a more Asian history from those with stronger ties to Australia.

Finding both the seemingly trivial species (a frog that barks like a dog), as well as some of the most spectacular creatures on Earth (the giant panda), was "of utmost importance," and in this cause David gladly risked poisoning, drowning, starvation, infestation, isolation, and all the other hazards of the species-seeking life, plus the likelihood of having his throat slit by brigands. Earlier martyrs had given up their lives for their faith. David was willing to die for what he regarded as an entirely natural combination of Christianity and Darwinian discovery.

"The Animals' Westminster Abbey"

Evolutionary thinking had changed a lot of things. "The wide question of the origin of species has been shaking the moral and intellectual world as by an earthquake," the president of the American Association for the Advancement of Science (AAAS) reminded his membership at about the same time Père David was contemplating his dots and commas. And that had only made the hunt for species more urgent: People wanted answers to the questions Darwin and Wallace had raised. "Never before has the study of animals been raised to so high a dignity as at present," another AAAS president declared in 1876. Naturalists, always insecure about their scientific standing, now had "universal laws" to match anything in physics or chemistry, he said. Zoology had become "the pivot on which the doctrine of man's origin hinges."

Thus the great age of biological discovery, the age of the species seekers, reached its apotheosis in the last three decades of the nineteenth century. Nations, museums, and universities all sent out collecting expe-

ditions, at times spending huge sums to keep teams of scientists conduct-
ing field surveys for months or years at a time. For the love of science
and the glory of having their names attached to newly discovered species, rich men and women also sponsored expeditions and sometimes
went along for the trip. Everyone bickered over priority. And natural history museums sprang up everywhere to sort through, and show off, the
booty; there were 150 of them in Germany alone, and 250 in the United
States. Nor was this phenomenon confined to the developed world: Cape
Town, Bombay, Calcutta, São Paulo, Buenos Aires, Shanghai, Sarawak,
Singapore, and Manila all had natural history museums of their own by
century's end.

At the top of this considerable heap, Richard Owen won permission for his natural history department to split off from the rest of the
British Museum. It reopened in 1881 in London's Kensington district,
in a new cathedral of a building celebrated by contemporaries as "a
true temple of nature" and "the animals' Westminster Abbey." Much
as Westminster Abbey was crowded with statues of war heroes, this
scientific cathedral also implicitly served as a showcase for the power
and reach of the British Empire.

European explorers were now intent on writing "the natural history of the globe with exhaustive comprehensiveness and precision,"
according to science historian Fa-Ti Fan, and they displayed an almost
religious conviction of their right to go anywhere in this cause. In the
minds of some explorers who misunderstood Darwinian survival of the
fittest, it was a right born of the natural superiority of their own more
highly evolved nations. They presumed that they could and should
bend inferior cultures to European purposes. The idea of a "right to
know" also derived its authority, according to Fan, "from a belief in
the universal validity of factual knowledge." The facts that explorers
were systematically gathering would yield useful knowledge, ultimately
bringing benefits to the inferior cultures, too. "This grand vision of scientific commonweal" was too important to be "restrained by human
boundaries—particularly, the boundaries drawn by the natives against
European scientific researchers."

Far Cathay

No boundaries were more vexing, or seemed to conceal more enticing wonders, than the ones that protected Imperial China. For more than a century, the Qing emperors had confined Western commerce to the southern port of Canton (now Quangzhou) and, even there, foreign merchants and their trading stations had to huddle along the riverbank outside the walled city. "Expeditions" by naturalists seldom went farther "than the gardens, nurseries, fish markets, drugstores, and curio-shops in the district," Fan writes in his book *British Naturalists in Qing China*. The Chinese themselves did the fieldwork, and took their finds to market for Westerners to discover. Thus Cantonese merchants imported live butterflies from the mountains 60 miles north and kept them "tethered by threads to thin bamboo sticks, so they could flutter and display their beautiful wings." Colorful beetles and dragonflies preserved in camphorwood boxes were also commonplace. Detailed drawings prepared by Chinese artists according to instructions from a British merchant-naturalist in Canton became the sole basis for naming 83 new species of fish.

But China itself remained a tantalizing blank. In the overheated Western imagination, it took on a dream life concocted from the fanciful imagery on the porcelains, textiles, paintings, and carved furniture that the Chinese also happily exported. China became Far Cathay, a mysterious land inhabited by "butterflies the size of puffins," as one connoisseur put it, and "gaudy birds with rainbow-hued plumage" perching in "forests of gnarled trees." Westerners imagined the Chinese themselves as "a serene, contemplative people," ornamented with "long curling fingernails" and "robes embroidered with gold," according to John R. Haddad in his book *The Romance of China*. China wasn't, as Fa-Ti Fan mordantly remarks, just "a big piece of land inhabited by many short tea drinkers."

Westerners desperately yearned to "unveil" the real China, and they got their first chance when trade expanded from Canton to other ports in 1842, following China's defeat by the British in the first Opium War.

Europeans cultivated a fanciful image of Far Cathay.

In 1860, a further humiliating defeat in the second Opium War forced the Qing government to allow "Big Nose" Westerners into the interior, with naturalists prominently and persistently among the nosiest. The agents of British Empire and British natural history were, as always, essentially identical. Diplomats, soldiers, and merchants had all grown accustomed to collecting specimens elsewhere in the world. In China, they quickly developed an informal national network for that purpose, operating out of the offices of the British Consular Services and the Chinese Imperial Maritime Customs Service, which, despite its name, was largely staffed with British administrators.

The motive was sometimes a practical matter of commerce. Silk-worms, for instance, were always of interest. According to legend, they had been the object of the earliest-known incident of industrial espionage in about 550 A.D., when two monks secreted the eggs of silkworms in their bamboo canes and carried them back to establish a competing silk industry in the West. Likewise in the nineteenth century, the British would spirit away Chinese tea plants, and tea planters, to establish a competing industry in India. When the Chinese tried to protect what they knew about local plants and animals, Fan writes, the British depicted them as "mean, jealous, and cunning." When the British used deception to circumvent the Chinese and get what they wanted "they were portrayed as smart, adventurous heroes outwitting the conceited natives."

Life in remote diplomatic outposts could be dull and lonely, so British functionaries also "collected for fun, for curiosity, for self-improvement, for ambition, for the intellectual satisfaction of making a contribution to science, for the good feeling of having a respectable hobby," Fan writes. "And sometimes they were requested by friends or ordered by superiors to investigate certain products of nature . . . Collecting specimens joined tennis, card games and hunting as regular pastimes." British missionaries could not do as much as they might have liked for the cause because they typically lived with their families in coastal cities tending to the urban poor. But French Catholic missionaries, unmarried and often traveling deep into the interior, took up the slack, none more boldly than Père Armand David.

"I Am Used to Everything"

David was relatively old for an explorer, in his midthirties when he first went to China in 1862, and almost 50 when he completed his last expedition and returned to France in 1874. Physically, he was unimposing, of medium height and lean build, with thin lips, deep-set eyes, and a broad, bony forehead. He seems to have been built for long-distance walking, with a quick stride and remarkable endurance developed growing up in a Basque family in the foothills of the Pyrenees. On expedition in China, he routinely walked 30 miles a day.

Like other naturalists, he needed to kill animals to get specimens for his natural history collection. But to the chagrin of his assistants, he generally refrained from filling his dinner pot with "these poor creatures, who revel in life so joyously." David and his assistants got by instead on rice, millet, and *tsamba*, flour made of cooked beans and barley. A handful of this "reddish dust," rolled with hot or cold water into a ball was "highly expeditious," but so tasteless it defied "all but the most ravenous hunger." He had to force himself to swallow enough to keep body and soul together. "When mixed with milk," he allowed, "it seems to be bearable." But milk was scarce.

For this, David had given up his original assignment as a science teacher at a Catholic college on the Italian Riviera, where he collected specimens along the shores of the Mediterranean and in the foothills of the Alps. He had requested transfer to China because the "privations of a missionary" suited his wish to do penance, and because he was "pursued with the thought of dying while working at the saving of infidels." In Beijing, he continued to collect specimens for the instruction of his new students. But he also sent some back to Paris, where Henri Milne-Edwards, director of the Muséum National d'Histoire Naturelle, was so impressed that he prevailed on the Superior General of the Vincentians to put David to work as a biological explorer.

Over the next decade, David undertook three major expeditions, first a seven-and-a-half-month journey northwest into Inner Mongolia, next for two years up the Yangtze River and overland deep into the southwestern province of Sichuan, and finally for 20 months through the Chin Ling (now Qin Ling) mountains of central China. Friends warned him repeatedly about bandits, armed rebellions, and anti-Christian persecutions, but David wrote that "it would be necessary to renounce all traveling to distant parts if one had to wait for peace in this Empire. "The wild places," reputed to be the haunts of thieves and malefactors, are precisely the ones that offer the most in the way of natural history in China."

Thus one morning in 1864, David was traveling by mule in a landscape of dark, forested hills when eight robbers on horseback ambled onto a lonely stretch of road and blocked his path. He was accompanied only by an assistant and a couple of timid cart-drivers. The robbers carried sabers, and had a reputation for hacking off the limbs of uncooperative victims. A few also had guns. But they were too confident in their numbers to bother drawing them.

Two faces of Père Armand David.

Their intended victim was, however, something the robbers had probably never seen before—a European. On first sight, David later noted in his diary, "We are readily believed to be endowed with extraordinary talents and superhuman power." He saw no reason to dispute this impression of "a universal and incontestable superiority." Combined with his faith in God, it gave him "strength and confidence." It may also have helped that he had his shotgun and revolver ready. He conveyed his disinclination "to let myself be despoiled gratuitously, and still less to let myself be killed by these prosaic cutthroats." The bandits went off instead to rob and burn down a nearby inn.

David recounted such episodes matter-of-factly, without swagger or braggadocio. He did not know how to spin a story, and his diary often omitted crucial details: Did he fire his gun at the feet of one of the bandit's horses? Did he even point it? The reader has to read between the lines of David's quietly purposeful narrative.

Another time, a bandit armed with sword and gun pretended to be a local hunter. David, still part schoolteacher, grilled him so closely about local wildlife that the flustered bandit backed off and went to rejoin his gang at their camp nearby. Not content, David took his shotgun and revolver and went out alone that night to visit, thinking his weapons would make "a salutary impression." But the bandits, perhaps fearful of further quizzing, had already extinguished their fire and fled, leaving behind a half-eaten meal of wild onions, which David scavenged for his companions.

The deep hostility to foreigners that David encountered almost everywhere was more dispiriting than the bandits. At times, officials suspected him of being a spy or a gold prospector. In Wanhsien (now Wanxian), an angry crowd surrounded his party, shouting out "vile epithets" and throwing stones. But David was not the sort to dwell on such episodes. After describing this incident in his diary, he began the next paragraph, a little oddly, "the most beautiful stones are abundant around Wanhsien," and proceeded to a quick travelogue about the ones fixed permanently in place as staircases, bridges, and tombs.

Because Europeans were "considered dangerous and evil beings," innkeepers often refused accommodation or demanded extortionate payments in advance "for the most deplorable hospitality." In one case, his party of six had to crowd into a dirty cabin that measured just six feet on a side—and the innkeeper thought that "the two old baldheaded nuns" already sleeping there should also stay. "With a little firmness," David overcame another innkeeper's resistance and won entry not only to the inn, but to the brick platform heated by smoke from the hearth, where "all the men of the house sleep wrapped up in their bed clothes." He came away infested with "several kinds of unpleasant little animals," probably lice, fleas, or both. The alternative was to camp out by a fire fueled with dung, which sometimes froze overnight. After one such night in his tent "on a bed of dried manure and stones that are by no means soft," he commented stoically, "I am used to everything."

"One Hears the Hatchet"

During his years in China, David collected hundreds of species previously unknown to science. Some of the plants he discovered are now commonplace in Western gardens, like *Buddleia davidii*, the butterfly bush, with its long, showy purple flowers, or many of the 52 previously unknown rhododendron species he collected. He also brought back a new clematis, a lily, a peach tree, roses, aconites, ranunculuses, primulas, gentians, oaks, magnolias, and fir trees. He discovered more than 100 new insect species, 65 new birds, 60 mammals, and numerous reptiles, amphibians, and fish. David seldom mentioned his religious feelings; that's not what the diary of his expeditions was about. But his plea-sure in the natural world clearly outweighed the hardships he endured en route: "I passionately love the beauties of nature," he wrote. "The marvels of the hand of God transport me with such admiration that in comparison the finest works of man seem only trivial."

One such marvel, the giant panda, would become one of the most beloved animals in the world. David first saw it in March 1869 as "a fine skin"—that is, a trophy or perhaps a sleeping mat—at a "pagan" land-owner's house in the Mu-Ping region of Sichuan province. The "famous white and black bear . . . appears to be fairly large. It is a remarkable species and I am delighted when I hear my hunters say that I shall certainly obtain the animal within a short time. They tell me they will go out tomorrow to kill this animal, which will provide an interesting novelty for science."

Less than two weeks later, shortly after David returned from his icy climb on Hong-chan-tin, his hunters presented him with his first panda, a juvenile, "which they sell to me very dearly." He considered it a new species (in fact, it was a new genus), "very remarkable not only because of its color, but also for its paws, which are hairy underneath, and for other characters." His hunters brought him a second specimen, live, a few weeks later. David noted that "it does not look very fierce" and "its stomach is full of leaves," though he omitted the painful but obvious intervening fact that the panda had not remained alive for long before being sacrificed to science.

The giant panda was of course hardly a new species to the Chinese. More than 1000 years earlier, an emperor in the Tang Dynasty had supposedly sent two live pandas to Japan as a token of friendship. A thousand years before that, during the Western Han dynasty, a panda had been the prize specimen in the Emperor's private zoo in Xi'an. But nobody beyond Asia—indeed, hardly anyone beyond central China—had ever heard of it, until David brought the species to the attention of the outside world. It became known to science as *Ailuropoda melanoleuca* David. Oddly, David himself never actually saw one in the wild. In fact, no Westerner would do so until 1916. But the Western interest in giant pandas that he instigated would eventually help save the species from almost certain extinction.

A few weeks after delivering the panda, David's hunters presented him with another spectacular species. The golden monkey was familiar to Europeans from images in Chinese paintings and porcelains, but according to David's translator Helen M. Fox, it was "so odd that it was thought to be an imaginary animal." It had a small bluish-white face surrounded by a fringe of flame-red hair, and whereas all other known primates were tropical, these lichen-eating monkeys lived, David wrote, "in trees in the highest mountains, now white with snow." The golden monkey became *Rhinopithecus roxellana*. In a multicultural mashup, the French naturalist Henri Milne-Edwards named the Chinese species after the Ukrainian wife of an Ottoman Turkish sultan, Suleiman the Magnificent, because monkey and wife both had distinctive up-turned noses.

To some modern readers, the discovery and naming of these species may seem like an act of cultural appropriation, and with no particularly heroic overtones to distract from the crime. It may have seemed that way at the time to the Chinese themselves. David wrote that when he was collecting in Muping province, the local prince tried to stop him "in a most severe fashion." David characterized the prince as a tyrant "detested by everyone because of his avarice and bad conduct," who had previously had Christian missionaries put to death by drowning, and, adding incest to murder, included his own sister in his harem. Concluding that the prince or his men merely wanted a share of the money he was spending, David brazened it out. "As for respecting the hunting laws, no one pays attention to them," he wrote. His hunters told him they had not killed the animals they brought him on the lands of the principality, and "so be it!"

As this passage suggests, David could at times express as much disdain for the Chinese people as any petulant colonial administrator. He was furious, for instance, when a former pupil, a Christian convert who had been assisting him, took another job at a higher salary. David figured that what he was paying was "entirely sufficient." He wrote (italics added), "I am annoyed at the widespread, sordid cupidity of these people, *who owe us everything* and instead of helping me voluntarily at *work which benefits the whole world,* try to exploit me and get as much cash as possible." The locals simply didn't get the "grand vision of scientific commonweal," to use the historian Fa-ti Fan's words.

But David was also a clear-eyed realist, even about his religion. "Above all do not think China will become Catholic," he wrote to his uncle and godfather. "At the pace things are going now, it will take forty to fifty thousand years before the whole Empire will become Christian." He also wrote that "The Chinese have a natural aptitude for business, and when they learn more European languages and customs they may gradually absorb the riches of the rest of the world . . . it will be difficult to contend against the shrewdness of the Chinese, their prudent composure, and their sense of thrift." But he warned both East and West that "a selfish and blind preoccupation with material interests" was reducing "this cosmos, so marvelous to him with eyes to see it, to a hard, matter-of-fact place."

David saw alarming changes happening in China long before the Chinese themselves, and before conservationists elsewhere, had even begun to contemplate the mass extinction of species. "From one year's end to another," he wrote, "one hears the hatchet and the axe cutting the most beautiful trees. The destruction of these primitive forests, of which there are only fragments in all of China, progresses with unfortunate speed. They will never be replaced. With the great trees will disappear a multitude of shrubs which cannot survive except in their shade; also all the animals small and large, which need the forest in order to live and perpetuate their species . . . it is unbelievable that the Creator could have placed so many diverse organisms on the earth, each one so admirable in its sphere, so perfect in its role only to permit man, his masterpiece, to destroy them forever." He foresaw a bleak future in which "hundreds of thousands of animals and plants given to us by God "would disappear and the Earth would be given over to horses, pigs, wheat, and potatoes."

So while David may well have been committing cultural appropriation, he was in some cases appropriating species only from likely extinction.

A Species Saved

The trend among modern critics is to focus on natural history in the nineteenth century mainly as an imperialist exercise, with great white hunters playing out their heroic exploits at the expense of local cultures. And there is certainly plenty of evidence that Western naturalists in China and elsewhere could at times be exploitative, disruptive, and dismissive of local knowledge. They assumed that their factual approach was universally valid, as critics like Fa-Ti Fan argue, and they believed in their unlimited right to impose it everywhere.

But though it is not fashionable to say so, the rest of the world had nothing to match the systematic approach to nature that Western science had been developing and refining since the time of Linnaeus and Buffon. The Chinese had a rich horticultural literature, according to Fan. Detailed information about plants and animals recorded in local gazetteers enabled visiting naturalists to track down species they might not otherwise have discovered. "In fact, however," he adds, "the Chinese did not have a discipline, a system of knowledge, or even a coherent scholarly tradition equivalent to Western notions of 'natural history,' 'botany,' or 'zoology.'" Phrases for the study of plants, *zhiwu xue*, and natural history, *bowu xue*, only appeared as mid-nineteenth-century translations of Western concepts.

The ability to distinguish one species from another, and to sort out the relationships among species, was—as David had suggested—a critical advance for understanding the nature of life on Earth. Thus every country from China to Gabon to Colombia would eventually adopt the Western system of classification. It would prove crucial even for local cultures wanting to understand their own plants and animals. What Winston Churchill said about democracy applied equally to the system of scientific classification—that it was the worst form of government except all the other forms that had been tried.

One of Armand David's least heroic discoveries was in some ways the best argument for the value of scientific methods. David first heard about the species that would become known as Père David's deer, or *Elaphurus davidianus,* soon after his arrival in China. It had long since been hunted out in the wild, but a herd of 120 animals survived in a deforested and overgrazed imperial hunting park a few miles south of Beijing. Locals called this deer *ssu-pu-hsiang,* "the four characters which do not match," because it supposedly had the tail of an ass, the hooves of a cow, the neck of a camel, and the antlers of a stag. David made the hike out to the imperial park and by befriending the guards or climbing on a wall he managed to get a glimpse of the herd. "All my attempts to secure a specimen, or even some remains, were unsuccessful," he wrote. "It is said that there is a death penalty for anyone who dares to kill one of these animals." But his sense of the "utmost importance" of each species—each "dot" or "comma"—made him persist, and in January 1866 he somehow got hold of two skins "in fairly good condition."

In the ensuing excitement about the "new" species, both French and British diplomats pressed the superintendent of the imperial estates— probably none too gently—to release live animals for shipment back to Europe. From these animals, a breeding population eventually became established at Woburn Abbey, the Duke of Bedford's estate north of London. In China meanwhile, soldiers in the Boxer Uprising of 1900 bivouacked on the old imperial hunting grounds, where they shot and ate the last remaining deer.

Père David's deer would thus have become one more species lost in the rush to a landscape of pigs and potatoes, except that it had flourished in captivity outside China. In 1985, Woburn Abbey shipped a breeding population back to Beijing, and the deer soon became established on the grounds of the same imperial park where Armand David first discovered them, now called the Beijing Milu Park ("milu" being the modern Chinese name for the species). The deer population rapidly expanded, as deer populations will do, and they have since been translocated to multiple protected areas around China. Instead of being extinct, the species now numbers close to 1000 animals in their native habitat.

INDUSTRIAL-SCALE NATURAL HISTORY

... and we may assert with strictly scientific
accuracy that the Rothschilds are the most
astonishing organisms the world has ever yet
seen.

—SAMUEL BUTLER

O F ALL THE ROTHSCHILDS, none was a stranger or more splendid
organism than Walter, born in 1868 in Tring, north of London,
and memorialized by his family as "a great stuttering bear of a man,"
6-foot-3, weighing more than 300 pounds, who "could keep the whole
house awake with his thundering snores" (and this in a house with
20-odd bedrooms). Everything about Walter was huge, including the
flaws—quirks, rather—in his personality. He was, to begin with, hope-
lessly tied to his mother and lived his entire life on the nursery floor,
a short stairway down from her apartment, at Tring Park, the family
home. No doubt as a result, he was alternately shy, socially awkward,
and arrogant. In *Dear Lord Rothschild*, a biography, the entomologist
Miriam Rothschild recalled her uncle's "almost infantile insouciance
regarding other peoples' problems." She may have been thinking about
the time when she was 13 and Walter, age 53, sized up her body type
in a booming foghorn voice, "Mama, isn't it strange that Miriam is completely *square* ?" Or maybe she had in mind one sunny afternoon when

Walter Rothschild failed in his family's banking empire but became the greatest species seeker of his time.

he received a nervous visitor to Tring Park while "swinging unselfconsciously in a hammock, 22 stone in weight and stark naked."

Flouting generations of Rothschild tradition, Walter was inept with money. To the deep dismay of his father, who had hoped that he would follow him as head of the British wing of the family's banking empire, Walter loathed everything to do with finance. He dutifully showed up to work for 18 years. But all concerned ultimately recognized it as a hopeless cause. Among other missteps, Walter allowed himself to be blackmailed to the point of financial ruin by one of his many mistresses. Dealing with the messy business of life caused him so much angst that for one two-year period, until his family found him out, he simply dumped his personal correspondence unopened into a series of 5-foot-tall wicker laundry baskets, which he then rammed shut with an iron bar, padlocked, and stacked up in a corner of his room. It is not a pretty image for a 300-pound man in a van Dyke beard, but Miriam described his life aptly as "a series of flying leaps from one sort of frying pan into yet another fire."

And yet Walter was a genius at one thing—the business of seeking species. At the age of 20, in his first year at university, he had already accumulated 46,000 specimens, mostly birds, butterflies, and moths. That same year, he sent out his first collecting expedition, to New Zealand and Hawaii. By 21, he had his own public museum at Tring, and in the first year alone 30,000 people showed up to see it. He went on, according to Miriam, to amass "the greatest collection of animals ever assembled by one man, ranging from starfish to gorillas. It included 2¼ million set [that is, pinned] butterflies and moths, 300,000 bird skins, 144 giant tortoises, 200,000 birds' eggs and 30,000 relevant scientific books. He, and the two collaborators he had selected as assistants, described between them 5000 new species and published over 1200 books and papers based on the collections."

Though the rest of his life was a shambles, Walter managed his natural history enterprises with a keen attention to detail and, in Miriam's phrase, "a streak of egotistical ruthlessness." Aided by two highly capable German naturalists, the entomologist Karl Jordan and the ornithologist Ernst Hartert, he ran Tring, in the words of one colleague, as "a model of how to collect and arrange and work in a museum—a model which the British Museum has signally failed to emulate."

For much of his adult life, Walter kept 400 collectors actively working in all corners of the Earth, according to Miriam Rothschild. That number does not include an additional 100 collectors specializing in fleas, because they worked mainly for Walter's younger brother Charles. (Charles almost matched Walter's passion for nature. Once, spotting a rare butterfly out the window of a train, he pulled the emergency cord, and then ran back down the tracks to collect his prize.) A map showing the whereabouts of Walter's various representatives looked, to one visitor at Tring, like "the world with a severe attack of measles."

Entirely apart from his family's tendency to do things on the grand scale, Walter Rothschild lived in an age of big biological surveys, with a new breed of collectors taking a more methodical approach to the habitats they studied. In place of spectacular novelties, naturalists now aimed to send back "crates of specimens," historian Robert E. Kohler writes. "Survey expeditions produced vast public collections, in the millions of specimens, all prepared and arranged according to standardized procedures," or at least that was the aim.

Early explorers had set out to dazzle the world with heroic tales and with strange new species, often collected higgledy-piggledy en route. The new species seekers meant to write monographs, not adventure stories. They wanted to have complete material for careful revisions of entire taxonomic groups, shifting species around based on better evidence and more advanced ideas about how different animals might be related. This industrial-scale natural history made for better science. It was more professional. But, except where Walter Rothschild was concerned, it could also at times seem just a little dull. In truth, naturalists now sometimes cultivated dullness, to separate themselves from the colorful amateurs who had preceded them. They wanted to be scientists, not showmen.

"What! . . . No Bottom!"

One of the most celebrated expeditions of the survey era traveled the world for three-and-a-half years (1873–1876) aboard the British Navy corvette *Challenger*, methodically dredging and trawling for specimens. T. H. Huxley, then head of the Royal Society, and Alfred Russel Wallace helped plan the research with the aim of advancing the Darwinian revolution. The expedition would eventually discover 4700 new species, some of them spectacularly beautiful. (The German biologist Ernst Haeckel produced a volume of the expedition's radiolarians, amoeboid protozoa which turned out, under the microscope, to resemble the most exquisite Christmas ornaments.) It would also open up a deep-sea world that until then had been thought to be utterly devoid of life. But the business of plodding across 68,890 nautical miles of the planet, stopping at regular intervals to conduct 492 deep-sea soundings, 133 bottom dredges, 151 open-water trawls, and 263 serial water-temperature observations was inevitably tedious.

Dropping and retrieving a dredge miles deep could consume an entire day. Even a parrot on board found it stupefying. Like the scientists, he had put the colorful heritage of his breed behind him, and in place of "bottles of rum" and "pieces of eight" now dwelt, as one zoologist put it, in the "rectified spirits-of-wine" world of the "scientifics." They evidently liked to invoke the name of William Carpenter, a vice-president and fellow of the Royal Society (abbreviated "F.R.S." after his name), who had been one of the chief promoters of the expedition. But in the parrot's rendition, it sounded something like a curse. As the dredge plunged endlessly downward, he liked to exclaim, "What! Two thousand fathoms and no bottom! Ah, Dr. Carpenter, F.R.S."

Getting to the bottom wasn't necessarily all that great either. A day's work often produced what looked to the sailors like nothing more than a bucket of muck. It probably didn't help that the "scientifics" tended to stay aloof from what the sailors called "drudging." They strolled around the deck of the 200-foot-long corvette improbably costumed in waistcoats and fob chains, as if waiting for tea to be served. Even George Campbell, a sublieutenant, found it irritating, though he surely knew something about staying aloof in his other life as a son of the Duke of Argyll. He complained in his journal about naturalists "to whom the whole cruise was a yachting expedition . . . to whom some new worm, coral, or echinoderm is a joy forever, who retires to a comfortable cabin to describe with enthusiasm this new animal, which we, without much enthusiasm, and with much weariness of spirit . . . had dragged up from the bottom of the sea."

The scientifics had reason to rejoice. Three days' work off the coast of New Zealand, for instance, revealed what an admiring zoologist still regarded 75 years later as "one of the most remarkable faunas ever discovered: Imagine . . . almost a mile below the surface, strange animals of bizarre shapes, swimming amongst waving fields of sea-lilies of intricate delicacy of pattern and brilliant hues. Think of submarine fields lit, not by any trace of daylight from above, but by the phosphorescence of the deep-sea corals."

To the Victorians, "the *Challenger* expedition was every bit as important as the *Apollo* moon landings would be to another great nation a century later," Richard Corfield writes in *The Silent Landscape*. The new species discovered by the "scientifics" mattered particularly because they seemed at the start to provide an important test of newfangled evolutionary theory. In the 1840s, naturalist Edward Forbes had proposed that the ocean below 1800 feet was an "azoic," or lifeless, zone. Wyville Thompson, the lead scientist on the *Challenger*, wanted to prove instead that the ocean depths were "the land of promise for the naturalist," where "endless novelties of extraordinary interest" could still be discovered. He thought the depths of the ocean would be inhabited by some of the same species that appear only as fossils on land.

The thinking was that land species had faced endless change, meaning that natural selection had forced them through one adaptation after another, with some species going extinct and leaving only fossil evidence behind. "But the Victorians believed," Corfield writes, "that on the unchanging ocean floor—the silent landscape—organisms would not have been forced to evolve . . . and, therefore, that the animals of the deep would be evolutionary throwbacks, living fossils." So the scientists rejoiced when one of their early dredges, a mile deep off the coast of Portugal, produced a living crinoid, or sea lily, whose flowerlike form was well known from the fossil record.

The expedition would turn up plenty of other "fossil" species and supposed missing links as it zigzagged across the planet. But by the nature of its methodical collecting, it also demonstrated the larger truth that the ocean depths are a living world, where natural selection operates much as it does on land. The *Challenger* collections were so rich that it took 19 years, and 50 volumes, to publish the complete reports. The expedition marked the birth of the science of oceanography.

Nature as Popular Attraction

The age of survey collecting coincided with an upsurge in public interest in biological discovery, partly as a result of a broad back-to-nature movement. Across the Atlantic, most Americans had started the century regarding nature only as an obstacle. "They may be said not to perceive the mighty forests that surround them till they fall beneath the hatchet," Alexis de Tocqueville remarked in 1831. "Their eyes are fixed upon another sight"—their "own march across these wilds, draining swamps, turning the course of rivers, peopling solitudes, and subduing nature."

But as the forests dwindled and abundant wildlife vanished, artists and essayists began to celebrate (or eulogize) nature. Loggers and hunters who had begun the century "as heroes in the battle against the dark wilderness," one historian writes, now got headlined in the *New York Times* as "pirates of the forest." During the Long Expedition of 1819, naturalist Thomas Say had argued for protective game laws to stop what was already an almost genocidal slaughter of bison. But by the end of the century, with the herd cut down from millions to hundreds, preservation of the species became a national movement. At about the same time, a new group calling itself the Audubon Society rose up to fight the killing of birds to provide feathers for women's hats.

People also took to the outdoors for their own health. A walking mania "swept the eastern seaboard like an epidemic in the 1870s," Robert E. Kohler writes, pioneered by clubs like the Pemigewasset Perambulators of Boston. Country driving with horse-and-buggy became the rage in the 1880s, and bicycle touring in the 1890s. The recreational camping movement sent city dwellers out in summertime droves to the Adirondacks or the seashore, and packing the kids off to summer camp became a new fashion. These trends weren't quite so single-mindedly focused on nature as the British manias for seaweeds, ferns, and infusoria. But natural history gave them an intellectual purpose.

Professional naturalists had ambivalent feelings about this newfound popularity, particularly as it showed up in their museums. The

stodgy old practice was to fill a display case with all the stuffed galli-
nules or fruit bats or weasels lined up in proper taxonomic order. The
approach was didactic, bettering visitors' minds with abstract scientific
principles. But taxidermists had rapidly become more skillful and they
now wanted not merely to educate but to excite the public with dioramas rendering dramatic scenes from the wild.

The French naturalist and specimen dealer Jules Verreaux depicted
a lion rearing up to claw an Arab courier down from the back of a camel
and in 1869 this exhibit drew crowds to the American Museum of
Natural History in New York. A few years later, the National Museum
of Natural History in Washington, D.C., displayed "Fight in the Tree-
Tops," a scene of "immense and hideously ugly male orang utans fighting furiously," with blood gushing from the wound where one sank his
fangs into the other. Critics worried about a certain lack of scientific
probity. But a defender responded, "If you cannot interest the visitor
you cannot instruct him; if he does not care to know what an animal is,
or what an object is used for, he will not read the label. "

Surprisingly, the dioramas actually helped turn museums into better collectors and protectors of specimens, according to Kohler. If the
old lineup-style displays no longer impressed the visiting public, they
didn't much please curators either. The standard practice had been
to put a museum's entire collection on public display, "and zoologists
watched with dismay as type specimens, the unique and indispensable reference specimens of taxonomic science, gradually faded in the
sunlight, lost their labels, and were eaten by moths and weevils." The
logical solution was to create separate exhibit and study collections,
with the important stuff carefully tucked away in insect-proof cabinets
in the back rooms.

Only the showier specimens now got put on public view, and since
taxidermists no longer had to stuff every specimen, "they could concentrate their efforts on a few elaborate and striking exhibits." Re-creating
a moment from nature in a good diorama required fresh skins of the
featured species, and also all the attendant details of its habitat, down
to rocks and grass. So instead of simply buying specimens, or accepting

them as gifts, museum curators and taxidermists now had to go out and become field collectors themselves. The comprehensive survey collections they brought back served the needs of showmen and scientists alike.

The "new museum" movement also encouraged a delicate shift in status, by literally putting a wall between professional naturalists and mere amateurs. The problem wasn't so much the amateurs themselves, at least not at first; the professionals still depended on them as an audience and also as collectors. But naturalists on staff at museums and universities suddenly felt themselves under siege from the other side: The rise of experimental laboratory science, starting with the medical use of improved microscopes to study cell function, made their work collecting specimens and mapping species distributions seem quaint.

"The glory of the field naturalist has departed," one taxonomist lamented in 1892. "The biologist or physiologist is the hero of the hour, and looks down with infinite contempt upon the luckless being who is still content to search for species." It probably didn't help the writer's case that his article carried the title, "The Happy Fungus-Hunter." But in the battle for status between the "worm slicer" physiologists and the "bug catcher" naturalists, creating a separate study area within museums allowed the naturalists to assert their status as bona fide scientists.

Bashing amateurs soon also became a way to reinforce the status of this new professional elite. In the 1870s, Yorkshire alone had 33 different natural history field clubs for enthusiastic amateur naturalists. But in that decade it also got a new Yorkshire College of Science, and the first lecturer in the new biology department, Louis Compton Miall, went on to denounce these clubs for "unintelligent collecting," long, dry, "wretchedly delivered papers," and recreation "of an instructive but by no means profound kind." Being profound was evidently a job for professionals. Miall was constructing an academic identity, historian Samuel J.M.M. Alberti writes, by contrasting himself with "a particular type of amateur straw man, the bumbling myopic list-maker, exhibiting a 'sordid craving' for selfish collecting, with no concern for any practices or techniques beyond capture and recording."

Big Was Good

To these insecure professionals, Walter Rothschild must have seemed like the worst sort of nightmare, not just an amateur naturalist, but a highly opinionated one, with enough money to dance rings around them, and plenty of "sordid cravings" to boot. Though he was serious about his taxonomy, Walter also wanted to wow the public and himself—not necessarily in that order. Big was good; bigger, better. "Walter Rothschild's ambitions were always aroused by large numbers and by objects of gigantic size, difficult and costly to obtain, on the eve of extinction or entirely extinct," one acquaintance recalled. His museum displayed the largest gorilla specimen known, the first white rhino ever captured, and a huge sea elephant 18 feet long and weighing 4 tons at death.

Numbers always mattered. "One of Walter's truly peculiar traits," Miriam Rothschild wrote, "was his meticulous recording in his letters of the number of specimens he had acquired and a list of the birds he had seen during his walks . . . He was like a schoolboy recording the runs he had scored in house matches." In his own mind, he was always competing, even against the great collectors of the age. After an expedition to La Grave, in the French Alps, he noted that his party had collected more than 5000 Lepidoptera specimens in 13 days, which he deemed "a record anywhere as Wallace's collections during his Malay Archipelago trip of several years was not 6000." Miriam commented, with a sigh, "Walter liked record bags. He never quite grew up."

He also loved anything odd or amazing from the animal world. When a shipment of live zebras arrived at Tring in 1894, he set out to train them as carriage animals. Zebras are headstrong, temperamental creatures. But so was Walter Rothschild. Since they would not otherwise submit, he devised a way to lasso them, in effect, by dropping their harnesses down over their heads from the stable ceiling. Then he trained them singly on a small trap, before moving them up to team work. A few months later, he thrilled London by driving a four-in-hand with three zebras and, for steadiness, one conventional carriage pony down Piccadilly and into the forecourt of Buckingham Palace. Later, he had his photograph taken dressed in a top hat riding a giant tortoise around the garden at Tring Park.

For insecure professionals, the worst thing about Walter Rothschild may have been that his amateur science was at least as good as their professional work, and often better. He had managed to complete only two lackluster years at Cambridge University, where he arrived with a flock of kiwis in tow. But from early childhood he had focused his attention on the animal world, filling up his "freakishly retentive" memory with everything he saw or read along the way. Once, much later in life, a visitor mentioned a rare New Guinea bird not in the Tring collection. Walter remarked, "That bird is illustrated on plate 87 of Gould's *Birds of New Guinea*," and it was.

As a teenager studying the displays at the new Natural History Museum in South Kensington, Walter had also attracted the attention of Albert Günther, keeper of zoology, who became his friend and trained him in sound Darwinian biology. Günther later helped Rothschild hire his curators Karl Jordan and Ernst Hartert, both also solid Darwinians. This trio would work together for almost 40 years, pursuing a clear and rigorous program to understand the proper classification of every specimen, by genus, species, and subspecies, the last of these a special obsession of the Tring scientists.

"I Have No Duplicates"

To outsiders, and particularly to professional rivals who could not hope to match his efforts, Walter's style of collecting must have seemed megalomaniacal, as if he meant to sweep up everything on earth where other collectors were content with a specimen or two. At one point, when ships visiting over the previous five years had taken and eaten 80 percent of the giant tortoise species on Duncan Island (now Isla Pinzón) in the Galapagos, he ordered his team to remove all 29 survivors "to save them for science" back at Tring. Rothschild meant to keep them alive, along with 30 giant tortoises from other Galapagos Islands. "I think 60 living Galapagos tortoises will make people stare," he wrote. (Happily, enough tortoises survived in the wild to provide the stock for a captive rearing program begun in 1965, with the result that roughly 350 tortoises now live on Pinzón. Rothschild's conservation efforts were more successful where he leased entire islands to protect wildlife on site—notably on Aldabra in the Indian Ocean.)

Other times, particularly with severely threatened island birds, Walter's idea of "saving species for science" meant only saving their skins in the collection drawers at Tring. "He has therefore been accused of hastening certain species into extinction," Barbara and Richard Mearns write in *The Bird Collectors*. "If these species had lasted a little longer perhaps some sort of captive-breeding programme could have rescued them, but perhaps not." Early on, Alfred Newton, a professor of his at Cambridge, warned Walter against overcollecting. "I can't agree with you in thinking that Zoology is best advanced by collectors of the kind you employ," he wrote. "No doubt they answer admirably the purpose of stocking a Museum; but they unstock the world—and that is a terrible consideration."

Visitors to Tring often worked up the nerve to wonder whether he didn't have an *awful* lot of the same kind of bird or butterfly. But Walter always replied, "I have no duplicates in my collection." What he meant was that understanding evolution meant knowing how different populations of a species vary from one place to another, or over a period of time. It would be absurd to expect one male and one female specimen to tell the whole story. Walter wanted a long series of specimens including all possible variations—the different wing patterns or colors or beak sizes that occur in separate populations or in different age groups, the hybrids, the albinos, the individuals with mixed male and female characteristics, even the teratological specimens—that is, the monsters. His characteristic cry of unmitigated delight on unpacking the latest collection to arrive at Tring was "Hartert! Hartert! *Schauen sie, Schauen sie* . . . COME AND SEE WHAT WE'VE GOT!"

Walter sometimes got too caught up in the minor variations. With cassowaries, big flightless birds from northern Australia and Papua New Guinea, he mistakenly proposed new species, according to Miriam, mainly because he was "bewitched by their multi-coloured wattles which proved in the long run rather capricious characters." But his photographic memory for the differences among specimens could also be of value in sorting out whether a species was really new. At one point, Walter dashed off a characteristically peremptory note to Hartert: "Please at once take one of the supposed new Petrels from the Galapagos Isles and relax the feet and legs and open the toes and make a note if the inside webs between the toes are yellow. I suddenly remembered either Beck or Harris in their diaries saying something about yellow webs. If this is so it is Wilson's petrel *Oceanites oceanicus* or *Oceanites gracilis.*"

In addition to his extraordinary memory, Walter also had a remarkably keen eye—a search image—for whatever species interested him, wherever it happened to turn up. Once driving across Hyde Park in the middle of London, he spotted a woman climbing into her car as her chauffeur cradled a rug in his arms to place over her legs. "Stop! Stop, Christopher!" Walter yelped to his own driver, "that rug is made out of the pelts of Tree Kangaroos," the topic of his current research. The woman went off colder but £30 richer, and the rug went to the Tring Museum.

His enthusiasm for anything to do with the animal world was highly contagious, and it motivated his collectors to travel in far more perilous regions and often pay a much higher price. Sadly, most of them seem to have been built on the heroic but understated model of Albert Stewart Meek. His autobiography carried the colorful title *A Naturalist in Cannibal Land* but mostly bore out Walter's assessment: "Meek is a man who faces a danger bravely and then forgets all about it." Among many other episodes, he worked with local hunters in Papua New Guinea who brought him butterflies collected with nets made of spider webs, or shot down with bow and arrow. One such find was the largest butterfly in the world, the Queen Alexandra's Birdwing (*Troides alexandrae*), with a 10-inch wingspan. To satisfy Tring's appetite for a complete series, Meek also found the pupae and bred up perfect specimens in captivity.

All of Walter's collectors routinely risked danger, suffering, and death. One collector wrote home poignantly about his solitary Christmas celebration in Gabon: "I was busy all day collecting butterflies which are now at Tring. I returned to my hut dead-tired & weary & feeling ill. I could not eat but made a feeble effort to swallow a mouthful of the small plum-pudding (a shilling tin which I had brought out with me from England 8 months previously). I felt fever coming on & went to bed, carefully blowing out my half-penny candle as candles are unobtainable . . . Then the fever took me, the shivering fit shaking my teeth & bones. I was delirious & moaned till dawn when the merciful perspiration broke out & the fever was broken, but it left me as weak as weak can be. That was my last Xmas 1907! How & where I shall spend this one, God alone knows." In Papua New Guinea, another Tring collector was less fortunate. For three nights, he ranted deliriously about seeing "motor cars, railway trains, children riding on clouds and the like," then died suddenly of "malarial fever and sleeplessness." The victim went unnamed in Meek's account, one of innumerable other collectors and assistants who died in the search for species.

This death, quickly passed over in the recounting, serves as a reminder of the subtext to all the overcrowded display cases at Tring ("I want to prove that one can put in one case 3 times as much as is usually done and *yet see all well*," Walter declared). It is the subtext to those endless drawers of carefully arranged specimens in museums around the world: Someone had collected each specimen; killed it; skinned it; stuffed it, set it, or put it in preservative; pencil-scratched a label for it; carried it cross-country; shipped it home; studied it; and classified it—and then repeated this ritual over and over, countless millions of times. For each specimen, someone had gone hungry and sleepless. Someone alone in a remote and hostile territory had wept. Someone had perhaps drowned, been murdered, suffered malaria, yellow fever, dysentery, or typhus. Someone had certainly cursed and complained, though not so much as we might expect. Someone had said, "Hunh!" And someone had rejoiced.

When a local assistant brought him a spectacular new butterfly species, *Ornithoptera chimaera*, Meek rewarded him extravagantly with two shillings, two tins of English bacon, and five sticks of tobacco. "I felt more pleased than if I had been left a fortune," he wrote. "A fine discovery of that sort stirs the heart of a collector. He forgets hardships and troubles, and remembers only that he has given something to science, taken from Nature one of her secrets. 'A little secret!' some may say, but naturalists do not think so."

Later naturalists would place high value on the work Walter Rothschild, his curators Hartert and Jordan, and their many collectors had done in putting together those lengthy series of specimens for each species. When a later ornithologist, David Lack, was working out how speciation had occurred in Darwin's finches, he had to spend months doing fieldwork in the Galapagos. But fieldwork wasn't enough. "You've just got to come back to the Rothschild collection and get down to measuring beaks," he wrote. He also noted that "The drawers of the Rothschild cabinet contain more representatives of some of the Hawaiian sicklebills than are alive in the Islands today."

Though Rothschild and his collectors were only dimly aware of how important genetics would become to the study of nature, their work preserved the broadest possible swath of genetic variation for each species. "It is only against the panorama of modern research that the full value of Walter's butterflies becomes apparent," Miriam Rothschild wrote. "Its importance lies in the unfolding and presentation before your eyes of a whole order—in all its variety and complexity, culled from continent to continent, from one far flung oceanic island to another, from desert and forest and prairie and mountain range. There is also an indefinable factor about these collections, a Walterian factor—call it what you will—a whiff of zest and wonder, which must somehow have been pinned in among the butterflies. Suddenly the outlook broadens, the horizon expands—a penny drops, new ideas materialize, the mind 'takes off.'"

Postscript

But a different Walterian factor shaped the eventual fate of the collection at Tring. In 1931, Walter's financial missteps caught up with him, possibly aggravated by his persistent blackmailer (a "charming, witty" aristocrat, according to Miriam, who never revealed her identity). With his characteristic secretiveness, Walter discussed his situation with no one. Instead, he made a sort of "Sophie's Choice" among his beloved children, and privately offered the entire bird collection, minus the cassowaries and a few other sentimental favorites, to the American Museum of Natural History in New York. Gertrude Vanderbilt Whitney made a gift to cover the $225,000 cost—less than a dollar apiece for specimens said at the time to have cost Walter altogether $1 million. The Rothschild family heard about the transaction first in the newspapers. At the Berlin Museum, to which Hartert had by then retired, a colleague recalled how he got the news: "'My collection! My collection!' he stammered out, his chest heaving and his clear eyes swimming with tears." Hartert died a year-and-a-half later, Walter Rothschild a few years after that, at age 69.

In her biography, Miriam Rothschild noted that Walter's portrait does not hang in the family bank and that "all traces of his activities have been expunged from the archives." But she added that, long after the family's many bankers have faded into pin-striped oblivion, Walter Rothschild will live on in "the giraffe, the bird of paradise, the scarlet climbing lily, the blue/mauve orchid, the silk moth with the mother-of-pearl 'windows' in its wings and the blind white intestinal worm," among scores of other new species that scientists have named in his honor.

Chapter Twenty-Two

"THE BLESSING OF A GOOD SKIRT"

> I am at a loss to understand on what grounds of
> justice or public policy a career which is open
> to the weakest and most foolish of the male sex
> should be forcibly closed to women of vigour
> and capacity.
>
> —T. H. HUXLEY

LATE ONE AFTERNOON in 1895, that rare animal for that era, a female species seeker, was hiking alone through a forest in the interior of Gabon, a few minutes ahead of the rest of her party. It was treacherous country. Her guides had pointed out the shredded bark of trees along the forest trail, meaning leopards in the neighborhood. The human inhabitants were also fearsome, said to be cannibals. But Mary Kingsley was in equal parts self-assured and self-deprecating, an attractive unmarried woman in her thirties with an independent income, roaming footloose over "the white man's grave." Where male explorers often resorted to the chest-thumping language of conquest, she relied instead on her undertstated wit, claiming, for instance, that she had "seen at close quarters specimens of the most important big game of Central Africa, and with the exception of snakes, I have run away from all of them." (She backed away slowly from snakes. Or ate them for dinner.) She was also unmistakably intrepid.

This time, though, going ahead on her own proved foolish. It was 5 p.m., and the path through the woods grew indistinct. If she paid close attention, she could pick it out from among the undergrowth. But then she came to a place where it vanished completely. She peered ahead, and thought she saw it resume again on the other side of a clump of brush. So she pushed on—and plunged to the bottom of a 15-foot-deep pit lined with sharpened spikes. In China, the naturalist Roger Fortune had stepped into such a trap and survived only by grabbing hold of a root as he went over the edge. In Hawaii, the botanist David Douglas, age 35, fell into a pit and was gored to death by a bull also trapped there.

Kingsley was saved by being female, or so she claimed. "It is at times like these that you realise the blessing of a good skirt," she wrote. "Had I paid heed to the advice of many people in England, who ought to have known better . . . and adopted masculine garments, I should have been spiked to the bone, and done for. Whereas, save for a good many bruises, here I was with the fulness of my skirt tucked under me, sitting on nine ebony spikes some twelve inches long, in comparative comfort, howling lustily to be hauled out." It was a curiously literal refutation of the charge that women were too frail for the ferocious challenges of fieldwork.

For much of the great age of discovery, women had been obliged to stay at home, doing little jobs about the house, and not much else. If they happened to marry naturalists, they could generally count on a lifetime of foul odors and playing second fiddle to species. Richard Owen's duties as an anatomist, for instance, included dissecting cadavers from the London Zoo, and he generally did this work at home. His wife Caroline once mildly noted that the aroma of rotting elephant was so overpowering on a summer day that she "got R. to smoke cigars all over the house." Another time, she came home to find a dead rhinoceros ("late of Wombwell's Menagerie") in her front hall. After five long days, she noted in her diary, "R. still at the rhinoceros."

Women who were scientifically inclined could perhaps provide illustrations for their husbands' books. Or they could become collectors, at least locally, and supply specimens to male scientists. Either way,

Mary Kingsley strode serenely across areas notorious as "the white man's grave."

they generally got no credit for their work. The omission could at times be spectacular. Popular books by marine biologist Philip Henry Gosse and by ornithologist John Gould owed much of their success to illustrations that were both accurate and beautiful, by Emily Bowes Gosse and Elizabeth Gould respectively, both uncredited.

In the clubby world of the naturalists, women were "left out, ignored, only brought in on festive occasions, because they could scarcely be seriously interested," David E. Allen writes, in his 1976 book *The Naturalist in Britain*, "or else they were deliberately kept out, because science was a man's business and the club a kind of intellectual stag-party where a male rattled his antlers: a place reserved apart for him, like his study, where women should never be allowed to intrude." Oddly, Allen put the blame partly on women themselves for wearing "impossible clothing," and for having "acquiesced in the exaggerated ideas about their fragility."

"The Morals of the Maids"

Women naturalists then suffered from "the crippling 'delicacy' of the age," though it was really more like "cultivated prurience," Lynn Barber writes, in *The Heyday of Natural History*. It obliged them "to shy in alarm at the least hint of sex or reproduction," or almost any other aspect of anatomy. Indeed, social convention required sparing them even the near occasion for such alarm. Though it began allowing women to attend its meetings in the 1830s, for instance, the British Association for the Advancement of Science barred them from a session where papers were to be read on the racy topic of reproduction in marsupials. The very word "mammal" was fraught with difficulty. In one public address, Richard Owen tactfully characterized the class Mammalia as "nourishing their young in a peculiar way" and thus avoided shocking women in his audience who might not otherwise know what their breasts were for.

Charles Darwin could muse over a "wonderfully developed" barnacle penis lying "coiled up, like a great worm," which when "fully extended . . . must equal between eight or nine times the length of the animal." But Victorian strictures seem to have left his daughter Henrietta freakishly unhinged on the topic of sex. *Period Piece*, a family history by Darwin's granddaughter Gwen Raverat, recounts how "Aunt Etty" used to make outings to eradicate the stinkhorn fungus, a rank-smelling toadstool whose scientific name, *Phallus impudicus*, or "shameless penis," aptly reflects its appearance.

"Armed with a basket and a pointed stick, and wearing a special hunting cloak and gloves, [Aunt Etty] would sniff her way round the wood, pausing here and there, her nostrils twitching, when she caught a whiff of her prey; then at last, with a deadly pounce, she would fall upon her victim, and poke his putrid carcase into her basket. At the end of the day's sport, the catch was brought back and burnt in the deepest secrecy on the drawing-room fire, with the door locked; *because of the morals of the maids.*"

When Beatrix Potter took a more serious scientific interest in fungi, as a young woman in the 1890s, her work was foredoomed by male prejudice. She conducted careful experiments, made significant discoveries her male counterparts had overlooked, and recorded them in detailed illustrations. But the director of Kew Gardens dismissed her and her work so curtly that she characterized him in her journal as a misogynist. The Linnean Society, which did not admit women, accepted a paper by Potter—to be read at one of its meetings by another male scientist who had rebuffed her. Potter took the hint, abandoning science to turn her attention to Mrs. Tiggy-Winkle and Peter Rabbit.

Women collected shells, from which the fleshy bits had been safely removed. Fossils, also notably unfleshy, were a safe, if not quite suitable, object for the most famous woman collector of the early nineteenth century. Mary Anning worked in Lyme Regis, on the southwest coast of England, and her finds included new species of ichthyosaur, plesiosaur, and pterosaur (for which male geologists took the credit). She appears in a painting from the period with her geological hammer in hand. But she is also covered head to toe in a voluminous cloak much like a chador.

The Trousers Beneath the Skirt

For all the force of her personality, even Mary Kingsley clearly struggled with how to present her gender in public. As she was preparing *Travels in West Africa*, an account of her expeditions, she became vexed with her publisher. "I do not understand quite what you mean by 'this story being told by a man,'" she wrote to Macmillan Brothers. "Where have I said it was? What would have been the good of my saying so." She added, "Of course I would rather not publish it under my own name and I really cannot draw the trail of the petticoat over the coast of these places. Neither can I have a picture of myself in trousers or any other such excitement of this sort added." (As a practical matter, she had in fact worn her brother's trousers in West Africa, but always kept them modestly covered beneath her skirts.) "I went out there as a naturalist not as a sort of circus," she wrote. "But if you would like my name will it not be sufficient to put M. H. Kingsley? It does not matter to the general public what I am as long as I tell the truth as well as I can." It did, however, matter to the public, and when Macmillan published the book under the name Mary H. Kingsley, it became a best-seller and she was the lion, or rather, as she put it, "the sea-serpent of the season." Even if scientists still resisted, it was no longer entirely taboo for a woman to join actively in the great cause of discovery.

Neither Kingsley, nor the other women species seekers who began to follow in her path, brought a markedly different sensibility to natural history. In truth, they often strove to enhance their scientific credibility by sounding like their male counterparts, at times seeming to distort their own perceptions. Thus Kingsley wrote about the "feeling of horrible disgust that an old gorilla gives on account of its hideousness of appearance." But she also sat and watched gorillas with evident delight, instead of simply shooting them for specimens. Like her male predecessors, she characterized her African hosts as intellectually and emotionally inferior. (Modern readers sometimes wishfully think Kingsley was being egalitarian when she wrote, "I feel certain that a black man is no more an undeveloped white man than a rabbit is an undeveloped hare." In fact, she was repeating the old canard that blacks and whites are separate species.) But she described African customs with respect and admiration. She also scandalized some readers when she attacked "the evil" being done in Africa "by misdirected missionary effort" to change those customs. Kinglsey even paid lip service to the idea that women are intellectually inferior to men, though everything in her own life testified to the opposite.

Glory and Dread

The combination of trousers worn beneath the skirt is, after all, a fair symbol for the ambivalence forced on early women naturalists. But when she could free herself from social strictures, Kingsley wrote in a voice that was far livelier and more engaging than that of almost any naturalist before her. Once, for instance, floundering helplessly through the "great grim twilight regions of the forest," she recorded the moment when the confusing shadows finally began to separate and sort themselves into recognizable species. An old hunter from the Fan tribe who had been guiding her recognized the moment of revelation, remarking, "Ah, you *see.*"

Kingsley was doubtless gratified, though she noted with chagrin that, even here, she did not get much credit for "the hard work I had been doing in order to see." On the contrary, the hunter attributed her developing awareness to "a little ivory half-moon" talisman, which a chief had given her with the magical power "to make man see bush." Whatever the source, it was a special power, and one to which every naturalist aspired (though no one else thought to write about it): the ability to go into a new habitat; start to piece together the individual species, however insignificant they might seem by themselves; and one day wake up to see order, a system, a web of mutual dependencies.

Like other naturalists, Kingsley collected reptiles, insects, and especially fish, the building blocks of such systems. (As the daughter of an amateur naturalist and travel writer, and as the niece of the writer, naturalist, and clergyman Charles Kingsley, she had grown up in the great tradition of discovery.) She wanted to find new species and have them approved by experts, if only to validate the scientific nature of her travels. In 1896, she eagerly reported the "verdict" on the haul from her second expedition to West Africa: "one absolutely new fish" and one equally new snake, among other treasures. Kingsley was relieved, "for I was beginning to fear I was an utter wind bag."

She was, in fact, anything but that. Her words were so deftly chosen that she seemed not merely to see life in the bush, but to bring it home to her readers intact and alive. In a letter to a friend, for instance, she gave a vivid sense of the shifting glory and dread entailed in penetrating upriver by canoe past the mangrove-lined West African coast: "You turn from the broad black silent main stream into" a creek, she wrote, "where the trees are interlaced so thickly overhead that the sunlight comes through the foliage, as though it had come through stained glass, a beautiful tender green. When there is a mist in the air, as at morning or evening, the very air itself seems green and you feel that if you could get some of it and bottle it up that you could shew it in England.

"The water in the creek soon turns from coffee colour to pure translucent green and breaks from the paddles in wreaths of foam like frosted silver and the air driven down into it by the stroke of the paddle comes softly up through the water gleaming like jewels. The mangroves leave off and the banks become covered with a dense and lovely growth of marsh Palm, oil palm, bamboo and thousands of other trees and shrubs, many of them bearing enormous sweet-scented flowers—parrots shriek and grand cranes—pelicans and flamingos get up off the bank where they have been busy fishing and go off with a swish and a rush, and you feel that you did well in coming to West Africa.

"But 5 or 6 hours later when it is getting dangerously near 6 o'clock when the night comes down from above (not like it does in England up from the ground), and you are cramped from sitting for hours and none too well fed, and the person steering begins to express a doubt about which creek is which, and the thick white wooly miasma-laden mist comes stealing out from among the tree stems floating in ghostly trails along the surface of the River, winding itself round the tree stems like a soft material—and now and then wrapping up the boat in its fold so that you can hardly see the bow oar—and the frogs start their hoarse croak and the crocodiles which swarm along the Rivers break the silence with a whining roar, or howl, you wish you had taken the advice of your friends and were well home at dinner in Cambridge, a place it is borne in upon you that it is extremely unlikely you will ever be privileged to see again."

In fact, Kingsley would see and enjoy Cambridge several more times in her life. She was well aware of the importance of disease prevention in her travels. Before her first expedition, a colleague had advised her that "nothing hinders a man, Miss Kingsley, half so much as dying," a handicap that also presumably applied to women. She seemed to think that her spare, energetic morphological type afforded her some protection. "The full-blooded, corpulent, and vigorous should avoid West Africa like the plague," she wrote. She had repeatedly seen such men and women "come out and die, and it gives one a sense of horror when they arrive at your West Coast station, for you feel a sort of accessory before the fact to murder." Malaria was the most common killer, and its cause remained unknown. Kingsley took a variety of precautions, both old-fashioned (avoiding the miasma-laden night air by dining before eight), and newfangled (sleeping under a mosquito net "whether or not there are mosquitoes in your district"). It worked until her third trip to Africa, when she contracted typhoid fever, probably from contaminated drinking water, while tending wounded soldiers in the Boer War. She died in 1900, age 37. At her request, the "sea serpent" of a few London seasons earlier was buried at sea.

An Elephantine Postscript

Despite the example of Mary Kingsley and other pioneering female biologists, women would only begin to take an independent role in the search for new species in the last decades of the twentieth century. Even then, when they were free to wear "bifurcated garments" and speak in their own voices, they did not bring a strikingly new perspective to the hunt for species. The differences between the sexes were never so great as the people who wanted to exclude women liked to pretend. (It wasn't Mars versus Venus, to use a modern construction, but more like men are from Minneapolis, women are from St. Paul.) And yet women seemed to see a lot. Maybe having been number two for so long simply made them try harder.

Today women biologists routinely work at the extremes of the planet, from Sylvia Earle exploring life at the bottom of the oceans, to Birgit Sattler revealing that the clouds, long thought sterile, are in fact a soup of microbial life. Some of them have discovered new primate species. Others have focused on understanding the behavior of old ones. Thus Jane Goodall described tool use for the first time in chimpanzees, causing her mentor Louis Leakey to reply by cable: "Now we must redefine 'tool,' redefine 'Man,' or accept chimpanzees as humans."

The belated acceptance of women as human, and as equal partners in the business of discovery, has also helped redefine evolutionary theory. In his 1871 book, *The Descent of Man*, Charles Darwin added the idea of sexual selection to his original theory of evolution by natural selection: It isn't enough, he argued, simply to avoid being killed by predators, famine, or disease. Species also evolve depending on which individuals do better at attracting members of the opposite sex.

We now recognize that sexual selection accounts for some of the most spectacular displays in the animal kingdom, from the peacock's tail to the courtship dancing of bowerbirds. It's also an important factor in producing the diversity of species. Males are continually competing against other males, and evolving lavish new displays to impress females. In most species, females sit back, survey the possibilities, and then do the choosing—and over time these choices can lead to splitting off as a new species. For a century after Darwin, male biologists managed to ignore this idea. They did so because it challenged an orthodoxy more sacrosanct than the biblical account of creation: the idea that males are in charge. But all that began to change with the arrival of women in science.

Back in the heyday of "cultivated prurience," a young woman visiting the London Zoological Gardens wrote home that the monkeys were "so very nasty" she did not even want to look at them. "Really, with a gentleman, I think it quite indelicate," she added. She preferred a "decent animal" like the elephant. One hundred and fifty years later, women field biologists, no longer constrained by false modesty, would take a closer look at elephants. Among other extraordinary behaviors, they discovered infrasonic communication, the ability of elephants to communicate across long distances using frequencies below the range of human hearing. They also discovered that male African elephants experience "musth," a recurring period of raging hormones when they belligerently compete for sexual opportunities. The phenomenon should really have been hard to miss. (The researchers named one of their study animals "Green Penis," for his outlandish appearance then.)

But generations of male scientists had simply failed to notice.

THE BEAST IN THE MOSQUITO

Pathology is, in the main, but the study of a
certain fauna and flora—a fauna and flora that
inhabits the human body.

—PATRICK MANSON

O NE NIGHT IN 1877, in a squalid port city on the southeastern coast of
China, a Scottish doctor named Patrick Manson performed a small
experiment with mosquitoes. The scope of the test was limited and the
design badly flawed. But it was the beginning of a spectacular quarter
century in which the work of the species seekers would bear fruit,
enabling humans for the first time to control diseases that had plagued
them forever. Much as uncovering the secret lives of gorillas and Galapagos finches reshaped how we think about the world and our place in
it, the work begun by Manson would soon revolutionize the ways we
live and die.

The experiment focused on elephantiasis, a common disorder in the
tropics. It does not kill its victims, but in its later stages can make them
wish it had: The limbs swell grotesquely and the sagging, thickened skin
becomes rough and fissured, like the hide of an elephant or an especially odious lizard. Male victims sometimes have to cart the enlarged
scrotum in front of them in a wheelbarrow.

The protocol for Manson's experiment was to have a servant named
Hin-Lo spend the night in the "mosquito house," a screened compart-

Though ridiculed by "sneering skeptics," Patrick Manson persistently advanced the revolutionary idea that mosquitoes spread disease.

ment in a crude shack. An attendant set a light beside the cot and left the door open for a half hour, to attract as many mosquitoes as possible. Then—sweet dreams—he snuffed out the light and shut the door, with Hin-Lo and the mosquitoes inside.

Most practitioners then categorized elephantiasis as just another symptom of malaria, the tropical diagnosis catchall. Whatever you called it, the cause was unknown and comparing notes or trying out theories was difficult at best for doctors treating the condition. The daily patient load was heavy and nobody was paying colonial doctors to waste their time on experiments. Places like Amoy (now Xiamen), where Manson worked for the Chinese Imperial Maritime Customs Service, were also thoroughly isolated. "I live in an out of the world place away from libraries, out of the run of what is going on," he wrote to a mentor in London, "so I do not know the value of my work or if it has been done before, or better." Manson felt "the need of books, criticism, and medical companionship."

He got the chance to catch up only during a furlough back home in 1875, spent partly courting his future wife—he was 31, she was the 17-year-old daughter of a ship captain—and partly in the library of the British Museum. A few years earlier, a British researcher in India and another in London had separately identified tiny, threadlike worms in the blood and urine of elephantiasis victims. They couldn't say whether these filarial worms (from the genus *Filaria*) actually caused the disease. But they were the first-known blood-borne parasites in humans, and the two researchers bickered in print over credit for the discovery.

Manson read their reports as a physician, of course. But he was also a naturalist, both by boyhood predilection and as part of the routine training of doctors at Scotland's Aberdeen University. So the filarial worms weren't just a medical problem to be treated (or a trophy to be claimed). They were a species, susceptible to investigation for all the naturalist's usual concerns with life history, behavior, home range, and habitat. That sensibility showed up back in Amoy, when Manson got to look at the living worms through a microscope and see for himself the "marvelous power and activity" of "this formidable looking animal."

Chinese tradition prevented him from working with human cadavers. So Manson turned first to comparative anatomy, familiarizing himself with filarial worms in dogs, where he could begin to see the damage they do in the arteries, heart, and lymphatic system. Then Manson and his assistants examined blood smears from 670 human patients. They found that the worms occurred at the highest rate in the sickest individuals—not a coincidence, he concluded, but "cause and consequence." Doctors now know that the accumulation of worms over time causes swellings and obstructions in the web of lymphatic vessels, trapping fluid in the lower extremities. But in the 1870s, "germ theory"—the idea that microorganisms cause many diseases—was still widely disputed.

Manson knew enough about parasites to think that the immature worms he was seeing in the blood couldn't complete their entire life cycle within a single human host. A few decades earlier, the Danish researcher Johann Steenstrup had revealed the bizarre complexity of parasitic life with an astonishing study of the trematode worms known as flukes. Many people were familiar with adult flukes, roughly the size and shape of a sage leaf, clamped onto the livers of sheep and other livestock. But nobody could figure out how they got there. Many scientists regarded them as proof that parasites could generate spontaneously from living tissue.

Steenstrup collected flukelike creatures from ponds and ditches, together with snails living there. By keeping them alive in the laboratory, he demonstrated that the snails serve as an intermediary host, or what he called a "nurse" species, for the flukes. It was the circle of life, but not quite as Disney might imagine it: A fluke produces 20,000 eggs per day, which get deposited in water along with the sheep's droppings. They hatch into a free-swimming form. It penetrates the body of a snail and develops through several bizarre stages before emerging again in a different free-swimming form. This form attaches itself to a blade of grass. And there, beneath a hardened cyst, it patiently waits to be eaten by a grazing sheep, and begin the cycle over again.

Steenstrup proposed his "alternation of generations" theory to describe this pattern of going back and forth between different forms and different hosts. His natural history perspective brought order to what had been "a complete chaos of seemingly irregular phenomena," according to one German scientist. It showed how parasites conformed to "the acknowledged laws of animal existence and propagation" and thus made nonsense of most arguments for spontaneous generation. Other researchers went on to describe parasites that developed by alternating between mice and mealworms, and between dogs and lice. The pairings of primary host and nurse species—like the connection between sheep and snails—were not always obvious. One early researcher tried to find the "nurse" species for canine filarial worms, for instance, by feeding their eggs to cockroaches and frogs.

But filarial worms are blood-borne parasites. So Manson hit on the idea that they might enter and exit the body by means of a blood-sucking animal. Because elephantiasis and other forms of filarial disease occur only in certain parts of the world, he also thought the "nurse" species should have "a corresponding and limited distribution." He ruled out fleas, lice, "bugs," and leeches as suspects, because "they are found pretty well all over the world." Instead, he settled on mosquitoes, though they also occur worldwide, and focused on *Culex fatigans*, a species commonly found in houses around Amoy. This stroke of insight set the stage for the long night in the mosquito house.

Hin-Lo, who was Manson's servant, was already infected with filarial worms. The question was whether a night of unlimited feeding would transfer them to the mosquitoes. In the morning, an attendant used puffs of tobacco smoke to immobilize the blood-swollen mosquitoes covering the walls. Then he trapped them beneath an inverted wine glass and transferred them to ventilated glass vials, to be kept alive in Manson's laboratory. The design of the experiment was flawed on several counts—and not just from Hin-Lo's itchy perspective. Manson accepted the common belief that mosquitoes took only a single blood meal in their brief lives. So it didn't occur to him that the mosquitoes collected that morning might already have been infected with worms from someone other than Hin-Lo. He also neglected to feed them in captivity, with the result that they died off over the course of a week.

Manson's pathological techniques were also crude. "I shall not easily forget the first mosquito I dissected," he wrote. "I tore off its abdomen and succeeded in expressing the blood the stomach contained." Under the microscope, he saw not just that the worms had entered the mosquito, but also that something in the mosquito's digestive tract had freed each immature worm, or microfiliaria, from the sheath in which it had been wrapped.

Somehow, from examining what was essentially a paste of mashed mosquito, Manson also discerned what happened next: The newly liberated microfilariae passed through the mosquito's abdominal lining and took up residence in the muscles of the thoracic cavity. There they continued to develop, "manifestly . . . on the road to a new human host." He had discovered "a new and revolutionary concept," according to medical historian Eli Chernin—"that certain bloodsucking arthropods can transmit human disease." It would eventually save millions of lives—arguably, millions every year in the modern era—and immortalize Manson as "the father of modern tropical medicine."

For the moment, Manson busied himself with another, disturbingly beautiful, behavior of his filarial worms: They were absent from human blood by day. (We now know that they retreat to the lungs.) But they reappeared each night, "the discharge commencing after sunset and continuing until near midnight." He had discovered a migration within the human body, like the nightly migration of plankton from the depths to the surface of the ocean. The human body was a habitat, a "peculiar territory" for its "peculiar fauna," as Manson's mentor in London put it. And the parts of the body were like districts, each with "its special attractions for particular parasitic forms" at particular times. (Evolutionary theory made this idea somewhat less distasteful: "It is only by accepting Mr. Darwin's hypothesis," his mentor concluded, "that we can escape the undignified conclusion that parasites were specially created to dwell in us, and consequently, also, that we are destined to entertain them.") Manson himself wrote, "It is marvelous how Nature has adapted the habits of the Filaria to those of the mosquito—the embryos are in the blood just at the time the mosquito selects for feeding." He wondered idly whether some similar pattern might not explain the intermittent fevers in malaria victims.

Because he thought mosquitoes bit only once, Manson missed the second half of the story. He concluded that, after biting a victim, mosquitoes normally go off to lay their eggs and die in a body of water. The "formidable looking, very active" worms escape from the dead body, he wrote, and make their way by means of drinking water "to the stomach of a human host." In fact, we now know that a mosquito typically bites multiple victims. The filarial worms line up like paratroopers in the infected mosquito's head and proboscis, and then escape onto the skin as the mosquito bites, finding their way into the bloodstream by way of the bite wound. But Manson would only come to accept the role of multiple biting years later, when he was building on his elephantiasis work to help solve the far more deadly puzzle called malaria. For the time being, the doctors on ships calling at Amoy mostly laughed at

"Crazy Pat Manson" for attributing so much evil to a threat as trivial as a mosquito.

But the idea that insects could cause disease must have been in the air then. In 1878, a Russian researcher named Grigory Nikolaevich Minkh theorized that lice, also a trivial nuisance for everyone from peasants to kings, might be the vector for typhus. Body lice are now so rare that we tend to think of "cooties" as an almost imaginary affliction. But among many other apocalyptic achievements, louse-borne typhus, together with plague and starvation, killed 10 million people in the Thirty Years War, versus 350,000 deaths in combat. It would kill millions more in World Wars I and II. (Anne Frank, imprisoned at Bergen-Belsen concentration camp, was among its many victims.)

At about the same time Manson and Minkh were theorizing about insect-borne disease, a Cuban researcher named Carlos Juan Finlay was testing the idea that mosquitoes transmit yellow fever, which still routinely terrorized communities as it swept across the landscape in deadly waves. Finlay focused on the mosquito species now called *Aedes aegypti*. He reared his experimental mosquitoes in captivity and allowed them to feed first on yellow fever victims and then on healthy volunteers, who soon also became sick with the disease. But for years after, Finlay's theory, like Manson's, mainly elicited incredulity and ridicule.

Chapter Twenty-Four

"WHY NOT TRY THE EXPERIMENT?"

There is no malaria without *Anopheles*.

—BATTISTA GRASSI

Historians tend to celebrate the last decades of the nineteenth century as a triumph of "hard" sciences like bacteriology not just over disease, but over the "soft" science of natural history. In fact, neither could have made much progress without the other. "In the age of the 'laboratory revolution,'" medical historian Shang-Jen Li writes, "natural history still provided templates for innovative research." It was a tradition with deep roots in British, and especially Scottish, medicine. Training young doctors to look closely at the natural world was a way to develop their observational skills. That's why ship surgeons and military doctors also often served as naturalists—while naturalists sometimes found themselves being dragooned as surgeons.

Moreover, it was a tradition that had already yielded extraordinary results, beginning with the eighteenth-century London surgeon and comparative anatomist John Hunter (the same John Hunter who was experimenting with electric eels at the outset of this book). One of Hunter's students was a country doctor and naturalist named Edward Jenner, who spent his leisure hours over a period of 15 years studying the behavior of cuckoos. Cuckoos are of course notorious for brood parasitism—the practice of laying their eggs in the nests of other birds. (It's the origin of the term "cuckold" for a man with an unfaithful wife.)

Jenner was the first to report that cuckoo chicks hatching in their adopted homes remove their native competition by heaving their eggs up out of the nest—an idea widely considered preposterous until photographic evidence proved it more than 130 years later. Under Hunter's direction, he also noted that they have a special egg-shaped groove in their backs for this murderous purpose.

Hunter drove Jenner to understand that close observation and careful experimentation mattered more than theory. In a famous letter about work Jenner was doing with hedgehogs, for instance, Hunter wrote, "I think your solution is just; but why think, why not try the Expt?" (He drove Jenner in other ways, too. Hearing that Lord Berkeley's collection included a toad specimen "of prodigious size," he wrote Jenner, "Let me know the truth of it, its dimensions, what bones are still in it, and if it can be stolen by some invisible being.") The science historian Lloyd Allan Wells argues that the discipline of the cuckoo study, combined "with Hunter's insistence on finely honed observation and cogent presentation, helped prepare Jenner's mind for his great work." That great work was of course to pioneer the development of the world's first vaccine, for the deadly and disfiguring disease, smallpox. It would lead in time not just to the eradication of smallpox, but to the development of vaccines against yellow fever, bubonic plague, cholera, polio, flu, diphtheria, measles, mumps, rubella, hepatitis A and B, tetanus, rabies, and countless other deadly diseases. Jenner thus gets credit for saving more lives than anyone in the history of medicine (though Hunter and the cuckoo surely also deserve honorable mention). Canadian Indians, who had suffered the genocidal effects of smallpox, wrote what Jenner regarded as his best thanks: "We shall not fail to teach our children the name of Jenner; and to thank the Great Spirit for bestowing on him so much wisdom, and so much benevolence."

That same marriage of natural history and medical research had also yielded countless smaller discoveries—for instance, revealing the cause of trichinosis. In 1835, during an autopsy on a 51-year-old bricklayer at St. Bartholomew's Hospital in London, the presiding doctors noted what they took to be "spiculae of bone" in the muscles, or "sandy diaphragm." But they thought about it merely as an inconvenience: It made their scalpels dull. Afterwards, James Paget, a first-year medical student, went back for a closer look with a hand lens at the tiny specks in the muscle. He took tissue slices to the British Museum, whose naturalists had the nearest available microscope. The specks turned out to be cysts and "nearly every cyst contained a small worm coiled up," a new species of roundworm, soon dubbed *Trichinella spiralis*. "All the men in the dissecting-rooms, teachers included, 'saw' the little specks in the muscle," Paget later recalled, "but I believe that I alone 'looked-at' them and 'observed' them: *no one trained in natural history could have failed to do so.*" (Italics added, and this postscript: Young Richard Owen, already making his reputation as a glory-hound, quickly got wind of the new species and rushed it into print, characteristically giving little credit to Paget.)

The discoveries made possible by the joined forces of natural history and medical research piled up through the nineteenth century. But they culminated in an almost miraculous rush in the 1890s, when diseases fell away in ranks, almost like their former victims. In a period of little more than five years, researchers suddenly linked cause and effect for epidemic killers that had routinely terrorized humans—yellow fever (1898–1901), plague (1894–1897), dysentery (1897), and above all malaria (1894–1898). In each case, the solution depended on having precise knowledge—both taxonomic and behavioral—of the species involved. Moreover, it often involved multiple species, including the bacterium or other organism that causes the disease, plus one or more host species that serve as a reservoir for this microbe, and a vector species to deliver it to the human victim. As Manson put it, "the etiology of disease"—that is, the study of its origins and causes—"is but a branch of natural history."

During an epidemic in Hong Kong in 1894, for instance, two bacteriologists working independently identified the bacillus that causes the plague. (The new species *Yersina pestis* got its name from the Swiss researcher Alexandre Yersin, but credit belongs equally to Japanese researcher Kitasato Shibasaburō. The ranks of the species seekers were no longer exclusively of European stock.) Researchers soon isolated rats as the reservoir of the disease and fleas as the means of transmission. But saying "fleas" wasn't nearly enough, according to medical historian J. R. Busvine. The European rat flea, *Nosopsyllus fasciatus*, rarely bites humans, while the human flea, *Pulex irritans*, does not normally deign to feed on rats. Defeating the disease depended on identifying a species that happily samples both and can thus function as a vector— the rat flea, *Xenopsylla cheopis* (discovered, incidentally, by Charles Rothschild).

The old miasma theory had been comparatively simple, blaming epidemic diseases on vague "noxious vapors" given off by even vaguer "putrefying matter." But now epidemic disease looked like an ecological puzzle, and public health workers were suddenly dependent on the work of the species seekers. In particular, the most widely ridiculed branch of that work, identifying obscure insect species, now gave medicine the essential tools for saving lives.

"I Have Formed a Theory"

Patrick Manson was back in England in 1894, age 50, when he belatedly took up the question of malaria. Gout and other medical issues had caused him to resign his colonial duties. He had planned to retire to Scotland with his wife and their three children. Instead, bad investments forced him to resume his medical practice in London.

At about the same time that Manson had been studying filarial worms in Amoy, Alphonse Laveran, a French military surgeon in Algeria, had discovered the organism in the blood that causes malaria. It would soon become an object of intense international interest under the name *Plasmodium malariae*, the first of the four *Plasmodium* species now known to cause malaria in humans. Laveran theorized that certain stages of this parasite's life cycle must occur outside the human body, as suggested by Steenstrup's "alternation of generations." Taking his lead from Manson, Laveran also proposed that mosquitoes might be the means of transmission. But he couldn't prove it.

He refused to test his theory by administering infected blood to a healthy human patient, Douglas Haynes writes, in his book *Impe-rial Medicine*, because "he was unwilling to cause a needless death," an unusual scruple in those days. (Manson would later deliberately infect his own son.) Nor did the ingenious bacteriological methods recently pioneered by Louis Pasteur prove helpful. Pasteur had made it a stan-dard technique to isolate the bacillus that causes a disease like anthrax or cholera and cultivate it in a test tube or petri dish. Many researchers now adamantly argued that bacteria also cause malaria. But Laveran's *Plasmodium* parasite was a single-celled protozoan, and in a petri dish it quickly died. "We are in the presence of microbes of a very special morphology," he wrote, "which we are unable to confuse with common species; these microbes are absolutely specific to malaria."

Laveran had seen the parasite in a flagellate stage with small whip-like appendages. Now other researchers began to track how it shape-shifts as it develops. When it first appears in the bloodstream, it has a rod or spindle shape. After it enters a red blood cell it takes on a ring form. Then it develops into a cluster of small disks, which accumu-late until they rupture the cell membrane. The timing of this rupture synchronized in blood cells throughout the body corresponds to the familiar cycle of fevers in malaria victims, a discovery that stirred up great excitement among researchers in the 1880s.

Manson became interested in malaria partly as a matter of national pride, especially after Laveran's work got translated into English in 1891. "It is little to our credit that continental nations, whose stake in tropical countries is infinitely smaller than ours, are nevertheless just as infinitely ahead of us in this matter," he declared, in a speech to the Hunterian Society. Manson made the analogy to his own research on filarial worms, and in an 1894 paper he declared that mosquitoes "or a similar suctorial insect" must be the means of transmitting malaria. He even applied for funding to test the idea with an expedition to the West Indies; but he was too old and gouty for the work. "Sneering skeptics" also dismissed his theory as too speculative, treating Manson, as he later put it, like "a sort of pathological Jules Verne."

That same year, another physician, on furlough home from India, literally came knocking at his door. Like the younger Manson, Ronald Ross had been putting in hours at the British Museum to catch up on his reading. Also like Manson, he had done little to distinguish himself during his first 15 years in medicine. When they first met, Ross still believed in the bacterial origin of malaria. But he was knowledgeable about tropical diseases and also a skilled microscopist, and the two men became friends. Soon after, as they were walking one day down Oxford Street, Manson confided, "I have formed a theory that mosquitoes carry malaria just as they carry *Filariae*." He wanted Ross to prove it.

"The Door Is Unlocked"

Over the next four years, Manson would steadily guide Ross's research, and his career, toward that end. The two men exchanged frequent letters about the direction of Ross's work, with Manson regularly posing new challenges and reporting the results to the scientific community in London. Ross quickly demonstrated that the malaria plasmodium doesn't get digested in the stomachs of certain mosquitoes fed on infected blood. Instead, he could see that it persists and develops in cells that take on a telltale pigmentation. The idea that the malaria plasmodium seems to thrive in the mosquito's stomach reminded Manson of the way his elephantiasis microfilariae lose their protective wrapping there. He thought it was the prelude to release of the organism and transmission back to humans via drinking water.

In 1897, Ross's careful microscopy revealed the existence of another stage in the plasmodium's development, a spore called the oocyst, embedded in the abdominal wall of "a grey mosquito, a dapple winged mosquito," species unknown. He was tracing the course of the disease by feeding his mosquitoes on birds now, at Manson's suggestion; human test subjects were surprisingly scarce in India, partly because patients feared that a British doctor with a syringe might be deliberately infecting them. Birds were easier to work with, in any case, and early in 1898, Ross reported to Manson that he could feed his mosquitoes on infected larks and sparrows and consistently produce healthy, developing plasmodium in their stomachs. "Dawn seems to be near," Manson replied.

Ronald Ross undertook his mosquito research in India under step-by-step guidance from Patrick Manson in London—but later disavowed his mentor.

Manson meanwhile lobbied successfully to get Ross assigned full time to the malaria study, urging his bosses to give him "the chance of striking a good stroke for England." He warned that "If we don't do it and do it soon, some Italians or Frenchmen or Americans will step in and show us how to do what we cant or wont do for ourselves. They are on track even now." Solving the puzzle of malaria was a chance to "rehabilitate our national character and to point out to the rest of the world how to deal with the most important disease."

In fact, a team led by the Italian zoologist Battista Grassi was pushing hard to solve the malaria puzzle first, and Ross knew it. Where Ross excelled at describing the inner workings of the plasmodium, the Italians applied what historian Ernesto Capanna calls a distinctly "zoological approach to the problem." In 1890, Grassi had coauthored a paper describing the malaria cycle in owls, pigeons, and sparrows, identifying a different protozoa species responsible for the disease in each—but not the means of transmission.

When he resumed work on malaria in the outskirts of Rome in 1895, Grassi's new colleagues persuaded him of the likely importance of mosquitoes. As an entomologist, Grassi knew the taxonomy of the mosquito family. He set out to map the incidence of malaria in local districts. Then he matched those data to the geographical distribution of different mosquito species, narrowing his study down to *Anopheles claviger* and two species in the genus *Culex*. As Grassi and his group were working, a British physician made a habit of dropping by on friendly visits. "The Italian scientists, flattered by the interest of an English colleague in their studies, greeted him without suspicion of his motives," Capanna writes. "He then reported to Ross the information obtained." Espionage, like medicine, was a common use of the naturalist's talent for close observation. But the Italians were at least cautious enough to beg off when the visitor asked to take away samples of the mosquito species they were studying.

Ross meanwhile was still pursuing Manson's theory that mosquitoes transmit the disease to humans by way of drinking water, rather than with their bite. But he could find no evidence that dead mosquitoes actually produce plasmodium spores. Instead, as he was tracking development of the plasmodium within the living mosquito, he noticed something unexpected. The cells containing plasmodium suddenly lost their pigmentation, and rod- or spindlelike structures once again appeared nearby, as if they had burst from their former containers. He cut the head off a mosquito, and "the rods were swarming here and were even *pouring out* from somewhere in streams."

Further dissection revealed that these rods had traveled from the thorax via a duct leading directly up between the mosquito's eyes. There, they crowded into a structure Ross correctly identified as the salivary glands, source of "the stinging fluid which the mosquito injects into the bite." Ross realized that, along with the saliva, the rods would probably be "poured out in vast numbers" under the skin of a bite victim, to be "swept away by the circulation of the blood." And there, they would "develop into malaria parasites, thus completing the cycle."

Manson's theory of water-borne transmission was suddenly history; it was the mosquito's bite that took the malaria plasmodium out of one victim and injected it into another. But his larger idea was also vindicated. "I think that I may now say Q.E.D. and congratulate you on the mosquito theory indeed," Ross wrote to his mentor. "The door is unlocked, and I am walking in and collecting the treasures." Manson was delighted. When he presented the latest developments to the British Medical Association meeting in Edinburgh that July 1898, he waited till the end of his talk, then milked the drama of the moment with a telegram from Ross breaking the news that the puzzle of malaria had been solved.

Or almost solved. Ross had demonstrated that mosquitoes can transmit malaria in birds, not humans. More seriously, he was still talking about his "grey" mosquito, without attempting to give it a proper Linnaean name. It was a peculiar omission for any researcher working in the British scientific tradition, and particularly on the eve of the twentieth century. It also meant that he had missed the single most important fact about transmission of malaria by mosquitoes. In Italy at about the same time, Grassi and his team had acquitted their two *Culex* species of any involvement in malaria. But with the bite of the *Anopheles* mosquito, they had successfully infected a healthy human volunteer with the disease. Ross and the Italians published their results within weeks of each other, and in the margin of Ross's article, Grassi jotted this terse judgment: "non dice che fosse *Anopheles*"—that is, "he doesn't say it was *Anopheles*." Grassi argued correctly that "there is no malaria without *Anopheles*." Mosquitoes of that genus alone are capable of spreading the disease, and targeting *Anopheles* is the only way to control it.

"Collecting the Treasures"

The aftermath of this triumph was, on one level, an appalling scramble for glory. Not only did Ross and Grassi wrestle for the laurels, but Grassi bickered in print with one of his own colleagues about which of them had first suggested the *Anopheles* connection. Ross soon also set out to minimize Patrick Manson's role in solving the malaria puzzle.

At first, Ross had given due credit, writing that Manson's "brilliant induction," his mosquito theory, had "so accurately indicated the true line of research that it has been my part merely to follow its direction." As he was about to head home from India, he added that he was "greatly looking forward to seeing Manson. To him belongs the solution of the malaria problem." But the story, like the plasmodium, began to shape-shift when Ross got back to England and had time to contemplate the treasures he might collect. A poem he later wrote about the great discovery captures his mood, in words reminiscent of Linnaeus at his most egomaniacal:

> *This day relenting God*
> *Hath placed within my hand*
> *A wondrous thing, and God*
> *Be praised. At his command,*
> *Seeking his secret deeds*
> *With tears and touring breath,*
> *I find thy cunning seed*
> *O million-murdering Death.*

Ross's attitude toward Manson seems to have changed for at least two reasons, beyond the normal tendency for a protégé to revolt against his mentor. He was disturbed that Manson soon began collaborating with Grassi, his rival, on experiments to firm up the role of mosquitoes in human malaria. (In July 1900, Grassi shipped live *Anopheles* mosquitoes infected with malaria to London, where Manson's son, a medical student, volunteered to be bitten and promptly came down

Focusing on the critical importance of species, Battista Grassi proved that "there is no malaria without Anopheles.*"*

with the disease. He recovered under treatment with quinine. But this was about as firmed-up a proof as anyone could desire.) Ross also began to distance himself, science historian Jeanne Guillemin suggests, because Manson's careful directions "raised uncomfortable questions" about Ross's "autonomy and originality." Ross found it advantageous to pretend that Laveran had been his intellectual godfather and denigrate Manson. After considerable political maneuvering to block out those who had done equal work in the cause of discovery, Ross alone won the 1902 Nobel Prize and declared in his acceptance speech, "I will begin with the great name Laveran . . ."

Later, Ross began to depict Manson as more a hindrance than a help to the discovery, because of his drinking-water theory of malaria transmission. "It is of little importance that a mosquito should carry the poison out of us," Ross wrote, "but one of *vast* importance that it should also put the poison back into us; *and this was no part of Manson's original teaching.*" A low point in this campaign occurred in 1912, when Ross seized on a harmless sentence in a letter completely unrelated to malaria and charged Manson with libel. The older man, having patiently endured ridicule for 20 years as "Mosquito Manson," was by now too secure, or too stoic, to rise to this bait. His attitude was that it takes two fools to make a quarrel, medical historian Eli Chernin writes, "and he did not mean to be one of them." He wrote a letter of apology and, at Ross's insistence, paid four pounds in legal costs. Ross had taken one of the most gloriously productive collaborations in the history of science and wreathed it in shame.

Escaping Natural Selection

It did not matter. Inspired by Grassi and his colleagues, Italy soon launched the world's first national campaign to eradicate malaria. It faced extraordinary obstacles. Malaria sinks its victims into a seemingly permanent state of torpor and stupidity, and in Italy they commonly suspected that the quinine being distributed by the state was a poison meant to rid the countryside of its excess poor. Rich and poor alike also objected that the screen doors and windows advocated by public health workers "transformed a house into a cage," science historian Frank Snowden writes in *Conquest of Malaria: Italy, 1900–1962*. They propped the screen doors open and sometimes stripped out the screening to use as a kitchen sieve. But by combining quinine treatment, window screens, mosquito control, and rural education, the campaign quickly demonstrated that malaria was not the inevitable "Italian national disease."

At the start in 1900, malaria was killing 16,000 people a year in Italy. By 1914, the death toll was down to just 2000. (The official tally counted only direct deaths from malaria. But since malaria mostly kills indirectly, by weakening the immune system and making victims more vulnerable to other illnesses, Grassi estimated that the real toll at the start was closer to 100,000 deaths a year.) The campaign faltered during World War I, but resumed soon after. It gradually transformed the nation, Snowden writes, not just saving lives, but lifting the poor, and particularly women, up out of an ancient enslavement, and enabling economic progress in place of chronic poverty.

The Roman Campagna, a marshy, malarious district around the capital, had been a notorious summertime deathtrap for thousands of years; Emile Zola had characterized it as "this vast cemetery." But by 1925, Grassi could write in the past tense of scenes that had already become unimaginable: "A quarter of a century ago, anyone who traveled in the Roman Campagna during the malarial season before the end of the agricultural year met at every step individuals stretched out on the ground wrapped in their coats and shivering from fever. Sometimes, if, moved to pity, one attempted to transport them to hospital, they died on the spot. This sad spectacle, which I shall have forever before my eyes, no longer occurs!" Grassi asked to be buried at Fiumicino, in the heart of the Roman Campagna, so he could be there to witness the eradication of malaria from Italy. Long before eradication finally occurred, in 1962, the campaign he had begun had provided a model for the world.

In the United States, malaria had been so routine in the late nineteenth century that it was the summertime regimen in Washington, D.C., "to take a mouthful of sulphate of quinine every morning before breakfast," an entomologist there later recalled. An 1881 cartoon by Thomas Nast depicted a gaunt figure in a top hat shivering with fever before a poster characterizing malaria as "a fashionable disease" in the nation's capital. Northern cities suffered, too. A map showing malaria incidence was red, as if with infected blood, from the Gulf of Mexico to the Great Lakes. But with the discovery of the transmission of malaria by *Anopheles* mosquitoes, the red zone drained away, till the disease was finally eradicated in 1949. That same transformation happened throughout the developed world. (A United Nations/World Bank "Roll Back Malaria" campaign is currently working to extend that record of success to tropical nations, where malaria continues to kill upwards of a million people a year, most of them young children in Africa.)

At about the same time that malaria was coming under control, scientists also began to limit the long, violent depredations of yellow fever. This disease, now largely forgotten in the developed world, appeared in unpredictable summer epidemics then, as far north as Boston. The symptoms began with muscle pain and rapidly advanced to intense fevers, chills, and vomiting. The skin turned yellow as the liver failed. Victims bled from the eyes. Their vomit turned black and had the consistency of used coffee grounds, from half-digested blood. Delirium and convulsions followed, typically leading to death. Adding to the horror of these symptoms was the utter inability to know when or why the disease would strike next within a family or a neighborhood.

Yellow fever seemed like a judgment from God, and the thick of an epidemic was a scene from hell. Cities fired cannons and sent up black clouds of smoke from burning tar to dispel the pestilential miasma. The dead accumulated faster than the living could bury them. An 1853 epidemic killed 8000 people in New Orleans, and the corpses at one cemetery were "piled on the ground, swollen, and bursting their coffins." A newspaper described "a feast of horrors" with "withered crones and fat huxter women" selling ice cream and other treats just outside the cemetery gates, as flies buzzed back and forth between their wares and "the green and festering corpses."

Many people panicked during epidemics, shutting down their businesses and fleeing the cities. Word of an epidemic also spread quickly, and even where official quarantines failed to limit travel, "shotgun quarantines" thrown up by nervous locals closed down most commerce. The worst American epidemic, in 1878, killed 4600 people in New Orleans and 5000, or one in every eight residents, in Memphis as it spread through much of the Mississippi River valley. A New York City merchant traveling the region that September wrote that "the country between Louisville, Kentucky and New Orleans is one entire scene of desolation and woe."

All that came to an end after members of the U.S. Yellow Fever Commission in Cuba went to visit Carlos Finlay at his home in Havana on August 1, 1900. He outlined his ideas about mosquito transmission and described the experiments he had conducted. Grateful at last to have a receptive scientific audience, he also presented his visitors with a porcelain dish of the black, seedlike eggs of the suspect mosquito, *Aedes aegypti*. American researchers led by U.S. Army Major Walter Reed then used mosquitoes hatched from those eggs to transmit yellow fever to human volunteers, confirming Finlay's theories. Mosquito control efforts began in Havana soon after, and the military governor there called it "the greatest step forward made by Medicine since Jenner's discovery of the vaccination." The U.S. Secretary of War, recalling the years when quarantines shut down the country "from the mouth of the Potomac to the mouth of the Rio Grande," predicted that it would never again "be possible for yellow fever to gain such headway."

But the real test came in 1905, when yellow fever once again struck New Orleans. Clergymen now preached the "mosquito doctrine" from the pulpit, and one medical authority urged that mosquito control "had to be taught and practiced as the very catechism and Bible of our entire conduct." This new religion involved identifying yellow fever patients quickly, screening and fumigating their rooms to prevent infected mosquitoes from spreading the disease, and, above all, screening the open cisterns that supplied water to every house—and which also served as mosquito breeding ponds. When some people continued to dismiss the mosquito connection, officials began fining and even jailing them for failing to take the necessary precautions. Early in August, President Teddy Roosevelt ordered the U.S. Public Health Service to take charge of the antimosquito campaign. A British observer reported: "Cleanly, swiftly, scientifically, the olive-green uniforms darted hither and thither . . . opening up every corner to the bright light of science . . . so swiftly that it was like watching a huge machine, well oiled and efficacious, performing a marvelous task with perfect show of ease."

The sort of horrid scenes familiar from past epidemics did not recur. Instead of continuing to climb through late summer, the number of new cases of yellow fever now began to drop off. By September, the local Elks Club felt confident enough to celebrate, with 700 jubilant marchers costumed in cheesecloth screening and hats resembling covered cisterns. (Knowing now that only female mosquitoes bite and thus

The knight-errant in triumph: the work of mosquito scientists like Alphonse Laveran saved millions of lives.

spread the disease, the marchers gleefully chanted, "The female of the species is more deadly than the male." But perhaps by way of gender balance, the festival also featured the opposite of a beauty contest, for ugly men.) In the end, yellow fever caused only 452 deaths in the city, and another 500 elsewhere in Louisiana. Officials estimated "that at least 3,500 lives were saved that summer in New Orleans alone." The lessons learned there quickly spread to other regions, with the result that 1905 was the last time the United States suffered the terror of a yellow fever epidemic.

The beast in the mosquito had been tamed.

Roughly 170 years earlier, a young medical student had qualified for his doctor's certification with an absurd thesis arguing that malaria was caused by particles of clay that got into drinking water, entered the bloodstream, and clogged the capillaries. It is probably just as well that he never practiced medicine. But by coincidence, that student, whose name was Carolus Linnaeus, would later provide the first scientific description of what we now think of as the yellow fever mosquito, *Aedes aegypti*. More important, he would devise the system of classification that enabled an army of his followers—among them Carlos Finlay, Alphonse Laveran, Ronald Ross, Battista Grassi, and especially Patrick "Mosquito" Manson—to identify and begin to understand the vast array of mosquitoes, lice, flies, worms, and other organisms that spread disease. Though Linnaeus certainly did not intend it, his system forced us to understand who we are as human beings, one primate among many. It helped us to see how we fit into the history of the planet and into the lives of other species. It also showed us, to our chagrin, how their lives sometimes fit into ours, turning the human body into a habitat.

It's possible to conclude that the Linnaean system—and the work of the species seekers—ultimately humbled the mighty *Homo sapiens*. But it also gave us life. At the dawn of the twentieth century, it allowed us, alone among Earth's creatures, not merely to understand natural selection, but to escape for a time from its terrible grasp, by conquering some of the deadliest diseases ever known.

THE NEW AGE
OF DISCOVERY

Come, my friends,
'Tis not too late to seek a newer world.
Push off, and sitting well in order smite
The sounding furrows; for my purpose holds
To sail beyond the sunset, and the baths
Of all the western stars, until I die.

—ALFRED TENNYSON, "ULYSSES"

OVER THE COURSE of the great age of discovery, the number of known animal species had grown almost a hundredfold, from the 4400 listed by Linnaeus in the 1758 edition of his *Systema Naturae* to 415,600 by the end of the nineteenth century. But even today, with the total of known species pushing 2 million, new species continue to turn up almost everywhere, at times much closer to home than most of us care to contemplate.

For instance, researchers have lately begun to sort through the wild kingdom of the human digestive tract—identifying 800 bacteria species in the colon, 700 in the mouth, 65 in the esophagus, and just 1 in the stomach. And it turns out that these fellow travelers matter. Researchers are discovering that the particular makeup of this symbiotic com-

373

munity helps determine how we absorb different foods, whether we stay thin or get fat, how the immune system functions, and in short how we live. Identifying the lone colonist of the acid world of the human stomach, *Helicobacter pylori*, has given us a cure for many ulcers. Before that, ulcers generally got blamed on the victim.

Modern discoveries are not, however, just about microfauna. Almost 200 years after Georges Cuvier discounted the likelihood "of discovering new species of large quadrupeds" alive in the modern world, such creatures continue to show up today, among them a beaked whale (*Mesoplodon perrini*) discovered off the California coast in 2002 and a 200-pound oxen (the saola, or *Pseudoryx nghetinhensis*) found on the Laos-Vietnam border in 1993. In our own taxonomic group, the primates, several new species typically turn up each year. Despite our romantic view of the past and our pessimistic ideas about the present, we are living in what some naturalists describe as "a new age of discovery," in which the number of new species "compares favorably with any time since the mid-1700s"—that is, since the beginning of scientific classification.

Why now? Researchers are finding significant new species in the wild partly because they can get to places where wars or political barriers once kept them out. New roads and rapid deforestation are also allowing them to reach places that were formerly too remote. So the new species are sometimes getting discovered and going extinct at about the same time. Modern species seekers also have the advantage of helicopters, submersibles, satellite mapping, species databases, and other modern tools for targeting their searches more methodically at understudied areas. Sometimes they also use old tools in clever new ways—for instance, collecting new butterfly species in Madagascar by using a slingshot to get traps baited with rotting fruit into the treetops. And sometimes researchers just get lucky, as when a new orchid species recently turned up in the viewfinder of a miniature camera researchers had strapped onto a tree kangaroo.

Are we "finding anything significantly different," authors Michael J. Donoghue and William S. Alverson ask in their paper, "A New Age of Discovery." Or is it just a matter of "filling gaps in an already well-known range of variation"? Then they go on for pages describing bizarre new species—some medically important, some noteworthy because they have turned up near major cities. (In one case, a tree 130 feet tall, representing not just a new species but a new genus, somehow went undetected until 1994, though it is an easy day-trip from Sydney, Australia's most populous city.) There is no shortage of discoveries, they conclude, capable of inducing "the sense of awe, amusement, and even befuddlement that remarkable new organisms inspired during the last great age of discovery."

In truth, many of the new species that have turned up in the 10 years since Donoghue and Alverson published their article seem almost surreal. Among them, for instance, are a striped rabbit in the Mekong Delta, and a colorfully painted Indonesian fish named *Histiophryne psychedelica*. (It swims by bouncing off the bottom so haphazardly that, according to researchers, it looks as if it "should be cited for DUI.") And this may be the single biggest discovery, both delightful and daunting, that the species seekers have given us: Nature is not just weird, but limitlessly so.

Modern species seekers still hold forth the goal of describing every species on Earth. In 2003, E. O. Wilson proposed the Encyclopedia of Life (http://www.eol.org) for that purpose, with the intent of having an Internet page for every species within 25 years. But as Wilson acknowledged at the time, "The truth is that we do not know how many species of organisms exist on Earth even to the nearest order of magnitude." He thought the ultimate figure would be about 10 million species. Other biologists say 50 or even 100 million. This is not quite infinite life, not quite variation forever and ever, amen. But it can feel that way to taxonomists working on a family of beetles, or land snails, or parasitic wasps. Not only do specimens from new collections continue to pour in much faster than anyone can catalogue them, but things long since pinned down and described often have a way of becoming something else on closer examination.

Genetic study is revealing that the conventional practice of separating species based on morphology—or as one lepidopterist recently put it, "how they look to a six-foot-tall diurnal mammal "—can be seriously misleading. For instance, it now appears that the giraffe, long thought to be one species, actually has a half-dozen "cryptic species" hidden beneath its dappled flesh, invisible to us but not to one another. And instead of one African elephant species, we now have a savanna species (*Loxodonta africana*) and a forest species (*L. cyclotis*), almost as different from each other as they are from the Indian elephant. Some of these "discoveries" may at first seem trivial. Who cares if genetic evidence reveals that one *Anopheles* mosquito species is actually 20 separate species, all of them completely identical to human eyes? But such fine distinctions now enable public health teams to target the species that actually carry malaria, and ignore the ones that don't. More efficient mosquito control means that children live who, just a year or two ago, would probably have died.

This new brand of discovery is also exciting because it gets us beyond things we can see with our own dim eyes and brings us closer to what the animals themselves clearly know. Behavioral cues that don't show up in a museum drawer—like scent, or vocalization, or auditory targeting on a particular frog song—can make a life-or-death, sex-or-solitude difference in the wild. As we open up this new view into the animal world, it is becoming evident that we live on what is still a little-known planet.

So biologists will continue to travel to the farthest corners of the Earth in pursuit of mysterious and elusive life-forms. In an age of mass extinctions, it may seem as if the discoveries will run out, probably sooner rather than later. The lemurs, songbirds, and butterflies are disappearing when we have hardly begun to know them and, in some areas, people have already been reduced to the distant memory of lost wildlife. This grim tide of extinction may eventually roll us up in its path, too.

But here is a consoling prospect: Plants and animals will continue to do what they have done for roughly the past 3 billion years—change, fall out of old niches and fill new ones, develop bright new colors, quirky shapes, and astonishing new behaviors. They will adapt. And out of our present apocalypse, strange new species will repopulate Planet Earth.

Perhaps, against all odds, we will adapt, too. And children not yet born will again grow up to know the incomparable delight of discovering new species.

NECROLOGY

This is a very preliminary list of those who died in the search for new species. See www.speciesseekers.com for more complete information.

Akeley, Carl (1864–1926), naturalist-taxidermist for the American Museum of Natural History, age 62, while collecting mammals in the eastern Congo, of dysentery.

Alexander, Capt. Boyd (1873–1910), explorer and ornithologist, age 47, murdered in what is now Chad.

Anderson, William (1750–1778), surgeon-naturalist on Cook's second and third voyages, at sea, age 28, possibly from scurvy.

Banister, John (1650–1692), British naturalist and clergyman, of a gunshot wound, age 42, while exploring in Virginia.

Batty, Joseph H. (?–1906), taxidermist and specimen hunter recently accused of fraudulent practices, "killed instantly by the accidental discharge of his gun," age unknown, in Mexico.

Bernstein, Heinrich Agathon (1828–1865), German physician and collector of birds and mammals, age 36, on the island of Batanta off New Guinea, cause unknown.

Biermann, Adolph (?–1880), curator of the Calcutta Botanical Garden, survived attack by tiger while walking in garden but succumbed a year later, age unknown, to cholera.

Boerlage, Jacob Gijsbert (1849–1900), Dutch botanist on his 51st birthday, on a botanical expedition to the Moluccas to identify plants described by Rumphius, cause unknown.

Boie, Heinrich (1784–1827), German ornithologist, age 43, of "gall fever," in Java, one of a long succession of naturalists to die in the service of the Dutch Natural History Commission to the East Indies.

Bowman, David (1838–1868), Scottish plant collector, robbed of his specimens in Colombia and said to have died of "mortification," but more likely from dysentery, age 30, in Bogota.

Buddingh, *Johan Adriaan* (1840–1870), Dutch civil servant and amateur collector for the Leiden Museum, age 30, in Batavia (Jakarta), Java, cause unknown.

Cahoon, *John Cyrus* (1863–1891), American ornithologist and field naturalist, fell off a sea cliff, age 28, in Newfoundland.

Cassin, *John* (1813–1869), American ornithologist who described 198 new species, age 55, apparently of arsenic poisoning.

Cralitz, *H.* (?–1637), German physician and naturalist on a Dutch West Indies Company expedition, soon after arrival in Brazil, cause unknown.

Craven, *Ian* (1962–1993), ornithologist, age 31, plane crash in Irian Jaya.

Dawson, *Elmer Yale* (1918–1966), Smithsonian Institution phycologist, age 48, drowned while diving for seaweeds in the Red Sea.

Doherty, *William* (1857–1901), American lepidopterist and specimen hunter for Walter Rothschild, age 44, of dysentery, in Kenya's Aberdare Mountains.

Douglas, *David* (1799–1834), Scottish botanist and explorer, age 35, fell into a pit trap already occupied by a bull, in Hawaii.

Dutreuil de Rhins, *Jules Léon* (1846–1894), French explorer, age 48, murdered in Eastern Tibet.

Eickwort, *George Campbell* (1949–1994), hymenopterist, age 45, car accident in Jamaica.

Feilner, *Sgt. John* (?–1864), ornithologist, age unknown, surprised and killed by Sioux while collecting ahead of his U.S. Army expedition in the Dakotas.

Fonseca, *Rene Marcelo* (1976–2004), Ecuadorian mammalogist, age 28, car accident.

Forsskål, *Pehr* (1732–1763), Helsinki-born "apostle" of Linnaeus's, age 31, of malaria in what is now Yemen.

Gambel, *William* (1823–1849), American naturalist, namesake of Gambel's quail, age 26, of typhoid fever in the Sierra Nevada.

Gentry, *Al* (1945–1993), botanist, age 48, plane crash in the mountains of Ecuador.

Gilbert, *John* (1810?–1845), British naturalist and explorer, collected Australian mammals and birds for John Gould until killed by a spear, age 35, during a nighttime raid on his camp by Aborigines.

Grant, *Harold J.* (1921–1966), American entomologist, age 45, drowned on an expedition collecting grasshoppers in Trinidad.

Griffith, *William* (1810–1845), British botanist in India and Afghanistan, age 34, of malaria.

Harrisson, Tom (1912–1976), British anthropologist and ornithologist, age 64, bus accident in Thailand.

Hasselquist, Fredric (1722–1752), Swedish "apostle" of Linnaeus's, made extensive collections in the Middle East, age 30, near Smyrna, cause unknown.

Hasselt, Johann Coenraad van (1797–1823), Dutch ornithologist, age 26, of an unknown tropical illness in Java.

Helfer, Johan Wilhelm (1810–1840), Austrian naturalist, age 29, murdered in the Andaman Islands.

Hemphill, Henry (1830–1914), American naturalist studying shells, age 84, of arsenic poisoning.

Hemprich, Wilhelm (1796–1825), surgeon in the Prussian army, naturalist, leader of a five-year expedition to Egypt and nearby countries, collecting 3000 plant and 4000 animal species, on which nine team members died, including Hemprich, age 28, probably of malaria, in Eritrea.

Hendee, Russell W. (1899–1929), mammalogist collecting for the Field Museum's Kelley-Roosevelts Expedition, age 30, of malaria in Vientiane.

Jacquemont, Victor (1801–1832), French botanist in India, age 31, of dysentery or malaria.

Kingsley, Mary H. (1862–1900), British explorer, ichthyologist, age 37, of typhoid fever in South Africa.

Köenig, Johann Gerhard (1728–1785), Polish-born physician and student of Linnaeus who introduced the Linnaean system to India, age 57, cause unknown.

Kramer, Gustav (1910–1959), German ornithologist, was attempting to capture young rock doves from a nest when he lost his footing and fell to his death, age 49, in southern Italy.

Kuhl, Heinrich (1797–1821), German ornithologist, age 23, in Java, of an unknown tropical disease.

Leitner, Edward F. (1812–1838), German-born physician and botanist, collector for Audubon and Bachman, shot by Indians, age 26, near Jupiter Inlet, Florida.

Macklot, Heinrich (1799–1832), naturalist, was so enraged when insurgents burned down his house, with all of his collections, that he organized a revenge attack and was speared to death, age 33, in Java.

Marcgraf, George (1610–1644), German physician and naturalist on a Dutch West Indies expedition, age 34, probably of malaria, in Brazil.

Meyer, Frank N. (1875–1918), American plant explorer, made four expeditions to China. Heading homeward down the Yangtze River at a time of political turmoil, he disappeared, age 43, from his ship and his body was recovered a week later.

Moorcroft, William (1765–1825), British veterinary surgeon and plant-collector in Tibet and Kashmir, also reputed to be a secret agent, age 60, murdered in Afghanistan.

Natterer, Johann (1787–1843), Vienna-born zoologist, survived 18 years collecting in Brazil, but died at home, age 56, of pulmonary hemorrhage, while working up his extensive collection.

Oort, Pieter van (1804–1834), artist who made numerous illustrations of landscapes, people, animals, and plants for the Dutch Natural History Commission in the East Indies, age 30, in Sumatra, of malaria.

Pambu (dates unknown), a Lepcha collector working with William Doherty, "murdered by the savages" in Papua New Guinea.

Parker, Ted (1953–1993), American ornithologist, age 40, killed in a plane crash in the mountains of Ecuador.

Parkinson, Sydney (1745–1771), artist on Cook's *Endeavour*, at sea, age 26, of dysentery.

Peale, Raphaelle (1774–1825), American artist and naturalist, age 51, of arsenic and mercury poisoning from his taxidermic work in the family museum.

Przhevalsky, Nikolai Mikhaylovich (1839–1888), Polish-Russian explorer, discoverer of only wild horse species, of typhus, age 49, in Kyrgyzstan.

Raalten, Gerrit van (1797–1829), Dutch artist with naturalists in Java, survived a rhino attack but succumbed, age 32, to fever.

Raddi, Giuseppe (1770–1829), Italian botanist, herpetologist in Brazil, age 59, of dysentery at Rhodes, during an expedition to the Nile.

Schlagintweit, Adolf (1829–1857), one of five German brothers who became naturalists and explorers, beheaded as a spy, age 28, in Kashgar.

Schweigger, August Friedrich (1783–1821), German naturalist, age 38, murdered by his guide on a research trip in Sicily.

Seetzen, Ulrich. J. (1767–1811), German explorer and naturalist specializing in snakes and frogs, traveled in the Middle East disguised as a beggar. Accused of stealing cultural treasures, he was poisoned to death, age 44, apparently on the order of an Imam in what is now Yemen.

Slowinski, Joseph (1962–2001), herpetologist, age 38, in northern Burma, snakebite.

Smithwick, Richard P. (1887–1909), American ornithologist, smothered to death while digging his way into a soft bank to raid a Belted Kingfisher nest, found "with his feet only projecting through the sand," age 22, in Virginia.

Stalker, Wilfred (1879–1910), British collector of natural history specimens in Southeast Asia, drowned, age 31, on the British Ornithological Union's 1909 expedition to New Guinea.

Stokes, William (?–1873), "sailor boy" on HMS *Challenger*, killed when block from oceanographers' dredge tore loose and hit him.

Stoliczka, Ferdinand (1838–1874), Czech paleontologist and naturalist, age 36, of altitude sickness while crossing the Himalayas in Ladakh, in India.

Suhm, Rudolf von Willemoes (1847–1875), German, the youngest of the "Scientifics," aboard HMS *Challenger*, dubbed "the baron" by crew, age 28, of erysipelas, an acute streptococcus infection.

Swain, Ralph B. (1912–1953), entomologist, ornithologist, botanist, age 41, murdered by bandits in Mexico.

Thorbjarnarson, John (1957–2010), American herpetologist specializing in crocodiles, age 52, of malaria, in India.

Townsend, John Kirk (1809–1851), American physician and naturalist, age 42, of arsenic poisoning.

Tungkyitbo (?–1891), Lepcha collector working with William Doherty, hospitalized in Java for unknown condition, then died at sea.

Wallace, Herbert (1828–1850), entomologist, age 22, of yellow fever in the Amazon.

Walsh, Benjamin (1808–1869), the first state entomologist in Illinois, age 61, after losing his foot in a train accident.

White, Samuel (1835–1880), Australian ornithologist, age 45, of pneumonia or fever during an expedition to the Aru Islands.

Whitehead, John (1860–1899), British collector of natural history specimens in Southeast Asia, age 39, of fever in Hainan, China.

ACKNOWLEDGMENTS

I'D LIKE to thank the John Simon Guggenheim Memorial Foundation, for the fellowship that made this book possible. For their generous recommendation of this project, thanks also to Chris Johns at *National Geographic*, historian Geoffrey Ward, David Wagner at the University of Connecticut, and Fred Strebeigh at Yale University. My editor Angela von der Lippe and my agent John Thornton invested their considerable faith in me when *The Species Seekers* was still just a vague idea and I am deeply grateful to them both.

At several publications where I am a contributor, editors and staff also provided valuable help. Thanks to Peter G. Brown, Dolly Setton, and Annie Gottlieb at *Natural History*; Carey Winfrey, Tom Frail, Marian Holmes, and Sarah Zielinski at *Smithsonian*; Kathrin Lassila and Thea Martin at the *Yale Alumni Magazine*; Laura Marmor at *The New York Times*; and James Bennet, Scott Stoessel, Amy Meeker, and Eleanor Smith at *Atlantic Monthly*.

For help with the research on this book, I am indebted to a small army of librarians and archivists, particularly those at Sterling Memorial Library and Kline Science Library at Yale University, the American Philosophical Society, the Phoebe Griffin Noyes Library, the iCONN interlibrary loan service, the University of North Carolina and the University of Virginia libraries, Gina Douglas and Lynda Brooks at the Linnean Society in London, Simon Chaplin at the Royal College of Surgeons, Dawn Dyer at the Bristol (England) Reference Library, David Denny of Icon Films, Bronwen Tomb of the University of Connecticut, and Clare Conniff of Bard College.

For generously offering their bright ideas (and answering my dumb questions), thanks to William L. Krinsky and Daniel Brinkman of the Peabody Museum of Natural History at Yale; Terri McFadden and Judy Chupasko of the Harvard Museum of Comparative Zoology; Don E. Wilson and Kristofer M. Helgen of the U.S. National Museum of Natural History; Joanne H. Cooper of the Natural History Museum, Tring, England; Frank Snowden at Yale University; Andrew Brower at Middle Tennessee State University; Daniel Margocsy at Harvard University; Chris Smeenk of the National Museum of Natural History, Leiden, the Netherlands; Linda Cayot and Johannah Barry of the Galapagos Conservancy; Paul Lawrence Farber of Oregon State University; and George Schaller of the Wildlife Conservation Society.

Thank you to James C. G. Conniff for his careful reading of the manuscript and to Karen Conniff for heroic work pulling together the many loose ends along the way. Thanks finally to my mother-in-law, the late Janice Ward Braeder. As a decoupage artist, she collected eighteenth- and nineteenth-century natural history illustrations. Many of those she left behind illustrate this book.

NOTES

INTRODUCTION: STRANGE THINGS, STRANGE LANDS

1 "What raptures": O'Brian 1997, p. 169. Rev. W. Sheffield, Keeper of the Ashmolean Museum, was writing to Gilbert White on the return of HMS *Endeavour.*

1 Dejean: Reitter 1960, pp. 188–189; Smith, Mittler, and Smith 1973, pp. 125–127.

2 The fossil skeleton of a giant salamander: Gette and Scherer 1982, p. 100.

2 Distinguishing plants from animals: Groner and Cornelius 1996, pp. 3, 4.

3 Linnaeus's 19 "apostles": Koerner 1999, p. 113.

5 Echidna: Gruber 1982.

5 Titian Peale: Philbrick 2004, p. 95.

6 Charles Waterton: Aldington 1949, p. 80.

6 Col. Howard Irby's retriever: Mearns and Mearns 1998, p. 55.

6 Darwin collecting bones from kitchen scraps: Farber 2000, p. 56.

6 "A great inconvenience": Bates 2002, p. 289.

6 Ants using leaves to thatch roofs: Bates 2002, p. 21.

7 Ants stealing food: Bates 2002, pp. 22, 23.

7 "Salt and sugar here liquefy": W. Doherty to T. G. Walsingham, October 2, 1889, Ms. Coll. of Lord Thomas de Grey Walsingham, Natural History Museum, London.

7 Doherty "Loss of all my collections": Hartert 1901.

8 "Insects are remarkably cheap": Letter from J. Thomson, October 2, 1848, J. L. LeConte Correspondence, American Philosophical Society, Philadelphia, PA.

8 "Burning & plundering": Letter from W. H. Pease, January 27, 1848, J. L. LeConte Correspondence, American Philosophical Society, Philadelphia, PA.

8 Dying ornithologists: Mearns and Mearns 1998, p. 40.

8 Benjamin Walsh: Berenbaum 1995, p. 280.

9 Edward Baker: *The Ibis*, 1944, pp. 413–415.

9 "The usual adventures": Hartert 1901.

9 Returned letter to Doherty: ? Janson to W. Doherty, June 14, 1901, Janson Family Archive, Natural History Museum, London.

9 Hugh Cuming's restlessness: Dance 1966, p. 156.

10 "Our perfect naturalist": Kingsley 1890, p. 46.

11 Professor Ptthmllnsprts: Kingsley 1917, p. 148.

11 Rafinesque's letter to Audubon: Audubon 1832, p. 455.

11 Mark Twain's Amazonian ambitions: Twain 1963, pp. 480, 481.

11 "Strange things of strange lands": Philbrick 2004, p. 57.

11 "Dead and living lizards": Philbrick 2004, p. 69.

12 Nabokov "On Discovering a Butterfly": *The New Yorker*, May 15, 1943.

12 "The truth is, the pleasure of finding new species": Kingsley 1890, p. 31.

13 Dall loses 34 pounds (April 27, 1865), has a rough boat ride (January 1, 1867), and shares the "singular delight" of discovery (August 1866): All in William H. Dall papers, 1865–1927, Smithsonian Institution Archives, Washington, D.C.

14 No hope of discovering large quadrupeds: Cuvier 1822, p. 61.

14 "Nothing remained for naturalists": *The New York Times*, March 28, 1861, p. 2.

14 "Diabolical caricature": Owen to Wyman, July 24, 1848, in Wyman 1866.

CHAPTER ONE: THAT GREAT BEAST OF A TOWN

15 Addison on beetles: J. Addison, *The Spectator*, August 26, 1710.

15 Electric eels: Moore 2005, pp. 31, 32, 35–42.

15 John Hunter "laying down the chissel": Foot 1794, p. 10.

16 Hunter's stint at Oxford: Wells 1974.

18 Royal menagerie: Dobson 1962.

18 Pope's "virtuoso class": Rousseau 1982, p. 200.

18 The universal sport: Rousseau 1982, p. 224.

18 "Refuse of Nature": J. Addison, *The Spectator*, August 26, 1710.

18 Johnson's Quisquilius: Johnson 1810, p. 67.

19 Ashton Lever: King 1996.

19 Newton, Bacon: Rousseau 1982, pp. 214, 215.

20 "A matter of little moment": Denny 1948.

20 "Seized with terror": Kalm 1772, p. 24.

20 Broad participation in natural history: Christie 1990.

20 Dr. Johnson vs. Dr. Hunter: Moore 2005, p. 153.

21 William Hunter's anatomy demonstrations and John "incapable of putting six lines together": Foot 1794, p. 60.

21 Stubbs working for Hunters: Egerton 1976, p. 11.

21 Solander "the wittiest pupil": Linnaeus to Abraham Bäck, March 30, 1759, L2509, The Linnaean Correspondence Online (http://linnaeus.c18.net).

21 Banks and Solander's specimens in 1771: Wheeler 1984.

22 "The living alligator": Waller 2002, p. 220.

23 Cartmen and porters: Porter 1994, p. 183.

23 "Wild Beasts": Greig 1926, p. 207.

23 Animal performing acts: Altick 1978, p. 40.

23 "That great beast of a town": Smith 1962.

24 Francis Burney describes Lever: King 1996.

24 "Embracing all of nature" and inducing "majestic awe": Smith 1962.

25 "A whale's pizzle": Prince 2003, p. 13.

25 Lever's collection: Smith 1962.

25 Banks on Lever: King 1996.

25 "THE FIRST MUSEUM IN THE UNIVERSE": King 1996.

25 Specimens from Banks, a traveler, the royal menagerie: Dobson 1962.

26 Stubbs and tiger: Altick 1978, p. 39.

26 John Hunter on the family farm: Moore 2005, pp. 16, 69.

27 "Are these the bones of Eskimaux": Altick 1978, p. 48; O'Brian 1997, p. 181.

27 Charles Byrne story: Altick 1978, p. 42; Moore 2005, pp. 199–211.

27 Dr. Dolittle, Dr. Jekyll and Mr. Hyde: Magyar 1994; Moore 2005, p. 218.

27 Hunter's dissections, contributions to modern science: Moore 2005, pp. 151–154.

28 Hunter as model for "Jack Tearguts": Moore 2005, p. 235.

29 "Immortal work": Linnaeus to John Ellis, September 4, 1770, L4410, The Linnaean Correspondence Online (http://linnaeus.c18.net).

29 Garden's letter to Ellis, instructions to mariner: Garden 1775.

30 Hunter and Walsh work on torpedo fish: Piccolino and Bresadola 2002.

30 "An electric spark," Galvani, and Volta: Piccolino and Bresadola 2002.

CHAPTER TWO: FINDING THE THREAD

33 "Naturalists, like men newly risen": Harvey 1857, p. 3.

33 Platypus history: Moyal 2004, pp. 3–18, Gruber 1991.

37 Possums on a London quay: Boorstin 1983, p. 461.

37 Monkeylike animal "called *Mongoose*": Sellers 1980, p. 35.

38 Footless bird of paradise specimens: Goldgar 2007, pp. 73, 74.

38 Linnaeus as a guide through "Nature's labyrinth": Kerr 1792, pp. 22, 23

39 Linnaeus's travels in Lapland: Koerner 1999, p. 56ff.

39 Linnaeus chosen by God: Koerner 1999, p. 23.

40 *Arum maculata* and flower petals as "bridal bed": Jenkins 1978, p. 32.

41 Importance of reproductive organs in taxonomy: Coleman 1964, pp. 20, 21.

41 Appreciation of the simplicity of Linnaeus's system: Swainson 1834, p. 37.

41 "Things not in Linnaeus": Anon. 1875.

42 Linnaeus's collecting parties: Koerner 1999, pp. 41, 42.

42 Nordic summer and "unsearchable perfection": Lindroth 1994, pp. 10–12.

42 "A compulsion of almost demonic intensity": Lindroth 1994, p. 22.

42 "Oh what kind of marvelous animals we are": Koerner 1999, p. 91.

43 Siegesbeck: Jönsson 2000.

43 Disapproval of Linnaeus's emphasis on sex: Boorstin 1983, p. 439.

44 "Paris is hell": Roger 1997, p. 30.

45 The *Histoire Naturelle* as a best seller: Roger 1997, p. 184.

45 "The way a beetle's wing should fold": Roger 1997, p. 241.

45 Buffon's "reprehensible statements": Roger 1997, p. 188.

45 Transformed by habitat and "American degeneracy": Roger 1997, p. 299; Semonin 2000, p. 6.

46 Buffon's "knack for connecting" observations to general theories: Roger 1997, p. 83.

46 Buffon as "Pliny and the Aristotle of France": Lyon and Sloan 1981, p. 8.

46 Tulips, barberries, and "imperceptible nuances": Roger 1997, p. 85.

46 Linnaeus learns about Buffon's attack: Abraham Bäck to C. Linnaeus, May 1, 1744, L0560, The Linnaean Correspondence Online (http://linnaeus.c18.net).

46 Linnaeus's dismissal of Buffon: Koerner 1999, p. 28.

47 A rudimentary hierarchy of classification: Gould 2000.

48 "Manifestly unfair" and "ignobly motivated": Lyon and Sloan 1981, p. 91.

48 Buffon and the age of the Earth: Roger 1997, pp. 409–413.

49 "The most important scientific document": Gould 2000, p. 87.

49 Hoquet's assessment of Buffon: personal interview.

49 Buffon's death and burial: Roger 1997, pp. 425, 426.

50 Linnaeus's advice to skeptic: Jönsson 2000.

50 Cuvier on needing both Linnaeus and Buffon: Coleman 1964, p. 24.

CHAPTER THREE: COLLECTING AND CONQUEST

53 "I shall give up the wife": Fornasiero, Monteath, and West-Sooby 2004, pp. 16, 17.

53 Stedman on Joanna, slavery: Stedman 1988, p. 88ff.

53 Educating Joanna: Lee 2004, p. 138.

53 Contemplating Joanna's possible fate: Stedman 1988, p. 90.

53 "Humanitarian pornography": Klarer 2005.

55 Stedman's intent to "purchase and educate" Joanna: Stedman 1988, p. 98.

55 Stedman's work modeled on *Tom Jones*: Stedman 1988, pp. lxvi–xix.

55 A planter's wife murdering a slave: Stedman 1988, p. 340.

55 A slave hanging from a hook in his ribs: Stedman 1988, p. 105.

55 "The Toucan and the Fly-catcher": Stedman 1988, p. 110.

55 "Flagellation of a Female Samboe Slave": Stedman 1988, p. 265.

55 "The Spur winged Water hen" and "the Red Curlew": Stedman 1988, p. 274.

55 Stedman's "incurable romanticism": Stedman 1988, p. xxvii.

55 Blake's poem and Stedman's description of a tiger: Stedman 1988, p. xlii.

55 Contributions to the Leverian and Leiden museums: Stedman 1988, p. xxix.

55 Science helps Stedman "distance himself": Stedman 1988, p. xxiv.

55 Stedman standing over dead slave: Stedman 1988, p. 2.

56 "My hands are guilty, but my heart is free": Stedman 1992, p. lxxii.

56 Trading by Dutch East India Company employees: Boxer 1977, p. 201.

57 Benjamin Franklin protects Capt. Cook: O'Brian 1997, p. 190.

57 George Washington aids British excavation: Semonin 2000, pp. 186, 191.

57 Jefferson president of American Philosophical Society: Cohen 2002, p. 86.

57 Mahlon Dickerson: Philbrick 2004, p. 32.

57 Joel Poinsett and the poinsettia: Philbrick 2004, pp. 40, 41.

57 "Useful knowledge": The full name of the organization Franklin founded was "The American Philosophical Society, Held at Philadelphia, for Promoting Useful Knowledge."

58 "Scientific gentlemen": Fornasiero, Monteath, and West-Sooby 2004, pp. 37, 73.

58 Joseph Banks's career: O'Brian 1997.

58 Banks advocated tea plantations in India: Jenkins 1978, p. 51.

58 Linnaeus and Queen Lovisa Ulrika: Frängsmyr 1994, pp. 54, 186.

58 The Queen commented on Linnaeus in a letter to her mother: August 6, 1751, included in *Bref och skrifvelser af och til Carl von Linné*, I (4): p. 281.

58 Linnaeus on fir-tree usefulness: Koerner 1999, p. 107.

59 Linnaeus's ideas on economy and imports: Koerner 1999, p. 109ff.

59 Loss of Sweden's colonies: Koerner 1999, p. 102.

59 Linnaeus on the famine of 1756: Koerner 1999, p. 92.

59 Linnaeus's idea to grow tropical plants in Sweden: Koerner 1999, p. 121ff.

59 Coconuts and fried Birds of Paradise: Koerner 1999, p. 123.

60 Increasing the wealth of those on unknown coasts and live specimens: Fornasiero, Monteath, and West-Sooby 2004, pp. 19, 20.

60 Flinders applies to Banks: Fornasiero, Monteath, and West-Sooby 2004, p. 35.

61 Dinner served in bowls used for sick-room filth: Stedman 1988, p. 33.

62 Banks sends his mistress ahead: Beaglehole 1974, pp. 306, 307.

62 Flinders had just wed: Fornasiero, Monteath, and West-Sooby 2004, pp. 36, 37.

62 The British haul of new genera and species: Fornasiero, Monteath, and West-Sooby 2004, p. 346.

63 The French haul: Farber 1982, p. 40.

63 The glory of dissecting a shark: Fornasiero, Monteath, and West-Sooby 2004, p. 290.

63 A new mollusk or a new landmass: Kingston 2007.

63 Catching butterflies at Van Diemen's Land: Fornasiero, Monteath, and West-Sooby 2004, p. 166.

63 "My heart is with thee": http://www.nmm.ac.uk/matthew-flinders, accessed February 9, 2010.

64 Prickly heat and other trials: Stedman 1988, p. 229.

64 Death rate in British Army, and at Cartagena: Curtin 1989, pp. 2–4.

64 "The white man's grave": Curtin 1961.

65 Lord Kames: Bewell 2004.

66 People murdered to give us coffee: Stedman 1988, p. lxv.

66 Sex with slaves: Stedman 1988, p. xxxii.

66 His esteem for Joanna: Stedman 1988, p. 249.

66 Suriname marriage: Stedman 1988, p. xxxv.

66 Raffles the child of a slaver: Wurtzburg 1954, p. 15.

66 Audubon's parentage, and as slave owner: Rhodes 2004, pp. 20, 115.

66 Du Chaillu's parentage: Bucher 1979.
67 Raffles loses his first wife: Wurtzburg 1954, p. 362.
67 Raffles's marriage to Sophia Hull: Wurtzburg 1954, p. 414.
67 Founding Singapore: Wurtzburg 1954, p. 486ff.
67 "Here all is life": Wurtzburg 1954, p. 606.
68 Raffles's "dear little rogues": Wurtzburg 1954, p. 597.
68 "The melancholy blank": Wurtzburg 1954, p. 664.
69 The sinking: Wurtzburg 1954, p. 675ff.
69 "My almost only child" and "Rob me not": Wurtzburg 1954, p. 606.
69 Stedman's marriage: Stedman 1988, p. lxxxvii.

CHAPTER FOUR: MAD ABOUT SHELLS

71 "The wealth and happiness of man": Poe 1840, p. 6.
71 Batavia's "glistening white coral": Rumphius 1999, p. lii.
71 Rumphius's life: Sarton 1937.
72 "He fled from chaos": Rumphius 1999, p. xxxviii.
72 "The Pliny of the Indies": Sarton 1937.
72 Rumphius in Batavia, "caritates": Nieuwenhuys 1982, p. 29; Rumphius 1999, p. cx.
72 "The water Indies": Rumphius 1999, p. xlix.
73 Eating shellfish made us human: Broadhurst et al. 2002.
73 Shell beads from 100,000 years ago: Vanhaeren et al. 2006.
73 Marine snails source of purple dye: Abbott 1982, p. 183.
74 Amsterdam shell collector: Goldgar 2007, p. 80.
74 "Shell-lunatics": Goldgar 2007, p. 84.
74 Vermeer's "Woman in Blue Reading a Letter" sold for 43 guilders: Van Seters 1962, p. 59.
74 *Conus gloriamaris* sells for 120 guilders: Van Seters 1962, p. 48.
75 Crushing *C. gloriamaris* underfoot: Abbott 1982, p. 136.
75 Vermeer sold for $30 million: *The New York Times*, July 7, 2004.
75 Dutch artists produced 5 million paintings: Prak 2005, p. 241.
76 Spiny oyster shell story: Dance 1966, p. 132.
76 Linnaeus and *Cypraea amethystea*: Dance 1966, p. 82.
77 John Marr "begging pardon for my Boldness": Wilkins 1957.
77 Shell specimens collected aboard the *Adventure*: Dance 1969, pp. 49, 50.
77 Sale of *Astraea heliotropium* to Lever: Dance 1969, pp. 49, 50.
77 *Cypraea aurantium*: Dance 1969, pp. 60, 61.
77 "The excellent artisan of the Universe": Spary 2004, p. 19.
77 The Precious Wentletrap a "work of art": Roemer 2004.
78 "Build thee more stately mansions": Holmes 1895.
78 A cross between antelope and leopard: Abbott 1982, p. 110.
79 Instructions to bird collectors: King 1996.
79 Shell collecting "particularly suited to ladies": Allen 1994, p. 113.
79 Pen shells: De Wit 1959, p. 187.

80 "His most poetic texts" and Portuguese man-of-war: Rumphius 1999, p. cx.

81 The starfish tree: Nieuwenhuys and Beekman 1982, p. 30.

81 *The Ambonese Curiosity Cabinet* not "a catalogue of lifeless oddities": Rumphius 1999, pp. civ and cv.

81 Poe gets Rumphius backwards: Poe 1840, p. 6.

81 Rumphius's library: De Wit 1959, p. 189.

82 Rumphius was struck blind in 1670: Rumpf 1981, pp. 33, 126.

82 Rumphius loses his wife: De Wit 1952.

82 Story of publication of the *Herbarium*: Sarton 1937.

83 Sale of Rumphius's collection and his response to it: Rumphius 1999, pp. civ–cvii.

CHAPTER FIVE: EXTINCT

85 "Why I insert the Mammoth": Jefferson 1853, p. 55.

85 Mastodon tooth discovery: Stanford 1959.

85 Lord Cornbury: Semonin 2000, pp. 15–17.

86 *Incognitum* as "the nation's first prehistoric monster": Semonin 2000, p. 1.

86 Edward Taylor's poem: Stanford 1959.

86 Cotton Mather on besting "Og and GOLIATH": Semonin 2000, p. 31.

86 A creature buried in Noah's flood: Stanford 1959.

86 Ichthyosaur fragments: Howe, Sharpe, and Torrens 1981, p. 6–7.

86 Robert Darwin's discovery of fossil plesiosaur in 1718: Stukely 1719.

87 Erasmus Darwin's "the Goddess of Minerals": Ruse 2003, p. 54.

87 "Numerous other petrified shells": Darwin 1803, vol. 1, p. 398.

87 Description of Erasmus Darwin, "Eat or be eaten": Butler 1879, p. 177.

87 All warm-blooded animals "from one living filament": Darwin 1803, vol. 1, p. 397.

87 Motto "E conchis omnia" and "Great wizard he": King-Hele and Darwin 2003, p. xiii.

87 "Darwinizing": Coleridge used the word in his "Notes on Stillingfleet," first published in *The Athenaeum*, March 27, 1875, 2474: 423.

88 "New vista of earth history": Rupke 2002.

89 "*Pseudelephant* . . . probably extinct": Hunter 1768.

89 Franklin on pursuing prey vs. grinding up tree branches: Cohen 2002, p. 94.

89 "An idea injurious to the Deity": Greene 1961, p. 100.

89 *Incognitum* "had been a monstrous carnivore": Semonin 2000, p. 8.

89 "Patriotism and prehistoric nature": Semonin 2000, p. 186.

89 The *incognitum* symbolizing "the immense power": Semonin 2000, p. 339.

90 "Such is the economy of Nature": Jefferson 1853, p. 56.

90 Buffon's "American degeneracy": Roger 1997, p. 299; Semonin 2000, p. 6.

90 Jefferson's animal tables: Jefferson 1853, p. 51.

90 Mammoth stifles Buffon's notion: Jefferson 1853, p. 47.

91 Jefferson's panther and moose skins: Semonin 2000, p. 220.

91 American need to disprove Buffon's theory: Patterson 2001, pp. 7–9.

91 "Let us try this question": Jefferson and Ford 1904, vol. 12, pp. 110, 111.
91 Indian stories: Jefferson 1853, p. 42.
93 Daniel Boone carries letter for Jefferson: Semonin 2000, p. 184.
93 Mastodon bones in the White House: Bedini 1985.
93 *Megalonyx* as antagonist to mammoth: Jefferson 1799.
93 "In the present interior of our continent": Jefferson 1799.
93 *Megatherium* discovery and identification: Rudwick 1997, pp. 27–34.
94 "But the questions are simple": Simpson 1942.
95 "The Ohio animal": Rudwick 1997, p. 22.
95 "The nomadic peoples": Cohen 2002, p. 111.
95 "The revolutions of this globe": Rudwick 1972, p. 109.
95 "A world anterior to ours": Rudwick 1972, p. 109.

CHAPTER SIX: THE RISING

97 "For the trumpet will sound": 1 Corinthians 15:52.
97 *Incognitum* bones in 1782 set him on "irresistible bewitching" quest: Burns 1932.
97 "Everything that walks": Burns 1932.
99 Peale hired neighborhood men to carry "a long string of Animals" to new museum in 1794: Sellers 1980, pp. 79, 80.
99 "The Birds and Beasts will teach thee!" and "we are proud": Brigham 1996.
99 Cuckoos "are faithful and constant to each other": Brigham 1996.
99 "An animal of uncommon magnitude": Sellers 1980, p. 113.
99 John Masten unearthed bones: Semonin 2000, p. 316.
99 Bought bones for $200: Sellers 1980, p. 128.
99 "My heart jumpt with joy" and public reaction: Sellers 1980, p. 129.
100 "Exhuming the First American Mastodon": Charles Willson Peale, 1806–1808, oil on canvas, 49 by 61½ inches, The Maryland Historical Society, in Sellers 1951.
100 "Provided I do not degrade": Sellers 1969, p. 305.
100 "Numberless tryals of puting first one piece, then another": Sellers 1969, p. 299.
100 "My Molatto Man Moses": Shaw 2005.
101 Achieved a "second creation": Sellers 1969, p. 299.
101 "TEN THOUSAND MOONS AGO": Semonin 2000, p. 329.
102 Eagerness to "see the great American wonder": Sellers 1969, p. 300.
102 "Mammoth" retailing: Sellers 1969, p. 300.
102 John Adams's warning: Sellers 1969, p. 300.
102 Rembrandt's publicity stunt: Sellers 1969, pp. 301–302.
103 Cuvier's "resurrection of such a spectacular zoo": Rudwick 1972, pp. 102–113.
103 "And the dead shall be raised imperishable": 1 Corinthians 15:52, see Rudwick 2005, pp. 423, 424.
104 Cuvier "As a new species of antiquarian": Adams 1969, p. 146.
105 "The form of the tooth leads to the form of the condyle": Coleman 1964, p. 120.

106 *Paleotherium, Anoplotherium,* and "this foot is made for this head": Coleman 1964, pp. 123–125.

106 Cuvier "departed significantly from the proclaimed theoretical principles": Coleman 1964, p. 125.

107 Cuvier's paper "On the great mastodon": Cohen 2002, p. 100.

107 Cuvier dismissed the idea that "all species may change into each other": Gould 2002, p. 491.

108 "Living organisms without number have been the victims": Cohen 2002, p. 122.

108 Cuvier like "the demiurge of a new creation": Cohen 2002, p. 122.

108 Audubon "Thus have I described Cuvier": Adams 1969, p. 159.

109 Honoré de Balzac, "a novelist trying to look like a scientist": Somerset 2002.

109 *La Peau de Chagrin*: Balzac 1891, p. 26.

CHAPTER SEVEN: THE RIVER ROLLING WESTWARD

111 "The time will arrive": Stroud 1992, p. 127.

112 Crossbills not a defect: Wilson 1839, p. 292.

112 Wilson's "brilliant eye": Burns 1909.

112 Wilson described less kindly, and their encounter: Audubon 1832, p. 438ff.

113 "If I have been mistaken": Allen 1951.

114 "I must trudge by myself": Stroud 2000, p. 104, 327n.

114 "Science or literature": Audubon 1832, p. 440.

114 "The Athens of America": Stroud 1992, p. 28.

115 *Western Engineer* dimensions: Bell 1957, p. 31.

115 "Some peculiarities": James 1823, p. 4.

115 Reporter's description of the vessel: Wood 1966, pp. 63, 64.

115 "Lawless and predatory bands of savages": Greenfield 1992, p. 106.

115 Description of artillery and two "propelling powers": T. R. Peale's "Long Expedition Journal" ms., American Philosophical Society, Philadelphia, PA.

116 Calhoun says "conciliate Indians," record "everything": Wood 1966, p. 74.

116 The "magnificence of the forests": Evans 1997, pp. 30–32.

117 *Triton lateralis*: James 1823, p. 301.

117 Rafinesque's *Necturus maculosus* prevailed: Evans 1997, p. 27.

117 Rafinesque as "lone, friendless": Boewe 2003, p. 19.

118 Nature as "a beautiful and modest woman": Rafinesque's "Lecture on Knowledge" ms., American Philosophical Society, Philadelphia, PA.

119 "His prurient desire": Agassiz in *American Journal of Science and Arts*, May 1854.

119 "The wild effusions": Boewe 2003, p. 145.

119 Audubon's account of Rafinesque's 1818 visit: Audubon, 1832, p. 455ff.

120 Sleeping beneath mastodon skeleton: Evans 1997, p. 21.

120 "The greatest botanist in the world": Duyker 1988, p. 66.

120 "The ridicule of the inconsiderate": Stroud 1992, pp. 35 and 36.

122 "Grisly bear": Stroud 1992, p. 29.

122 "Messrs. T. Peale": Evans 1997, p. 31.

123 "Proper rules, principles": Rafinesque's "Lecture on Knowledge" ms., American Philosophical Society, Philadelphia, PA.

123 Rafinesque "danced, hugged": Audubon 1832, p. 456.

123 Say on splitters: Stroud 1992, pp. 52, 274.

123 British naturalist admonished Audubon: Herrick 1917, vol. 2, p. 107.

124 Audubon's bullet-proof Devil-Jack Diamond fish: Herrick 1917, vol. 1, p. 293.

124 *Journal*'s apology and "indispensable" rejection: Stroud 1992, p. 53.

125 A style best left to the "lords of Europe": Stroud 1992, p. 63.

125 "Not even a loophole": Stroud 1992, p. 205.

125 "In many respects the most gifted": Fitzpatrick 1911, p. 52.

125 "The best field botanist of his time": Boewe 2003, p. 47.

125 "Very acceptable" descriptions of bats: Boewe 2003, p. 147.

126 "a fine illustration of incipient species": Boewe 2005, p. 246.

126 Rafinesque and Darwin: Boewe 2003, p. 118.

126 Darwin calls Rafinesque "a poor naturalist": C. Darwin to J. Hooker, December 29, 1860, Letter 3034, Darwin Correspondence Project.

127 "Takes botany from the multitude": Daniels 1967.

CHAPTER EIGHT: "IF THEY LOST THEIR SKALPS"

129 "The object of the expedition": Wood 1966, p. 73.

129 "If they lost their skalps": Stroud 1992, p. 75.

130 Oto chief "I am not so silly": James 1823, vol. 1, p. 212.

130 Tolerance of homosexuality: James 1823, vol. 1, p. 129.

130 Bison slaughter: James 1823, vol. II, p. 168.

130 Biddle calls expedition "chimerical & impossible" and criticizes Long: Wood 1966, p. 90.

131 Bell refuses to obey Long: Wood 1966, p. 111.

131 "Contemptible knavery": Rotella 1991.

131 Long ordered to be quick, and James fumes about it: Rotella 1991.

131 "Detestable parsimony": Wood 1966, p. 119.

131 "The Great American Desert": Stroud 1992, p. 122.

132 "The greatest of all privations": James 1823, vol. 3, p. 98.

132 Racing down "bends and hills," eating skunkish soup: Evans 1997, p. 174.

132 Robbed of notebooks, salvages iconic animals: Evans 1997, pp. 178, 179.

133 Chasing beetle before crowd of Kansa Indians: Stroud 1992, p. 87.

133 "boatload of knowledge": Stroud 1992, p. 175.

133 Seeing no benefit in science: Stroud 1992, p. 193.

134 Say careful to avoid making synonyms: Stroud 1992, p. 127.

134 "I made a narrow escape": T. Say to C. Wilkins, January 14, 1832, Thomas Say Collection, American Philosophical Society, Philadelphia, PA.

134 Mule deer "without doubt a new species": James 1823, vol. 2, p. 89.

137 "Silly Sanchos": Porter 1979.

137 "Say's heart will throb": Stroud 1992, p. 57.

137 "A real and palpable vision": Herrick 1917, vol. 1, p. 360.

137 "Basely traduced the character" and "wretched collection of engraved birds":
     G. Ord to C. Waterton, July 20, 1831, George Ord Collection, American Philo-
     sophical Society, Philadelphia, PA.

138 Ord blackballed Audubon: Rhodes 2004, p. 221.

138 Ord's slanderous eulogy: Lucy W. Say (Say's widow) to John L. LeConte, M.D.,
     June 25, 1860, J. L. LeConte Correspondence, American Philosophical Society,
     Philadelphia, PA.

139 Credit for *Cyttaria darwinii*: Boewe 2005, pp. 120, 124.

139 Sale of Rafinesque's estate as trash: Haldeman 1842.

139 "Absently endangering himself": Thoreau 1980.

139 "Quadruped; seen by starlight": Cooper 1881, p. 85.

141 Say's description of *Canis latrans*: James 1823, vol. 1, p. 332.

CHAPTER NINE: THE BURDEN OF SPECIMENS

143 "A science of dead things": Gosse 1851, p. v.

143 Leach's eccentricities: Edwards 1969, pp. 575, 576.

143 "He valued it at fourscore thousand": MacGregor 1994, p. 48.

145 In the crypts of Montagu House: Günther 1975, p. 59.

146 "A world in miniature" and "ARSNIC POISON": Sellers 1980, pp. 26–28.

146 "Righteously elated": Elman 1977, pp. 65, 66.

146 Loading a double-barreled shotgun: Hollerbach 1996.

147 "The animal is immediately freed from pain": Graves 1818, p. 128.

147 "Steeping seeds" and horsehair nets: Graves 1818, p. 135.

147 Audubon on getting birds intoxicated: Rhodes 2004, p. 191.

147 "Bat fowling": Graves 1818, p. 21.

147 Audubon kills a golden eagle: St. John 1864, pp. 197–199.

148 "The familiar stomp and swat": Hollerbach 1996.

148 "And the poor beetle," with methods for killing insects: Lettsom 1779,
     pp. 7, 8.

148 "Because it bleached bright colors": Hollerbach 1996.

149 "Never put away a bird unlabelled": Coues 1890, p. 51.

149 Destroyed by "ravenous Insects": Farber 1977.

149 A field station in Jamaica: Gosse 1851, pp. 235–239.

150 Packing done, a naturalist "throws himself on his bed": Mearns and Mearns
     1998, p. 40.

150 "Camphorated spirit of wine" and "work of devastation": Prince 2003, p. 31.

151 "Our progress was often held up": Humboldt 1996, p. 8.

CHAPTER TEN: ARSENIC AND IMMORTALITY

153 "When a visitor shall gaze upon it": Lloyd 1985, p. 35.

153 Father of modern taxidermy: Lloyd 1985, p. 39.

153 A "unique dodo": Aldington 1949, p. 65.

153 First nature preserve: Lloyd 1985, p 32.

153 Anti-pollution action: Edginton 1996, p. 167.

154 Waterton's austere life: Waterton 1909, pp. xiv, 131; Hobson 1867, p. 155; Edginton 1996, p. 12.

155 The woodnymph hummingbird, *Thalurania watertonii*: Edginton 1996, p. 58.

155 Experiments with curare: Aldington 1949, pp. 70, 71.

155 Waterton's criticism of taxidermy, his own sacramental approach, and a sparrow's wonted pertness: Waterton 1909, pp. 315–320.

156 Retaining its pristine form: Waterton 1909, p. 331.

156 Mercury on the top hats and pants: Lloyd 1985, p. 35.

156 Waterton's mercurial behaviors: Edginton 1996, p. 66.

156 The "Nondescript" and monkeys stepping into our shoes: Hobson 1867, pp. 266–368.

157 Burchell's wagon travels in South Africa: Burchell 1822, pp. 149–152; Pickering 1998.

157 Hunters who added to his collection: Davies and Hull 1983.

157 Burchell's collections packed up in chests: Burchell to Swainson, September 27, 1819, Swainson Correspondence, Linnean Society of London.

158 "Do contradict so groundless a story": Burchell to Swainson, August 31, 1825, Swainson Correspondence, Linnean Society of London.

158 Government would not take him by the hand: Burchell to Swainson, February 28, 1831, Swainson Correspondence, Linnean Society of London.

158 John E. Gray visits: Davies and Hull 1983.

159 Other collections also remained locked up, undescribed: Koerner 1999, p. 169; Philbrick 2004, p. 17; C. Darwin to C. S. Darwin, October 24, 1836, Letter 313, Darwin Correspondence Project.

159 Grieving at the sight of unpacked boxes: Burchell to Swainson, October 3, 1839, Swainson Correspondence, Linnean Society of London.

160 French physician warns of "a dreadful poison": Farber 1977.

160 "Bloated piñatas": Prince 2003, p. 15.

160 Camper complains about chimps: A translation of Camper in the Richard Owen Papers, Royal College of Surgeons, London.

162 Taxidermy "wrong at every point": Prince 2003, p. 18.

162 The bird "wildly speaking": J. Nuttall to W. Swainson, August 4, 1819, Swainson Correspondence, Linnean Society of London.

162 Bécoeur: Rookmaaker 2006.

163 Taxidermy no longer a problem and its effect on science: Farber 1977.

164 Waterton against arsenic: Aldington 1949, p. 76.

164 Arguing that arsenic not so dangerous: *Random Notes on Natural History*, vol. 1, no. X (Providence, RI, Southwick & Jencks, October 1884), p. 3.

165 Taxidermy as magic and the hummingbird ecstacy: Pascoe 2005, pp. 40–44.

166 "Promethean boldness": Waterton 1909, p. 316.

CHAPTER ELEVEN: "AM I NOT A MAN AND A BROTHER?"

167 "I am haunted by the human chimpanzees I saw [in Ireland]": Curtis 1968, p. 84.

167 Darwins in antislavery movement: Desmond and Moore 2009, p. 2.

167 Visit to Tierra del Fuego: Darwin 1909, pp. 218, 219.

168 "Could our progenitors": Darwin 1909, p. 507.

168 Darwin 1838 Notebook C: [Transmutation of species (1838.02-1838.07) CUL-DAR122, The Complete Works of Darwin Online.

168 On the color of skin, shape of nose: Park 1878, p. 50.

169 "Connexions" between species: J. Children to W. Swainson, July 11, 1831, Swainson Correspondence, Linnean Society of London.

170 One of first apes in European literature: Schwartz 1988, p. 7.

171 Tulp's illustration of ape: Schiebinger 1993, p. 100.

171 "Dedicatory epistle": Tyson 1699.

171 Firsthand knowledge and *Homo troglodytes*: Schiebinger 1993, pp. 78–80.

172 "Scarcely any mark" to distinguish man from ape: Schiebinger 1993, p. 80.

173 "Broad assertion" based on seeing a single ape: Greene 1961, p. 184.

173 "By day hides; by night it sees": Greene 1961, pp. 184, 185.

173 The "remarkably adhesive term *Caucasian*": Jordan 1974, p. 101.

173 "An universal freckle": Greene 1961, pp. 223–224.

173 Kant on dephlogisticating: Greene 1961, p. 234.

174 "Foxes, bears, hares" turn white: Smith 1810, p. 78.

174 "The peculiar deformities of the African race": Smith 1810, p. 145.

174 Becoming white: Fredrickson 1987, p. 72.

174 "The influence of the state of society": Stanton 1960, p. 12.

174 Henry Moss: Stanton 1960, pp. 6, 7.

175 James Cowles Prichard: Greene 1961, p. 242.

176 Camper and skull shape: Meijer 1999, p. 167ff.

176 Comparing skulls to Dutch and Flemish art: Greene 1961, p. 190.

176 Measured facial angles: Schiebinger 2004, p. 149.

176 "I advance it therefore": Jefferson 1853, p. 155.

177 David Hume quote: Jordan 1974, pp. 109, 110.

177 Charles White quote: Jordan 1974, p. 200.

177 Robert Schomburgk: Riviere 1998.

CHAPTER TWELVE: CRANIOLOGICAL LONGINGS

179 "You have often asked for an Ishogo skull": Du Chaillu 1871, p. 67.

180 Owen's anatomical magic with moa: Gruber 1987.

180 Owen's tower story: Owen 1845, p. 300ff.

182 "Some Sculls of Negroes": Irmscher 1999, p. 303.

182 A barrel of skulls: J. Cassin to T. Savage, June 18, 1843, in Savage 1843.

182 Paying $3 for skulls: Vaucaire 1930, p. 160.

182 Banks on Maori face tattooing: Banks 1962, pp. 13, 14.

182 Moko tattoo: Robley 2003, pp. 169, 170.

183 Skulls "in a highly unsavory" condition: Dain 2002, pp. 199, 200.

183 Morton "of no arrogant pretensions": Humboldt and Alexander 1852, p. 153.

183 Measuring skulls with mustard seed and No. 8 shot: Wyman 1868.

183 No scientific man had a higher reputation: Humboldt and Alexander 1852, p. 153.

183 "He didn't openly . . . species": Michael 1988, pp. 349–354.

184 In his 1844 book: Stanton 1960, p. 99.

184 Gould reanalyzed the data: Gould 1978, p. 503.

185 A more recent study: Michael 1988.

185 Scholars have found: Fredrickson 1987, p. 77.

185 Elevating racial repugnance: Jordan 1974, p. 187.

185 Jefferson quote: Jordan 1974, p. 180.

185 Audubon: Rhodes 2004, pp. 4–6, 21.

186 Disgusted by "the Citron hue": Nobles 2005, p. 8.

186 Samuel Morton quotes: Morton 1847.

186 Morton's separate species conclusion: Morton 1847, p. 211.

186 He accepted a Russian account: Stephens 2000, p. 168.

187 William Stanton on Morton: Stanton 1960, p. 115, 116.

187 Pickering's travels: Stanton 1960, p. 93.

187 U.S. Ex Ex brought back: Philbrick 2004, p. 332.

187 James Dwight Dana: Philbrick 2004, p. 343.

188 Pickering was impressed: Stanton 1960, p. 93.

188 Pickering counted not 5: Pickering 1854, p. 2.

188 Nothing had been "modified or moulded by Climate": Stanton 1960, p. 93.

188 "Amorphous as a fog": Desmond and Moore, 2009, p. 120.

188 "Niggerology": Fredrickson 1987, p. 78.

189 "Their highest civilization" and "the most contented population": Nott 1849, pp. 42, 20.

189 Morton's "great pleasure": Stanton 1960, p. 121.

189 "Parson skinning": Gossett 1997, p. 64.

189 "The whole field is open": Stanton 1960, p. 118.

189 Cutting natural history loose from the Bible: Nott 1849, p. 7.

189 Nott depicts self in "the last great battle": Dain 2002, p. 221.

189 Bachman's studies: Stephens 2000, p. 176.

189 His species discoveries: Wayne 1906, p. 227.

190 Bachman on swans, human skeletons, nightingales, domestic cattle: Bachman 1850, pp. 21, 22, 25, 126, 127.

190 Nott wants to kill Bachman as warning against "all such blasphemies": Stephens 2000, pp. 186, 192.

191 "Majestic in his wrath": Douglass 1897, pp. 44, 45.

191 Douglass's commencement address: Brotz 1966, p. 226ff.

192 Spy's faint praise and "It could speak": Chesebrough 1998, p. 46.

CHAPTER THIRTEEN: "A FOOL TO NATURE"

193 "The enthusiasm of naturalists": Swainson 1834, pp. 119, 120.

193 Thomas Edward's early life, love of "beasts": Smiles 1879, p 1ff.; for perspective on how Smiles idealized Edward, see Secord 2003.

193 Horseleech, jackdaw, and moles: Smiles 1879, pp. 28, 25, 58.

194 Meeting badger, cats: Smiles 1879, pp. 96, 100.

195 The persistent weasel: Smiles 1879, pp. 111–113.

195 Knowing nothing about Malthus: Smiles 1879, pp. 88, 89.

195 Books promised, but not sent: Smiles 1879, p. 293.

195 Suicide attempt: Secord 2003.

196 "A fool to nature": Smiles 1879, p. 383.

196 Mosses, madrepores, limpet, and alligator fevers: Barber 1980, p. 13.

197 "Wandering in Epping Forest at dead of night": Kingsley 1890, p. 5.

197 A stonemason "fit company for dukes": Kingsley 1890, pp. 6–7.

197 Affluent Victorians and "that vampire, *ennui*": Barber 1980, pp. 19, 20.

198 The role of specimen dealers, and growth of museums: Barrow 2000.

199 "An English fool, named Smithson": G. Ord to C. Waterton, November 21, 1842, George Ord Collection, American Philosophical Society, Philadelphia, PA.

199 "Some of my Botanic enthusiasm": J. Barratt to T. Savage, March 31, 1841, in Savage 1843.

200 "The value of Turpentine": J. Barratt to T. Savage, April 18, 1841, in Savage 1843.

200 "The preservation of your health": A. Gould to T. Savage, January 8, 1842, in Savage 1843.

200 "All I should ask in the way of trouble": J. Wyman to T. Savage, June 25, 1850, in Savage 1843.

201 Edward Forbes on excursion: Wilson and Geikie 1861, pp. 389–391.

202 The iguanodon's encore: Lyell 1842, vol. 1, p. 193.

203 Edward Forbes's Valentine: Wilson and Geikie 1861, p. 387.

204 Natural history in Lear and Carroll, whimsy in Darwin, and nonsense as part of craze for discovery: Lecercle 1994, pp. 202, 203.

206 "Soulless utilitarians": Allen 1994, p. 74.

206 Birds to free us from insects: Barber 1980, p. 76.

206 Charles Babbage on "extensive manufactures": Allen 1994, p. 75.

207 A glassy window furnished by gracious Creator: Gosse 1851, p. 364.

207 Margaret Gatty's background: Sheffield 2004, p. 31.

207 The dragonfly ascends: Gatty 1908, p. 128.

207 "Why so much beauty": Bayne 1871, vol. 1, p. 393.

208 Darwin on Euclid: Dear 2008, p. 93.

CHAPTER FOURTEEN: THE WORLD TURNED UPSIDE DOWN

209 "What an old hen wd know": Desmond and Moore 1991, p. 94.

209 Life had "taken forms suitable to contemporary circumstances": Schwartz 1999.

209 "Individuals who possess not the requisite strength": Matthew 1831, p. 365.

210 Darwin on "not having discovered" the precedent: C. Darwin to C. Lyell, April 10, 1860, Letter 2754, Darwin Correspondence Project.

210 Matthew's calling card: http://www.ucmp.berkeley.edu/history/matthew.html, accessed March 23, 2010.

210 *Vestiges* as a *Victorian Sensation* full of "dangerous" ideas: Secord 2000, p. 1.

210 Cosmos began in primordial "fire mist": Secord 2000, p. 1.

210 "Ridiculous" notion of the Almighty having "to interfere personally": Chambers 1845, p. 116.

211 "Darwin had been scooped": Secord 2000, p. 429.

211 "Wondrous" constitution of the human brain: Chambers 1845, p. 232.

211 "Well, son of a cabbage": Secord 2000, p. 261.

211 Might we "actually *get our tails again?*": Secord 2000, p. 164.

212 Nightingale "as *Vestiges* would have it": Secord 2000, p. 444.

212 "The world cannot bear to be turned upside down": *Edinburgh Review*, July 1845.

212 "An iron heel upon the head of the filthy abortion": Secord 2000, p. 242.

212 "And the serpent coils a false philosophy": *Edinburgh Review*, July 1845.

212 "In the footprints of death": Chambers 1845, p. xv.

212 "A thunder-stroke of annihilation": Chambers 1845, p. xxvi.

213 A power "which might be turned to better purposes": Mills 1984.

213 Darwin disliked "that strange unphilosophical, but capitally-written book": C. Darwin to W. Fox, April 24, 1845, Letter 859, Darwin Correspondence Project.

213 "Geology strikes me as bad, & his zoology far worse": C. Darwin to J. Hooker, January 7, 1845, Letter 814, Darwin Correspondence Project.

213 A meal even the poorest laborer could afford: Schwartz 1999.

215 Chambers cited the example of bees manipulating their larvae: Chambers 1845, pp. 161, 162.

215 "From year to year, and from age to age,": Schwartz 1999.

216 Simpson credited *Vestiges* with "a contribution to the history of opinion, as distinct from that of ideas": Simpson 1959.

216 He felt "like a peacock": Desmond and Moore 1991, p. 208.

216 Darwin felt "like a duke": C. Darwin to C. Lyell, August 9, 1838, Letter 424, Darwin Correspondence Project.

216 "Complete scribbler; God help the Public": Darwin to C. Whitley, May 8, 1838, Letter 411a, Darwin Correspondence Project.

216 "As to a wife": C. Darwin to C. Whitley, May 8, 1838, Letter 411a, Darwin Correspondence Project.

217 "the Horneritas": E. Wedgewood to C. Darwin, November 30, 1838, Letter 447, Darwin Correspondence Project.

217 "For such facts": Barlow 1963.

218 John Herschel publicly endorsed natural process of creation, "in contradistinction to a miraculous": Cannon 1961.

218 "Checks" on population could provide "a force like a hundred thousand wedges": Darwin 1838, Notebook D, Complete Works of Darwin Online.

218 "If one is quiet in London": Darwin 1904, p. 269.

219 Anglican dons and the collapse of civilization: Desmond and Moore 1991, p. 321.

219 Undermining the stability of species would "undermine the whole moral and social fabric": Desmond and Moore 1991, p. 321.

220 No one has "a right to examine the question of species who has not minutely described many," "I will, life serving, attempt my work," and "My only comfort": C. Darwin to J. Hooker, September 10, 1845, Letter 915, Darwin Correspondence Project.

220 Difficult to tell apart, leaving him in a state of "inexplicable confusion": http://www.sulloway.org/Finches.pdf, accessed February 12, 2010.

221 "My life goes on like Clockwork and I am fixed on the spot where I shall end it": C. Darwin to R. FitzRoy, October 1, 1846, Letter 1002, Darwin Correspondence Project.

222 "Where does [your father] do his barnacles?": Desmond and Moore 1991, p. 407.

222 "I have been getting on well with my beloved cirripedia": C. Darwin to H. Hooker, May 10, 1848, Letter 1174, Darwin Correspondence Project.

222 "I hate a Barnacle as no man ever did": C. Darwin to W. D. Fox, October 24, 1852, Letter 1489, Darwin Correspondence Project.

222 "I have gnashed my teeth": C. Darwin to J. D. Hooker, September 25, 1853, Letter 1532, Darwin Correspondence Project.

223 Billiards "does me a deal of good, & drives the horrid species out of my head": C. Darwin to W. D. Fox, March 24, 1852, Letter 2436, Darwin Correspondence Project.

224 "I do not consider it a hasty generalization but rather as an ingenious hypothesis": Wallace 1905, p. 254.

224 "It furnishes a subject for every observer of nature to attend to": Wallace 1905, p. 254.

224 "Nothing ever stimulated my zeal so much": C. Darwin to A. R. Wallace, September 22, 1865, Letter 4896, Darwin Correspondence Project.

224 "Principally with a view to the theory of the origin of species": Wallace 1905, p. 256.

CHAPTER FIFTEEN: A PRIMATE NAMED SAVAGE

227 "The hideous monkey": Greene 1961, p. 174.

227 Savage letter of July 16, 1847: Wyman 1866.

227 Savage and Wyman present gorilla: Savage and Wyman 1847.

228 Wagstaff credited with gorilla discovery: http://www.bbc.co.uk/bristol/content/features/2001/01/05/bristol_treasures/episode1/bristol_treasures_fis-handeggs.shtml, accessed January 30, 2010.

229 "I . . . very much regret": Savage to Wyman, September 17, 1847, in Wyman 1866.

229 "Diabolical caricature": Owen to Wyman, July 24,1848, in Wyman 1866.

229 Savage background: Dunn 1992.

229 "The African's mind": Savage to Wyman, July 29, 1847, in Wyman 1866.

230 Driver ants: Savage 1847.

230 Susan's death, "blood of the martyrs": Library of Virginia Digital Collection, http://files.usgwarchives.net/va/fredericksburg/cemeteries/masonic.txt, accessed February 6, 2010.

230 Wyman background: Appel 1988.

230 The chimpanzee collaboration: Savage and Wyman 1844.

231 "Unexpectedly detained": Savage and Wyman 1847.

231 Wilson background: Dubose 1895.

231 "A monkey-like animal": Savage and Wyman 1847.

231 "An animal of an extraordinary character": Savage to Stutchbury, April 24, 1847, City of Bristol (England) Museum & Art Gallery.

232 "Marvellous accounts," description of gait, etc.: Savage and Wyman 1847.

232 "Three competitors" and the price he paid: Savage to Wyman, July 29, 1847, in Wyman 1866.

233 Hanno's voyage: http://www.metrum.org/mapping/hanno.htm, accessed February 1, 2010.

233 Establishing priority with brief notice: Wyman to Savage, July 30, 1847, in Savage 1843; Savage to Wyman, August 3, 1847, in Wyman 1866.

233 "Not a little surprised": Wyman to Savage, June 26, 1848, in Savage 1843.

234 "New Orang-Outang": Westwood 1847.

234 Stutchbury's gratitude: Stutchbury to Savage, September 10, 1842, in Savage 1843.

234 "Had I not taken": Savage to Wyman, May 22, 1848, in Wyman 1866.

234 Stutchbury asking Bristol captains for specimens: Owen 1848.

235 Owen "a gentleman": Savage to Wyman, May 22, 1848, in Wyman 1866.

235 *Troglogdyes savagei*: Owen 1848.

235 "A joint memoir": Wyman 1866.

235 "Flesh-creeping suggestiveness": Allen 1994.

236 "We had anticipated him": Wyman 1866.

236 "Clearly to light": Owen 1855.

236 Creating the genus *Gorilla*: Geoffroy Saint-Hilaire 1853.

236 Adding Wyman as co-describer: Coolidge 1929.

236 "Disagreeably human": Raby 1997, p. 182.

237 Comparing Negro and Orang: Savage and Wyman 1847.

237 "Mr. O'Rangoutang": Noakes 2002.

237 Owen's "archetypal light": Wilson 1996.

238 Gorilla ballet, quadrille: Kennedy and Whittaker 1976.

238 "Simple, modest, manly, true": Smallwood and Smallwood 1941, p. 346.

CHAPTER SIXTEEN: "SPECIES MEN"

241 Landslips (*terras cahidas*) on the Amazon: Bates 2002, p. 281.

241 The "thundering peal" of falling trees and "a horrible uproar": Bates 2002, p. 281.

242 Bates's family background: Anon. 1892.

242 Body covered with insect bites: Bates 2002, p. 278.

243 "Eleven of the best years of my life": Bates 2002, p. 10.

243 Death of Herbert Wallace: Raby 2002, p. 76.

243 Bates's parents visit Samuel Stevens: Raby 2002, p. 56.

244 Temperamental differences in Wallace-Bates split: Raby 2002, p. 45.

244 A profit of £27: Raby 1997, p. 81.

245 A child mimics flailing entomologist: Bates 2002, p. 301.

245 "A solitary stranger on a strange errand": Bates 2002, p. 283.

245 Bathing with alligators: Bates 2002, p. 288.

245 Snake encounters: Bates 2002, p. 113.

245 No danger from natives: Bates 2002, p. 287.

246 Saving collections before saving his own life: Bates 2002, pp. 99 and 100.

247 His shipments awaited in England with "intense interest": Anon. 1892.

247 "The iconographer of remarkable forms": Anon. 1892.

247 "Species men" vs. "species grubbers": H. Bates to C. Darwin, November 24, 1862, Letter 4138, Darwin Correspondence Project.

247 Alarming village with snake-mimicking caterpillar: Bates 1861, p. 509.

248 The Heliconids: Bates 1861, pp. 495–561.

249 Developing idea of Batesian mimicry: Bates 1861, p. 502.

249 "One of the most beautiful phenomena in Nature": Woodcock 1969, pp. 231, 232.

250 Starved for reading material: Bates 2002, p. 288.

250 River dolphins, bird calls, sudden screams, howler monkeys: Bates 2002, pp. 47, 90.

250 "Coyness is not always a sign of innocence": Bates 2002, p. 341.

250 "Mingled squalor, luxuriance, and beauty": Bates 2002, p. 12.

251 A "little career of looseness": Woodcock 1969, p. 157.

251 Showing wildlife images: Bates 2002, p. 256.

251 "Coldness of desire and deadness of feeling": Bates 2002, p. 275.

251 "The contemplation of nature": Bates 2002, pp. 288–289.

251 Eating smoked spider monkey: Bates 2002, p. 250.

251 Becoming an "oldish, yellow-faced man": Obituary 1892.

251 Anticipating dreary England: Bates 2002, p. 419.

252 Gray accuses him of "idleness": J. Hooker to C. Darwin, May 13, 1863, Letter 4165, Darwin Correspondence Project.

252 A poet with better social connections beats Bates for zoological job: Günther 1975, p. 177.

CHAPTER SEVENTEEN: "LABOURER IN THE FIELD"

253 "I'd be an Indian here, and live content": Wallace 1895, p. 180.

253 Sinking of the *Helen*: Wallace 1895, pp. 271–279; Wallace 1905, pp. 302–307.

254 "Now everything was gone": Wallace 1895, p. 278.

254 Wallace arrives back in England: Wallace 1905, p. 305ff.

255 His insurance payout and his sensible resolutions: Wallace 1905, p. 309.

255 "The young man in a hurry": Interview, *The Pall Mall Magazine*, March 1909.

255 His Zoological Society talk: Wallace 1852, full text at http://people.wku.edu/charles.smith/wallace/S008.htm, accessed February 13, 2010.

257 Gray's comment on Wallace's inadequate labels: Wallace and Berry 2003, p. 112.

257 Closely allied species in adjoining districts: Wallace 1855.

258 The connoisseurs of the country houses and the "labourer in the field": Raby 2002, p. 85.

259 Losing specimens when exact species is everything: Wallace 1895, p. 329.

259 "The most difficult and . . . interesting problem in the natural history of the earth": Wallace 1855.

259 Malay Archipelago as finest field for naturalist: Wallace 1905, p. 326; Wallace and Berry 2003, p. 113.

259 The need to visit the largest number of islands: Raby 2002, p. 144.

259 Holed up during the Borneo monsoon: Wallace 1905, p. 354.

260 Wallace proposes role of closely allied species in species creation: Wallace 1855.

260 Darwin sees "nothing very new": Desmond and Moore 1991, p. 438.

260 The country house connoisseurs grumble: Wallace 1905, p. 355.

261 Capturing an *Ornithoptera*: Wallace 1886, pp. 217, 218.

261 Aphids like white rabbits nibbling: *A. R. Wallace's Malay Archipelago Journals and Notebook*, Linnean Society of London.

261 "Wherever I go dogs bark," pack animals "stick out their necks": *A. R. Wallace's Malay Archipelago Journals and Notebook*, Linnean Society of London.

262 An islander indignant that Wallace comes from "Ung-lung": *A. R. Wallace's Malay Archipelago Journals and Notebook*, Linnean Society of London.

262 Wallace's reading: *A. R. Wallace's Malay Archipelago Journals and Notebook*, Linnean Society of London.

262 Scientific rules require listing errors: *A. R. Wallace's Malay Archipelago Journals and Notebook*, Linnean Society of London.

263 Running amok: *A. R. Wallace's Malay Archipelago Journals and Notebook*, Linnean Society of London.

CHAPTER EIGHTEEN: THE SLOW POWER OF NATURAL FORCES

265 "A struggle for existence," from Wallace's "On the Tendency of Varieties to Depart Indefinitely from the Original type."

265 Notebooks on species question: Desmond and Moore 1991, p. 438.

265 Lyell's antievolutionary convictions: Desmond and Moore 1991, p. 131.

265 Lyell's first notebook entry after Wallace article: Gould 1970, pp. 663, 664.

265 Wallace on whales: *A. R. Wallace's Malay Archipelago Journals and Notebook*, Linnean Society of London.

266 Lyell's "Lord Chancellor Manner": Gosse 2005, p. 90.

266 Description of Lyell: Desmond 1998, p. 153.

266 Lyell's view of evolution: Gould 1970, pp. 663, 664.

266 Wallace's thoughts on slow power of natural forces: Wallace 1855.

266 Darwin takes idea from Lyell: Recker 1990.

266 "In a small group of islands": *A. R. Wallace's Malay Archipelago Journals and Notebook*, Linnean Society of London.

267 Wallace's notes on natural selection: *A. R. Wallace's Malay Archipelago Journals and Notebook*, Linnean Society of London.

267 Lyell's visit to Down House: Lyell 1970, p. xliii.

267 Darwin's pigeons: Desmond and Moore 1991, p. 438.

267 His motives were the same: C. Darwin to P. Gosse, September 22, 1856, Letter 1958, Darwin Correspondence Project.

267 Pigeons as "most wonderful case of variation" and being "hand & glove" with . . . Spital-field weavers": C. Darwin to J. D. Dana, September 29, 1856, Letter 1964, Darwin Correspondence Project.

267 Pigeons as "the greatest treat": C. Darwin to C. Lyell, November 4, 1855, Letter 1772, Darwin Correspondence Project.

268 "Dreadful work" with pigeons: C. Darwin to T. C. Eyton, November 26, 1855, Letter 1784, Darwin Correspondence Project.

268 Fifteen species of pigeon: Lyell 1881, p. 213.

269 "An ourang": Lyell 1881, p. 213.

269 Gathering at Down House, and Lyell soon hears of it: Desmond and Moore 1991, pp. 436–439.

269 "More & more unorthodox": C. Lyell to C. Darwin, May 1–2, 1856, Letter 1862, Darwin Correspondence Project (with note on mosses and aldermen).

270 "The indefinite modifiability doctrine": Lyell 1881, p. 214.

270 Wallace sent Darwin specimens: Desmond and Moore 1991, p. 454.

270 Darwin warns Wallace: Desmond and Moore 1991, p. 455.

270 "We have thought much alike": C. Darwin to A. R. Wallace, May 1, 1857: Letter 2086, Darwin Correspondence Project.

270 "The mere description of species": C. Darwin to A. R. Wallace, May 1, 1857, Letter 2192, Darwin Correspondence Project.

270 Owen on anatomical evidence agreeing with what God has revealed: *The Methodist Quarterly Review*, August 1855, p. 168.

270 Owen's lectures at Linnean Society: Wilson 1996.

271 Huxley description: Desmond 1998, p. xiv.

271 Bringing down Owen's scheme: Desmond 1998, p. 240.

271 Birth of only child: Gosse 2005, p. 15.

271 Gosse prying shells from rocks: Gosse 1851, p. 33.

272 "You who have so watched all sea-creatures": C. Darwin to P. H. Gosse, April 27, 1857, Letter 2082, Darwin Correspondence Project.

272 Gosse's experience of species modification: Gosse 2005, pp. 85, 86, 76.

272 Great anticipations of success: Gosse 2005, p. 91.

272 God hid fossils: Gosse 1857, p. 335.

272   The moment of creation: Gosse 1857, p. 351.

273   Representation of Earth's past: Gosse 1857, pp. 45, 353, 337.

273   "One enormous and superfluous lie": Gosse 2005, p. 92.

273   Gosse as the "spoiled darling" of the public: Gosse 2005, p. 92.

273   The Gosses' "grim season" in Devon: Gosse 2005, p. 93ff.

274   Submarine gardens of fabulous beauty: Gosse 2005, p. 114.

275   Wallace in Gilolo reading Malthus: Raby 2002, p. 130ff.

275   Wallace's conception that "*the fittest would survive*": Wallace 1905, p. 362.

275   Wallace in Ternate planning trip to New Guinea: *A. R. Wallace's Malay Archipelago Journals and Notebook*, Linnean Society of London.

276   Birds of paradise quote: *A. R. Wallace's Malay Archipelago Journals and Notebook*, Linnean Society of London.

276   Darwin gets Wallace's 20-page letter: Desmond and Moore 1991, p. 466.

276   Text of Wallace's letter, and the entire Darwin-Wallace presentation: http://www.linnean.org/index.php?id=380, accessed February 12, 2010.

277   "Grand enough soul not to care": C. Darwin to J. Hooker, July 13, 1858, Letter 2306, Darwin Correspondence Project.

277   "All my originality": C. Darwin to C. Lyell, June 18, 1858, Letter 2285, Darwin Correspondence Project.

277   Darwin's daughter ill: Desmond and Moore 1991, pp. 470, 471.

277   Darwin's son dies of scarlet fever: Desmond and Moore 1991, p. 469.

277   Lyell and Hooker took the matter in hand: Raby 2002, p. 287–289.

277   Wallace's letter to his mother: Raby 2002, p. 141.

277   Wallace's "sudden intuition" and Darwin's "prolonged labours": Wallace 1905, p. 193.

278   "The accumulative action of Natural selection": C. Darwin to A. Gray, September 5, 1857, Letter 2136, Darwin Correspondence Project.

278   "Two indefatigable naturalists . . . conceived the same very ingenious theory": J. D. Hooker and C Lyell to the Linnean Society, June 30, 1858, Letter 2299, Darwin Correspondence Project.

279   Darwin buried his son: Desmond and Moore 1991, p. 470.

279   "Terrible wet weather": *A. R. Wallace's Malay Archipelago Journals and Notebook*, Linnean Society of London.

279   "I feel it & shall not forget it": C. Darwin to J. Hooker, July 13, 1858, Letter 2306, Darwin Correspondence Project.

279   Linnean Society's President's comments: Desmond and Moore 1991, p. 470.

280   "Just as noble a conception of Deity": C. Kingsley to C. Darwin, November 18, 1859, Letter 2534, Darwin Correspondence Project.

280   "The way a beetle's wing should fold": Roger 1997, p. 241.

280   A visit with the queen and the removal of coffins from crypt: R. Owen to C. Owen, February 25, 1859, Owen Papers, Royal College of Surgeons, London.

281   Darwin has "given the world *a new science*": Raby 2002, p. 151.

281   Bates's encounter with Lyell: Recounted by Edward Clodd in an 1893 edition of *Naturalist on the River Amazons*, pp. lxxi, lxxiii.

281 Darwin praises Bates's book: C. Darwin to J. Hooker, April 17, 1863, Letter 4103, Darwin Correspondence Project.

282 Frederick Jackson credits helper and Wallace "claimed the glory": Mearns and Mearns 1998, pp. 58, 59.

282 Discovering Wallace's Standardwing: Wallace 1886, p. 329.

282 "Discovered by myself": Wallace 1886, p. 563.

282 "Ingenious speculation": A. R. Wallace to C. Darwin, May 29, 1864, Letter 4514, Darwin Correspondence Project.

283 Naturalists do not own their discoveries: Kingsley 1855, p. 46.

CHAPTER NINETEEN: THE GORILLA WAR

285 "Suspicion of negro sympathies": *The New York Times*, July 6, 1861.

285 Talk at the Metropolitan Tabernacle: *The Times* [London], October 3, 1861.

287 Lincoln as "the original gorilla": Sears 1988, p. 132.

287 The "gorilla war": *The Athenaeum*, July 13, 1861, p. 51.

288 "Inexpedient" facts: Bucher 1979.

288 A biography that would have "blown the roof off": Bucher 1979.

288 Du Chaillu's background in Gabon: Bucher 1979.

288 John Cassin and endorsement of the Academy of Natural Sciences: Bucher 1979.

289 Du Chaillu's 8000 miles and 2000 birds: Du Chaillu 1868, p. x.

289 Du Chaillu as "their white man": Du Chaillu 1868, p. 286.

289 Mpongwe election rituals: Du Chaillu 1868, p. 43.

289 "Just as she would go to market and carry thence a roast or steak": Du Chaillu 1868, p. 74.

289 Bee-eaters and manatees: Du Chaillu 1868, p. 262, 451.

290 "My nerves firm as a rock": Du Chaillu 1868, p. 158.

290 Gorilla as "hellish dream creature": Du Chaillu 1868, p. 101.

290 Dispelling myths and gorilla as strict vegetarian: Du Chaillu 1868, p. 394.

290 "A very amusing book": *The Athenaeum*, May 11, 1861.

291 Gray attacks: *The Athenaeum*, May 18, 1861.

291 Gray as an "old malignant fool": C. Darwin to J. Hooker, May 15, 1863, Letter 4167, Darwin Correspondence Project.

291 And a slanderer: J. Hooker to C. Darwin, May 24, 1863, Letter 4169, Darwin Correspondence Project.

291 "You feed that dog too much": Günther 1975, p. 184.

291 Murchison praises Du Chaillu: Mandelstam 1994.

291 Owen on "The Gorilla and The Negro": *The Athenaeum*, March 23, 1861.

292 Huxley predicts "heart-burnings and jealousies": Mandelstam 1994.

292 Illustrations copied from other naturalists and other "falsifications": *The Athenaeum*, June 1, 1861.

293 Gray "luxuriously lodged in his museum": *The Athenaeum*, May 25, 1861.

293 The "gross personal attack": *The New York Times*, July 6, 1861.

293 The fight over Owen's purchase of Du Chaillu's specimens: Mandelstam
     1994.

293 An attempt "to humbug the scientific world": McMillan 1996.

295 Du Chaillu's origins: Vaucaire 1930; Bucher 1979.

295 Publisher obliged Du Chaillu to sensationalize: Akeley 1924, pp. 238, 239.

295 "Exceedingly queer English": Smith 1903.

295 Kneeland's slapdash style: Stratton and Mannix 2005, pp. 598, 599.

296 Du Chaillu's consciousness about race: Du Chaillu 1868, pp. 188, 303, 439.

296 Eating monkey "too much like roast-baby": Du Chaillu 1868, p. 80.

297 "Look at your white friend": Du Chaillu 1868, p. 330.

297 Waterton's suspicions: McCook 1996.

297 Ord blames African blood, "spurious origin" for "wondrous narratives": G. Ord
     to C. Waterton, October 20, 1861. George Ord Collection, American Philo-
     sopical Society, Philadelphia, PA.

297 Du Chaillu's breach with the Academy of Natural Sciences: McCook 1996.

298 Clodd on "Negroid face": Clodd 1926, p. 72.

298 The Octoroon: *The Athenaeum*, December 7, 1861.

298 Du Chaillu as a socializer, womanizer: S. Kneeland to M. de Guichainville,
     January 21, 1886, Samuel Kneeland Correspondence, American Philosophical
     Society, Philadelphia, PA.

299 De Morgan's ghastly joke: A. de Morgan to W. H. Dixon, October 1, 1861,
     Letters of Augustus de Morgan, American Philosophical Society, Philadelphia,
     PA.

299 Du Chaillu "hurt to the quick": Du Chaillu 1871, p. vi.

299 Du Chaillu back in Gabon: Du Chaillu 1871, p. 1.

299 Gorillas to British Museum: Du Chaillu 1871, pp. 65, 66.

300 Keeping his journal in triplicate: Du Chaillu 1871, p. xi.

300 Du Chaillu blamed for smallpox epidemic: Patterson 1974.

300 The final confrontation: Du Chaillu 1871, pp. 351–365.

301 Retreat to the coast: Du Chaillu 1871, pp. 397, 398.

301 Renewing the fight with Gray: Du Chaillu 1871, p. 117.

302 Du Chaillu's name prevails: Allman 1866.

302 Du Chaillu's discoveries: K. Helgen and R. Dowsett, personal communication.

302 Richard Burton's endorsement: Burton 1876, p. vii.

302 Mary Kingsley's tart advice: Kingsley 1897, p. 369.

302 Du Chaillu becomes "the ideal bachelor": Vaucaire 1930, p. 291.

303 Abbé André Raponda Walker: Walker 1960, pp. 85, 108, 132, 167.

CHAPTER TWENTY: BIG NOSES AND SHORT TEA-DRINKERS

305 "Thefts and murders": David 1949, p. 253.

305 Climbing Hong-chan-tin: David 1949, pp. 278–281.

306 The value of "the most minute objects": David 1949, xxi.

307 "Shaking the moral and intellectual world" and the study of animals "raised to so high a dignity: Blum 1993, p. 237.

308 Natural history museums everywhere: Mearns and Mearns 1998, p. 65; Farber 2000, p. 91; Fan 2004, p. 224; and Ripley 1970, pp. 149–152.

308 "The animals' Westminster Abbey": Wheeler 1993, p. 49.

308 "Exhaustive comprehensiveness" and the "grand vision of scientific commonweal": Fan 2004, pp. 89, 90.

309 Collecting in the marketplace: Fan 2004, pp. 26–29.

309 Drawings as basis for 83 new fish species: Fan 2004, p. 56.

309 Far Cathay: Haddad 2008, p. 51.

309 "Many short tea drinkers": Fan 2004, p. 13.

310 The British natural history network in China: Fan 2004, p. 67ff.

311 Chinese depicted as "mean, jealous, and cunning": Fan 2004, p. 86.

311 Natural history as "a respectable hobby": Fan 2004, p. 74.

311 David's dates in China: David 1949, pp. xii, xv.

311 "These poor creatures, who revel in life so joyously": David 1949, p. 177.

311 Eating tsamba: David 1949, pp. 84, 85, 104, 175.

312 "The saving of infidels": David 1949, p. xviii.

312 Put to work as a biological explorer: David 1949, p. xx.

312 His expeditions: David 1949, pp. vii, viii.

312 "It would be necessary to renounce all traveling": David 1949, p. 6.

312 "The haunts of thieves and malefactors": David 1949, p. 253.

313 "Endowed with extraordinary talents": David 1949, pp. 6–8.

314 Making "a salutary impression": David 1949, p. 77.

314 Hostility to foreigners, suspected of being a spy: David 1949, pp. 278, 49.

314 The stones of Wanhsien: David 1949, p. 228.

314 Europeans "dangerous and evil beings"and "the most deplorable hospitality": David 1949, pp. 58, 42, 43.

314 Frozen dung and "I am used to everything": David 1949, pp. 67, 84.

314 His species discoveries: David 1949, p. xxx; and Mearns and Mearns 1998, pp. 275, 282.

315 "The marvels of the hand of God": David 1949, p. 29.

315 The "famous white and black bear": David 1949, p. 276.

315 "Its stomach is full of leaves": David 1949, pp. 283, 284.

315 Early Chinese knowledge of pandas: Schaller 1994, pp. 61, 62.

316 Seeing pandas in the wild: Wildt 2006, p. 8.

316 Golden monkey "thought to be an imaginary animal": David 1949, xxix.

316 Muping prince as tyrant: David 1949, p. 285.

317 Petulance toward his former pupil: David 1949, p. 290.

317 Till China becomes Christian: David 1949, p. xvi.

317 A natural aptitude for business: David 1949, pp. 182, 183.

317 "One hears the hatchet": Schaller 1994, p. ix.

318 The Earth given over to horses, pigs, potatoes: David 1949, xxii.

318 What the Chinese knew about natural history: Fan 2004, pp. 105, 106.

319 Père David's deer: David 1949, p. 6.

319 Recovery of deer: Jiang & Harris 2008.

CHAPTER TWENTY-ONE: INDUSTRIAL-SCALE NATURAL HISTORY

321 The Rothschilds as astonishing organisms: Butler 1817, p. 52.

321 "A great stuttering bear": Rothschild 2009.

321 Walter Rothschild's height and weight: Rothschild 1983, p. 53.

321 His snoring, the 20-odd bedrooms: Rothschild 2009; Rothschild 1983, p. 288.

321 Walter's insouciance about other peoples' problems: Rothschild 1983, p. 87.

321 Walter on Miriam's being "completely *square*": Rothschild 1983, p. 290.

322 Walter receiving his visitor stark naked: Rothschild 1983, p. 101.

322 Financial ineptness, blackmail, laundry baskets: Rothschild 1983, p. 92.

323 Walter's life "a series of flying leaps": Rothschild 1983, p. 224.

323 Walter at university, his travels, visitors to his museum: Rothschild 1983, p. 101.

323 Description of Walter's collection: Rothschild 1983, pp. 2, 121.

323 His "egotistical ruthlessness": Rothschild 1983, p. 110.

323 Tring as "a model of how to collect": Rothschild 1983, p. 120. (But the quote requires a caveat. It came from Richard Meinertzhagen, who turned out to be an ornithological hoax artist.)

323 Tring's 400 collectors: Rothschild 1983, p. 155.

323 Charles's 100 flea collectors: Rothschild 1983, p. 336.

323 Charles stops train for butterfly: Rothschild 1983, p. 174.

323 Map of collectors like "severe attack of measles": Rothschild 1983, p. 155.

324 Survey expeditions want "crates of specimens": Kohler 2006, pp. 8, 110, 111.

325 The parrot's commentary: Fell 1950.

325 "Drudging": Campbell 1877, p. 482.

325 Campbell's complaints about naturalists: Campbell 1877, p. 494.

325 "One of the most remarkable faunas": Fell 1950, p. 78.

326 *Challenger* expedition and *Apollo* moon landing: Corfield 2003, p. 7.

326 Forbes's proposed "azoic" zone: Corfield 2003, p. 2.

326 "The land of promise" and "endless novelties": Thomson 1873, p. 49.

326 Victorian belief in unchanging seafloor: Corfield 2003, p. 28.

326 Crinoids off Portugal: Corfield 2003, p. 27.

326 Publishing 50 volumes: Hedgpeth 1946.

326 Birth of oceanography: Hedgpeth 1946.

327 Americans treating nature as an obstacle: De Tocqueville 1904, p. 559.

327 Loggers and hunters from "heroes" to "pirates": B. Edmonson, "Environmental Affairs in New York State: An Historical Overview," http://www.archives .nysed.gov/a/research/res_topics_env_hist_nature.shtml, accessed January 31, 2010; *The New York Times*, September 25, 1889.

327 Thomas Say's early push to protect bison: James 1823, vol. 2, p. 482.

327 Audubon Society and killing birds for feathers: Allen 1886.

327 Walking mania pioneered by clubs: Kohler 2006, pp. 48, 52.

327 Camping and summer camps: Kohler 2006, pp. 50, 51.

328 "If you cannot interest the visitor": Lucas 1914.

328 Deterioration of type specimens: Kohler 2006, p. 108.

328 Concentrating on a few striking exhibits: Kohler 2006, pp. 107–110.

329 "New museum" movement: Alberti 2001.

329 "The glory of the field naturalist has departed": Allen 1998; Grove 1892.

329 Yorkshire's 33 field clubs and Louis Compton Miall: Alberti 2001.

330 "Walter Rothschild's ambitions": Günther 1975, p. 426.

330 Rothschild's large museum displays: Gray 2006.

330 Walter's fixation on numbers, "record bags": Rothschild 1983, pp. 71, 163.

330 Walter training zebras: Rothschild 1983, p. 108.

331 Walter takes kiwis to college: Rothschild 1983, p. 70.

331 His "freakishly retentive" memory: Rothschild 1983, p. 122.

331 "That bird is illustrated": Rothschild 1983, p. 158.

331 Walter's 60 living Galapagos tortoises: Günther 1975, p. 426, 427.

332 "Saving species for science" or "hastening" into extinction: Mearns and Mearns 1998, p. 124.

332 Newton warns against unstocking the world: Rothschild 1983, p. 77.

332 "I have no duplicates in my collection": Rothschild 1983, p. 138.

332 Walter wanted a long series of specimens: Rothschild 1983, p. 151.

332 "COME AND SEE WHAT WE'VE GOT": Rothschild 1983, p. 132.

333 "Bewitched" by cassowaries: Rothschild 1983, p. 291.

333 Galapagos petrels: Rothschild 1983, p. 158.

333 "Tree kangaroos in Hyde Park: Rothschild 1983, p. 74.

333 Facing danger and forgetting about it: Meek 1913, p. x.

333 Collecting butterflies with spider webs and Queen Alexandra's Birdwing: Meek 1913, pp. 172–174.

334 Solitary Christmas celebration: Letter from W. I. Anrope (?) to a Miss Bowdler Sharpe, November 21, 1908, M. H. Kingsley papers, British Library.

334 "Motor cars" delirium: Meek 1913, p. 172.

334 Packing three times as much into a display: Rothschild 1983, p. 108.

335 Preferring a new butterfly to great wealth: Meek 1913, p. 142.

335 Needing to come back to Rothschild's collection: Rothschild 1983, pp. 138, 139.

335 "The panorama of modern research": Rothschild 1983, p. 152.

336 The blackmailing aristocrat: Rothschild 1983, p. 42.

336 " 'My collection!' " : Stresemann 1975, p. 268.

336 Walter "expunged from the archives": Rothschild 1983, p. 316.

336 Walter lives on in species named for him: Rothschild 1983, p. 217.

CHAPTER TWENTY-TWO: "THE BLESSING OF A GOOD SKIRT"

337 T. H. Huxley in an egalitarian mood: *The Times* [London], July 8, 1874.

337 Seeing big game at close quarters: Kingsley 1897, p. 268.

337 Kingsley hiking ahead of her party: Kingsley 1897, p. 209.

338 Roger Fortune's pit trap: Bishop 1996, p. 75.

338 Kingsley on "the blessing of a good skirt": Kingsley 1897, pp. 269, 270.

338 Caroline Owen quotes: Owen 1894, pp. 121, 296.

339 A kind of "intellectual stag-party": Allen 1994, pp. 150, 151.

340 "Cultivated prurience": Barber 1980, pp. 132, 133.

340 Women, British science, and marsupial sex: Sheffield 2001, pp. 67, 79.

340 Owen skittish about Mammalia: Barber 1980, p. 133.

340 Darwin's barnacle penis: Stott 2003, p. 220.

340 "Aunt Etty" eradicating *Phallus impudicus*: Raverat 1952, p. 135.

341 Beatrix Potter's experiences as a scientist: Sheffield 2004, p. 87; Gates 1999, pp. 84, 85.

341 Mary Anning: Barber 1980, pp. 126, 127.

341 "This story being told by a man": M. Kingsley to A. Macmillan, December 18, 1894, Macmillan Archive, British Library.

342 Kingsley wore brother's trousers: Lloyd 1985, p. 22.

342 "The sea-serpent of the season": Kingsley 1897, p. xi.

342 Watching gorillas instead of shooting them: Kingsley 1897, p. 268.

342 Kingsley's opinion of African hosts: Kingsley 1897, pp. 488, 500.

342 "The evil" done "by misdirected missionary effort": Kingsley 1897, p. 508.

343 "Make man see bush": Kingsley 1897, pp. 101, 102.

343 "One absolutely new fish" testifies she is not a windbag: M. Kingsley to A. Macmillan, February 18, 1896, Macmillan Archive, British Library.

344 "The very air itself seems green": M. Kingsley to B. C. Skeat, October 25, 1893, M. H. Kingsley Papers, British Library.

344 "Nothing hinders a man half so much as dying": Kingsley 1897, p. 515.

345 Feeling like an accessory to murder: Kingsley 1897, pp. 519, 520.

345 Precautions against disease: Kingsley 1897, pp. 519, 520.

345 Kingsley's death: Lloyd 1985, p. 146.

345 Leakey's cable to Goodall: Peterson 2006, p. 212.

346 Young woman's letter about monkeys: Rolt 1957, p. 93.

346 Elephant researchers: Joyce Poole, personal communication.

CHAPTER TWENTY-THREE: THE BEAST IN THE MOSQUITO

347 "The study of a certain fauna and flora": Li 2002.

347 A small experiment with mosquitoes: Haynes 2000, pp. 51, 52.

348 Malaria, the catchall: Haynes 2000, p. 43.

348 "An out of the world place": Cook 1992, p. 70.

348 Feeling "the need of books": Chernin 1983.

349 Manson's marriage: Cook 1992, p. 71.

349 British researcher identified worms: Haynes 2001, p. 9, 54, 55.

349 Manson's medical training: Li 2002.

349 "This formidable" animal: Haynes 2001, pp. 54, 55.

349 Manson working with filarial worms in dogs: Li 2002.

349 "Cause and consequence": Haynes 2001, p. 50.

349 Johann Steenstrup's work: Farley 1972.

350 "A complete chaos": Farley 1972.

350 Dogs and lice, sheep and snails: Kershaw 1963.

350 Manson's decision to work with mosquitoes, and flaws in his approach: Haynes 2001, pp. 51, 52.

351 "I tore off its abdomen": Cook 1992, p. 72.

351 Manson discerns what happens to microfilariae: Haynes 2001, p. 62.

351 "On the road to a new human host": Cook 1992, p. 74.

351 "A new and revolutionary concept": Chernin 1983.

352 "Father of modern tropical medicine": Li 2002.

352 A migration within the human body: Chernin 1983.

352 The body's fauna, each in its "peculiar territory": Li 2002.

352 Darwinism makes these discoveries more tolerable: Li 2002.

352 "It is marvelous" how filaria are adapted to mosquitoes: Chernin 1983.

352 The drinking water hypothesis: Chernin 1983, Kershaw 1963.

353 "Crazy Pat Manson": L. O. Howard article for *Engineering News Record*, L. O. Howard Papers, American Philosophical Society, Philadelphia, PA.

353 Minkh's lice theory: Busvine 1976, p. 237.

353 Death statistics during Thirty Years War: J. M. Conlon, "The Historical Impact of Epidemic Typhus," http://entomology.montana.edu/historybug/TYPHUS-Conlon.pdf, accessed February 12, 2010.

353 Carlos Juan Finley: Del Regato 2001.

CHAPTER TWENTY-FOUR: "WHY NOT TRY THE EXPERIMENT?"

355 "No malaria without Anopheles": Capanna 2006.

355 Natural history provides templates for innovative research: Li 2002.

355 Edward Jenner and the cuckoo: Scott 1974.

356 "Why not try the Expt?": J. Hunter to E. Jenner, August 2, 1775, Hunter-Jenner Letters, Royal College of Surgeons, London.

356 A toad specimen: Wells 1974.

356 Preparing Jenner's mind: Wells 1974.

356 Thanks from Canadian Indians: Wells 1974.

356 James Paget's discovery at 1835 autopsy: Cobbold 1879, pp. 6–8.

357 "A small worm coiled up": Paget 1866.

357 Richard Owen stealing glory: http://www.trichinella.org/history_1.htm, accessed March 23, 2010.

357 Plague: Busvine 1976, pp. 230, 231.

357 Manson quote: Li 2002.

357 Plague discovery: Busvine 1976, pp. 230–233; see also http://www.ncbi.nlm.nih.gov/pmc/articles/PMC1124411.

358 Manson takes up malaria question in 1894: Haynes 2001, p. 83.

358 Laveran's discovery of *Plasmodium malariae:* Haynes 2001, p. 87–91.

359 "Unwilling to cause a needless death": Haynes 2001, p. 90.

359 Manson would infect son: Guilleman 2002.

359 "Microbes of a very special morphology": Haynes 2001, p. 199.

359 *Plasmodium malariae* shape-shifts: Haynes 2001, pp. 88–92.

359 Manson on continental competitors: Haynes 2001, p. 95.

359 Mosquitoes transmitting malaria: Manson 1894.

359 Manson applied for expedition: Haynes 2001, p. 101.

359 "Sneering skeptics" view him as a "pathological Jules Verne": Haynes 2001, pp. 108, 121.

360 Ronald Ross begins to work with Manson: Haynes 2001, pp. 100–104.

360 Parasites don't die in mosquitoes' stomachs: Haynes 2001, p. 106.

360 Ross's discovery of oocyst in stomach of unknown mosquito species: Capanna 2006.

360 Ross reported feeding mosquitoes on birds: Haynes 2001, p. 118.

361 Manson got Ross assigned full time to malaria study: Haynes 2001, p. 112.

361 Battista Grassi's work: Capanna 2006.

362 Grassi discovers *Anopheles* mosquito: Capanna, 2006.

363 Ross discovers transmission by bite: Haynes 2001, p. 119–123.

363 Grassi bickered in print with colleague: Grassi 1899.

364 Ross's poem: Ross 1923, p. 226.

364 Ross's change in attitude toward Manson: Guillemin 2002.

366 Ross charged Manson with libel: Chernin 1988.

366 Manson's attitude toward Ross: Chernin 1988.

366 Italy's campaign to eradicate malaria: Snowden 2006, pp. 4, 72, 75, 89, 231.

367 Zola quote: Snowden 2006, p. 402.

367 Grassi writing in 1925: Snowden 2006, p. 91.

367 Grassi's desire to be buried in the Roman Campagna: Snowden 2006, p. 101.

367 Washingtonians take "a mouthful": L.O. Howard article for *Engineering News Record*, L. O. Howard Papers, American Philosophical Society, Philadelphia, PA.

367 Map of malaria incidence: "Statistical atlas of the United States based on the results of the ninth census 1870," available at http://memory.loc.gov/ammem/browse/ListSome.php?category=Maps., accessed March 23, 2010.

367 Malaria eradicated in U.S.: http://www.cdc.gov/malaria/history/eradication_us.htm, accessed February 15, 2010.

367 Current malaria statistics: http://www.rollbackmalaria.org/keyfacts.html, accessed February 15, 2010.

368 Yellow fever statistics and descriptions in 1853 New Orleans: Carrigan 1963.

368 1878 epidemic: Blum 2003.

369 Cuba's military governor's quote: Del Regato 2001.

369 U.S. Secretary of War on the end of quarantines: L.O. Howard Papers, American Philosophical Society, Philadelphia, PA.

369 1905 New Orleans mosquito campaign: Carrigan 1988.

370 Estimate that "3,500 lives were saved": "Reminiscences of Early Work," L. O. Howard Papers, American Philosophical Society, Philadelphia, PA.

370 Linnaeus theorized on the cause of malaria: Frängsmyr 1994, p. 46.

371 Linnaeus described *Aedes aegypti*: http://www.itis.gov, accessed March 23, 2010.

EPILOGUE: THE NEW AGE OF DISCOVERY

373 Number of species: Barber 1980, p. 65.

373 Digestive tract biodiversity: Dethlefsen et al. 2006.

374 No hope of discovering large quadrupeds: Cuvier, 1822, p. 61.

374 New beaked whale: Dalebout et al. 2002.

374 The 200-pound oxen: Dung 1993.

374 A new age of discovery: Donoghue and Alverson 2000.

374 New orchid: Informally reported by Woodland Park Zoo, Seattle, WA.

375 New tree genus 130-feet tall: Donoghue and Alverson 2000.

375 Striped rabbit: Reported in *Nature*, August 19, 1999.

375 *Histiophryne psychedelica*: reported in *Copeia*, February 2009.

375 "We do not know how many species of organisms exist on Earth": Wilson 2003.

376 "How they look to a six-foot-tall diurnal mammal": Dan Janzen, personal communication.

376 Giraffe as six species: D. M. Brown, et al, "Extensive population genetic structure in the giraffe," *BMC Biology* 5 (2007): 57.

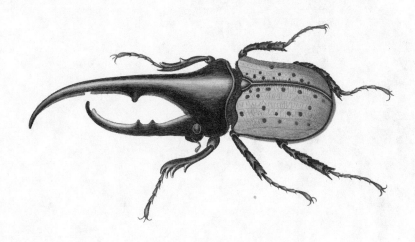

BIBLIOGRAPHY

Abbott, R. T. 1982. *Kingdom of the Seashell*. New York: Crown.

Adams, A. B. 1969. *Eternal Quest: The Story of the Great Naturalists*. New York: G. P. Putnam's Sons.

Akeley, C. E. 1924. *In Brightest Africa*. Garden City, NY: Doubleday.

Alberti, S. 2001. "Amateurs and Professionals in One County: Biology and Natural History in Late Victorian Yorkshire." *Journal of the History of Biology*, 34 (1): 115–47.

Aldington, R. 1949. *The Strange Life of Charles Waterton, 1782–1865*. London: Evans Bros.

Allen, D. E. 1994. *The Naturalist in Britain*. Princeton, NJ: Princeton University Press.

———. 1998. "On Parallel Lines: Natural History and Biology from the Late Victorian Period." *Archives of Natural History*, 25(3): 361–71.

Allen, E. G. 1951. "The History of American Ornithology before Audubon." *Transactions of the American Philosophical Society* (New Ser.), 41(3): 387–591.

Allen, J. A. 1886. "Bird Destruction." *Science*, 8(183): 118 and 19.

Allman, G. J. 1866. "On the Characters and Affinities of Potamogale." *Transactions of the Royal Society of London*, vi:1-16.

Altick, R. D. 1978. *The Shows of London*. Cambridge, MA: Harvard University Press.

Anon. 1875. "Biographical Note of the Late Dr. John Edward Gray." *The Annals and Magazine of Natural History*, Ser. 4, 15: 281-85.

Anon. 1892. "Obituary: Henry Walter Bates." *Proceedings of the Royal Geographical Society* (New Ser.), 14(4): 245–57.

Appel, T. A. 1988. "Jeffries Wyman, Philosophical Anatomy, and the Scientific Reception of Darwin in America." *Journal of the History of Biology*, 21: 69–94.

Audubon, J. J. 1832. *Ornithological Biography: Or, An Account of the Habits of the Birds of the United States of America*. Philadelphia, PA: Carey & Hart.

———. 1834. Audubon Papers, Houghton Library, Harvard University, Cambridge, MA.

Bachman, J. 1850. *The Doctrine of the Unity of the Human Race Examined on the Principles of Science*. Charleston, SC: C. Canning.

Balzac, H. de 1891. *La Peau de Chagrin*. Paris: Ancienne Maison, Michel Lévy Frère.

Banks, J. 1962. *The Endeavour Journal of Joseph Banks*, ed. J. C. Beaglehole. New South Wales: Angus & Robertson.

Barber, L. 1980. *The Heyday of Natural History, 1820–1870*. Garden City, NY: Doubleday.

Barlow, N. 1963. "Darwin's Ornithological Notes." *Bulletin of the British Museum (Natural History) Historical Series*, 2(7): 262.

Barrow, M. 2000, "The Specimen Dealer: Entrepreneurial Natural History in America's Gilded Age," *Journal of the History of Biology*, 33: 493–534.

Bates, H. W. 1861. "Contributions to an Insect Fauna of the Amazon Valley. Lepidoptera: Heliconidae." *Transactions of the Linnean Society*, 23: 495–566.

———. 2002. *The Naturalist on the River Amazons*. Santa Barbara, CA: The Narrative Press.

Bayne, P. 1871. *The Life and Letters of Hugh Miller*. Boston: Gould and Lincoln.

Beaglehole, J. C. 1974. *The Journals of Captain James Cook on His Voyages of Discovery*. London: Hakluyt Society at the University Press.

Bedini, S. A. 1985. "Thomas Jefferson and American Vertebrate Paleontology," Virginia Division of Mineral Resources Publication 61. Charlottesville, VA: Commonwealth of Virginia.

Bell, J. R. 1957. *The Journal of Captain John R. Bell, Official Journalist for the Stephen H. Long Expedition to the Rocky Mountains, 1820*, eds. H. M. Fuller and L. R. Hafen. Glendale, CA: A. H. Clark Co.

Berenbaum, M. R. 1995. *Bugs in the System: Insects and Their Impact on Human Affairs.* New York: Addison Wesley.

Bewell, A. 2004."Romanticism and Colonial Natural History." *Studies in Romanticism*, 43(4).

Bishop, G. 1996. *Travels in Imperial China: The Explorations and Discoveries of Père David*. London: Cassell Publishers.

Blum, A. S. 1993. *Picturing Nature: American Nineteenth-Century Zoological Illustration*. Princeton, NJ: Princeton University Press.

Blum, E. J. 2003. "The Crucible of Disease: Trauma, Memory, and Reconciliation During the Yellow Fever Epidemic." *The Journal of Southern History*, 69(4): 791–820.

Blunt, W. 2002. *Linnaeus: The Compleat Naturalist*. Princeton, NJ: Princeton University Press.

Boewe, C. 2003. *Profiles of Rafinesque*. Knoxville, TN: University of Tennessee Press.

———. 2005. *A C.S. Rafinesque Anthology: Constantine Samuel Rafinesque*, Jefferson, NC: McFarland & Co.

Boorstin, D. 1983. *The Discoverers: A History of Man's Search to Know His World and Himself*. New York: Random House.

Boxer, C. R. 1977. *The Dutch Seaborne Empire, 1600–1800*. London: Taylor & Francis.

Brigham, D. 1996. "Ask the Beasts and They Shall Teach Thee: The Human Lessons of Charles Willson Peale's Natural History Displays." *The Huntington Library Quarterly*, 59(2/3): 183–206.

Broadhurst, C. L., et al. 2002. "Brain-Specific Lipids from Marine, Lacustrine, or Terrestrial Food Resources: Potential Impact on Early African *Homo sapiens*." *Comparative Biochemistry & Physiology* (B) 131: 653–73.

Broberg, G. 1994. "*Homo sapiens*: Linnaeus's Classification of Man." In *Linnaeus: The Man and His Work*, ed. Tore Frängsmyr. Canton, MA: Science History Publications.

Brotz, H. (ed.) 1966. *Negro Social and Political Thought, 1850–1920*. New York: Basic Books.

Bucher, Jr., H. H. 1979. "Canonization by Repetition: Paul Du Chaillu in Historiography." *Revue Francaise d'Histoire d'Outre-Mer*, LXVI: 15–32.

Burchell, W. J. 1822. *Travels in the Interior of Southern Africa*. London: Longman, Hurst.

Burns, F. L. 1909. "Alexander Wilson. V: The Completion of the American Ornithology." *The Wilson Bulletin*, 1(21): 16–35.

———. 1932. "Charles W. and Titian R. Peale and the Ornithological Section of the Old Philadelphia Museum." *The Wilson Bulletin*, 44(1): 23–35.

Burton, R. 1876. *Two Trips to Gorilla Land and the Cataracts of the Congo*, Vol. 1, London: Sampson Low.

Busvine, J. R. 1976. *Insects, Hygiene and History*. London: The Athlone Press.

Butler, S. 1917. "Lucubratio Ebria." In *The Note-Books of Samuel Butler*. New York: E. P. Dutton.

———. 1879. *Evolution, Old and New*. London: Harwicke & Bogue.

Bynum, W., and C. Overy. 1998. *The Beast in the Mosquito: The Correspondence of Ronald Ross & Patrick Manson*. Amsterdam: Rodopi.

Campbell, G. G. 1877. *Log Letters from "The Challenger."* London: Macmillan.

Cannon, W. F. 1961. "The Impact of Uniformitarianism: Two Letters from John Herschel to Charles Lyell." *Proceedings of the American Philosophical Society*, 105(3): 301–14.

Capanna, E. 2006. "Grassi versus Ross: Who Solved the Riddle of Malaria?" *International Microbiology*, 9(1): 69–74.

Carrigan, J. A. 1963. "Impact of Epidemic Yellow Fever on Life in Louisiana." *Louisiana History: The Journal of the Louisiana Historical Association*, 4(1).

———. 1988. "Mass Communication and Public Health: The 1905 Campaign Against Yellow Fever in New Orleans." *Louisiana History: The Journal of the Louisiana Historical Association*, 29(1).

Chambers, R. 1845. *Vestiges of the Natural History of Creation*. New York: Wiley & Putnam.

Chernin, E. 1983. "Sir Patrick Manson's Studies on the Transmission and Biology of Filariasis." *Reviews of Infectious Diseases*, 5(1): 148–66.

———. 1988. "Sir Ronald Ross vs. Sir Patrick Manson: A Matter of Libel." *Journal of the History of Medicine and Allied Sciences*, 43(3): 262–74.

Chesebrough, D. B. 1998. *Fredrick Douglass: Oratory from Slavery*. Westport, CT: Greenwood Publishing Group.

Christie, J.R.R. 1990. "Ideology and Representation in Eighteenth-Century Natural History." *Oxford Art Journal*, 13(1): 3–10.

Clodd, E. 1926. *Memories*. London: Watts & Co.

Clutton-Brock, J. 1995. "Aristotle, the Scale of Nature, and Modern Attitudes to Animals." *Social Research*, 62(3).

Cobbold, T. S. 1879. *Entozoa: An Introduction to the Study of Helminthology*. London: Groombridge & Sons.

Cohen, C. 2002. *The Fate of the Mammoth: Fossils, Myths, and History*. Trans. W. Rodarmor. Chicago, IL: University of Chicago Press.

Coleman, W. 1964. *Georges Cuvier, Zoologist, a Study in the History of Evolution Theory*. Cambridge, MA: Harvard University Press.

Cook, G. C. 1992. *From the Greenwich Hulks to Old St Pancreas: A History of Tropical Disease in London*. London: Athlone Press.

Coolidge, H. J. 1929. "A Revision of the Genus *Gorilla*." *Memoirs of the Museum of Comparative Zoology at Harvard*, 50:295–381.

Cooper, J. F. 1881. *The Prairie: A Tale*. New York: D. Appleton & Co.

Corfield, R. 2003. *The Silent Landscape: The Scientific Voyage of HMS Challenger*. Washington, DC: Joseph Henry Press.

Coues, E. 1890. *Handbook of Field and General Ornithology*. London: Macmillan.

Curtin, P. D. 1961. "The White Man's Grave: Image and Reality, 1780–1850." *The Journal of British Studies*, 1(1): 94–110.

———. 1989. *Death by Migration: Europe's Encounter with the Tropical World in the Nineteenth Century*. New York: Cambridge University Press.

Curtis, L. P. 1968. *Anglo Saxons and Celts: A Study of Anti-Irish Prejudice in Victorian England*. Bridgeport, CT: University of Bridgeport.

Cuvier, G. 1822. *Essays on the Theory of the Earth*, 4th ed. Edinburgh: William Blackwood.

Dain, B. 2002. *A Hideous Monster of the Mind: American Race Theory in the Early Republic*. Cambridge, MA: Harvard University Press.

Dalebout, M. L., J. G. Mead, C. S. Baker, A. N. Baker and A. L. Van Helden. 2002. "A New Species of Beaked Whale, *Mesoplodon perrini* sp. n. (Cetacea: Ziphiidae), Discovered Through Phylogenic Analysis of Mitochondrial DNA Sequences." *Marine Mammal Science*, 18(3): 577–608.

Dance, S. P. 1966. *Shell Collecting: An Illustrated History*. Berkeley, CA: University of California Press.

———. 1969. *Rare Shells*. London: Faber and Faber.

Daniels, G. H. 1967. "The Process of Professionalization in American Science: The Emergent Period: 1820–1860." *Isis*, 58(2): 150–66.

Darwin, C. 1882. *On the Origin of Species by Natural Selection, or the Preservation of Favored Races in the Struggle for Life*. New York: D. Appleton and Company.

———. 1904. *The Life and Letters of Charles Darwin Including an Autobiographical Chapter*. Ed. F. Darwin. New York: D. Appelton.

———. 1909. *The Voyage of the Beagle*. New York: P. F. Collier & Son.

Darwin Correspondence Project, http://www.darwinproject.ac.uk/home.

"The Complete Works of Darwin Online," http://darwin-online.org.uk.

Darwin, E. 1803. *Zoonomia; or, the Laws of Organic Life*. Boston: Thomas and Andrews.

David, A. 1949. *Abbé David's Diary*. Trans. H. M. Fox. Cambridge, MA: Harvard University Press.

Davies, K. C., and J. Hull. 1983. "Burchell's South African Bird Collection (1810–1815)." *Archives of Natural History*, 11(2): 317–42.

Dear, P. 2008. *The Intelligibility of Nature: How Science Makes Sense of the World*. Chicago, IL: University of Chicago Press.

Del Regato, J. A. 2001. "Carlos Juan Finlay (1833–1915)." *Journal of Public Health Policy* 22(1): 98–104.

Denny, M. 1948. "Linnaeus and His Disciple in Carolina: Alexander Garden." *Isis*, 38(3/4): 161–74.

Desmond, A., 1998. *Huxley: From Devil's Disciple to Evolution's High Priest*. Reading, MA: Addison-Wesley.

Desmond, A., and J. R. Moore, 1991. *Darwin*. London: Michael Joseph.

Desmond, A., and J. R. Moore. 2009. *Darwin's Sacred Cause: Race, Slavery and the Quest for Human Origins*. London: Allen Lane.

Dethlefsen, L., et al. 2006. "Assembly of the Human Intestinal Microbiota." *Trends in Ecology & Evolution*, 21(9): 517–23.

De Tocqueville, A., 1904. *Democracy in America, Volume II*. Trans. H. Reeve. New York: D. Appleton & Co.

De Wit, H. C. D. 1952. "In Memory of G. E. Rumphius." *Taxon*, 1(7): 101–10.

———. 1959. *Rumphius Memorial Volume*. Amsterdam: Uitgeverij En Drukkerij Hollandia N.V.

Dobson, J. 1962. "John Hunter's Animals." *Journal of the History of Medicine and Allied Sciences*, 17(4): 479–86.

Donoghue, M. J., and W. S. Alverson. 2000. "A New Age of Discovery." *Annals of the Missouri Botanical Garden*, 87: 110–26.

Donovan, E. 1805. *Instructions for Collecting and Preserving Various Subjects of Natural History*. London: F. C. and J. Rivington.

Douglass, H. 1897. *In Memoriam: Frederick Douglass*. Philadelphia, PA: John C. Yorston & Co.

Dubose, H. C. 1895. *Memoirs of Rev. John Leighton Wilson*. Richmond, VA: Presbyterian Committee of Publication.

Du Chaillu, P. B. 1868. *Exploration & Adventures in Equatorial Africa*. New York: Harper & Brothers.

———. 1871. *A Journey to Ashango-land, and Further Penetration into Equatorial Africa*. New York: Harper & Brothers.

Dung, V. V., et al. 1993. "A New Species of Living Bovid from Vietnam." *Nature* 363: 443–45.

Dunn, D. E. 1992. *A History of the Episcopal Church in Liberia, 1821–1980*. Toronto: Scarecrow Press.

Duyker, E. 1988, *Nature's Argonaut: Daniel Solander 1733–1782*, Melbourne, AU: Miegunyah Press.

Edginton, B. W. 1996. *Charles Waterton*. Cambridge, UK: Lutterworth Press.

Edwards, E. 1969. *Lives of the Founders of the British Museum: With Notices of its Chief Augmentors and Other Benefactors*. New York: Burt Franklin.

Egerton, J. 1976. *George Stubbs, Anatomist and Animal Painter*. London: Tate Gallery.

Elman, R. 1977. *First in the Field*. New York: Mason/Charter.

Estensen, M. 2002. *The Life of Matthew Flinders*. Sydney: Allen & Unwin.

Evans, H. E. 1997. *The Natural History of the Long Expedition to the Rocky Mountains, 1819–1820*, New York: Oxford University Press.

Fan, F. 2004. *British Naturalists in Qing China: Science, Empire, and Cultural Encounter*. Cambridge, MA: Harvard University Press.

Farber, P. L. 1977. "The Development of Taxidermy and the History of Ornitholoy." *Isis*, 68(4): 550–66.

———. 1982. *Discovering Birds: The Emergence of Ornithology as a Scientific Discipline, 1760–1850*. Baltimore, MD: Johns Hopkins University Press.

———. 2000. *Finding Order in Nature: The Naturalist Tradition from Linnaeus to E. O. Wilson*. Baltimore, MD: Johns Hopkins University Press.

Farley, J. 1972. "The Spontaneous Generation Controversy (1700–1860): The Origin of Parasitic Worms." *Journal of the History of Biology*, 5: 95–125.

Fell, H. B. 1950. "New Zealand Crinoids." *Tuatara: Journal of the Biological Society*, 3(2): 78–85.

Fitzpatrick, T. J. 1911. *Rafinesque: A Sketch of His Life*. Des Moines, IA: The Historical Department of Iowa.

Foot, J. 1794. *The Life of John Hunter*. London: T. Becket.

Fornasiero, J., P. Monteath, and J. West-Sooby. 2004. *Encountering Terra Australis: The Australian Voyages of Nicolas Baudin and Matthew Flinders*. Kent Town, AU: Wakefield Press.

Frängsmyr, T. 1994. *Linnaeus: The Man and His Work*. Canton, MA: Science History Publications.

Fredrickson, G. M. 1987. *The Black Image in the White Mind: The Debate on Afro-American Character and Destiny, 1817–1914*. New York: Harper & Row.

Garden, A. 1775. "An Account of the Gymnotus Electricus, or Electrical Eel." *Philosophical Transactions (1683–1775)*, 65: 102–10.

Gates, B. 1999. *Kindred Nature: Victorian and Edwardian Women Embrace the Living World*. Chicago, IL: University of Chicago Press.

Gatty, A. 1908. *Parables from Nature*. London: J. M. Dent.

Geoffroy Saint-Hilaire, I. 1853. "Sur les rapports naturels du Gorille; remarques faites à la suite de la lecture de M. Duvernoy." *Compte rendu des séances de l'Academie des Sciences Mai* 36: 933–36.

Gette, P. A., and G. Scherer. 1982. *Insects Etc: An Anthology of Arthropods Featuring a Bounty of Beetles*. Trans. G. Zappler. New York: Hudson Hills Press.

Goldgar, A. 2007. *Tulipmania: Money, Honor, and Knowledge in the Dutch Golden Age*. Chicago, IL: University of Chicago Press.

Gosse, E. 2005. *Father and Son*. Gloucestershire, UK: Nonsuch Publishing.

Gosse, P. H. 1851. *A Naturalist's Sojourn in Jamaica*. London: Longman.

———. 1857. *Omphalos: An Attempt to Untie the Geological Knot*. London: John Van Voorst.

Gossett, T. F. 1997. *Race: The History of an Idea in America*. New York: Oxford University Press.

Gould, S. J. 1970. "Private Thoughts of Lyell on Progression and Evolution." *Science* (New Ser.), 169(3946): 663–64.

———. 1978. "Morton's Ranking of Races by Cranial Capacity." *Science* (New Ser.), 200 (4341): 503–9.

———. 2000. "Linnaeus's Luck?" *Natural History*, 109: 18–25, 66–76.

———. 2002. *The Structure of Evolutionary Theory*. Cambridge, MA: Harvard University Press.

Grassi, B. 1899. "Mosquitoes and Malaria." *The British Medical Journal*, 2 (2020): 748, 749.

Graves, G. 1818. *Naturalist's Pocketbook or Tourist's Companion*. London: Sherwood, Neely and Jones.

Gray, V. 2006. "Something in the Genes: Walter Rothschild, Zoological Collector Extraordinaire." A lecture delivered at Royal College of Surgeons, London, October 25, 2006.

Greene, J. C. 1961. *The Death of Adam: Evolution and Its Impact on Western Thought*. Ames, IA: The Iowa State University Press.

Greenfield, B. R. 1992. *Narrating Discovery: The Romantic Explorer in American Literature, 1790–1855*. New York: Columbia University Press.

Greig, J. (ed). 1926. *The Diaries of a Duchess: Extracts from the Diaries of the First Duchess of Northumberland (1716–1776)*. London: Hodder and Stoughton.

Groner, J., and P. F. Cornelius. 1996. *John Ellis: Merchant, Microscopist, Naturalist, and King's Agent: A Biologist of His Times*. Pacific Grove, CA: Boxwood Press.

Grove, W. B. 1892. "The Happy Fungus-Hunter." *Midland Naturalist*, 15: 158–61.

Gruber, J. 1982. "What Is It? The Echidna Comes to England." *Archives of Natural History*, 11(1): 1–15.

———. 1987. "From Myth to Reality: The Case of the Moa." *Archives of Natural History*, 14(3): 339–52.

———. 1991. "Does the Platypus Lay Eggs? The History of an Event in Science." *Archives of Natural History*, 18(1): 51–123.

Guilleman, J. 2002. "Choosing Scientific Patrimony: Sir Ronald Ross, Alphonse Laveran, and the Mosquito-Vector Hypothesis for Malaria," *Journal of the History of Medicine and Allied Sciences*, 57(4): 385–409.

Günther, A. E. 1975. *A Century of Zoology at the British Museum*. London: Dawson & Sons.

Haddad, J. R. 2008. *The Romance of China: Excursions to China in U.S. Culture, 1776–1876*. New York: Columbia University Press.

Haldeman, S. S. 1842. "Notice of the Zoological Writings of the Late C. S. Rafinesque." *American Journal of Science and the Arts*, 42: 280–91.

Hartert, E. 1901. "William Doherty, Obituary." *Novitates Zoologicae*, VIII (4).

Harvey, W. H. 1857. *The Sea-Side Book*. London: John Van Voorst.

Haynes, D. M. 2001. *Imperial Medicine: Patrick Manson and the Conquest of Tropical Disease*. Philadelphia, PA: University of Pennsylvania Press.

Hedeen, S. 2008. *Big Bone Lick: The Cradle of American Paleontology*. Lexington, KY: University Press of Kentucky.

Hedgepeth, J. W. 1946. "The Voyage of the Challenger." *The Scientific Monthly*, 63 (3): 194–202.

Herrick, F. H. 1917. *Audubon the Naturalist*. New York: D. Appleton.

Hobson, R. 1867. *Charles Waterton: His Home, Habits, and Handiwork*. London: Whittaker & Co.

Hollerbach, A. L. 1996. "Of Sangfroid and Sphinx Moths: Cruelty, Public Relations, and the Growth of Entomology in England, 1800–1840." *Osiris* (2nd Ser.), 11: 201–20.

Holmes, O. W. 1895. *The Complete Poetical Works of Oliver Wendell Holmes*. Boston: Houghton, Mifflin.

Howard, L. O. *Engineering News Record* article, L. O. Howard Papers, American Philosophical Society, Philadelphia, PA.

Humboldt, A. Von. 1996. *Personal Narrative of a Journey to the Equinoctial Regions of the New Continent: Abridged Edition*. London: Penguin Classics.

Humboldt, C. D., and M. Alexander. 1852. "Extracts from a Memoir of Samuel George Morton, M.D., Late President of the Academy of Natural Sciences of Philadelphia." *American Journal of Science and Arts (1820–1879)*, 13(38): 153.

Hunter, W. 1768. "Observations on the Bones, Commonly Supposed to Be Elephant Bones, Which Have Been Found Near the River Ohio in America." *Philosophical Transactions (1683–1775)*, 58: 34–45.

Irmscher, C. 1999. *The Poetics of Natural History: From John Bartram to William James*. New Brunswick, NJ: Rutgers University Press.

James, E. 1823. *Account of an Expedition from Pittsburgh to the Rocky Mountains Performed in the Years 1819, 1820*. London: Longman, Hurst, Rees, Orme, and Brown.

Jefferson, T. 1799. "A Memoir of the Discovery of Certain Bones of a Quadruped of the Clawed Kind in the Western Parts of Virginia." *Transactions of the American Philosophical Society*, 4: 246–60.

———. 1853. *Notes on the State of Virginia*, Richmond, VA: J. W. Randolph.

Jefferson. T. 1904. *The Works of Thomas Jefferson*. Ed. P. L. Ford. New York: G.P. Putnam's Sons.

Jenkins, A. C. 1978. *The Naturalists: Pioneers of Natural History*. London: Hamish Hamilton.

Jiang, Z. and R. B. Harris. 2008. *Elaphurus davidianus*. IUCN Red List of Threatened Species, version 2009.1 at www.iucnredlist.org, accessed October 30, 2009.

Johnson, S. 1810. *The Rambler*. London: Luke Hansard & Sons.

Jönsson, A. 2000. "Odium Botanicorum: The Polemics between Carl Linnaeus and Johann Georg Siegesbeck." In Språkets speglingar. Festskrift till Birger Bergh, ed. A. Jönsson and A. Piltz. Ängelholm, Sweden: Skåneförlaget, 555–66.

Jordan, W. D. 1974. *The White Man's Burden: Historical Origins of Racism in the United States*. New York: Oxford University Press.

Kalm, P. 1772. *Travels into North America*. London: Lowndes. www.americanjourneys.org/aj-117a, accessed July 23, 2010.

Kennedy, K.A.R., and J. Whittaker. 1976. "The Ape in Stateroom 10." *Natural History*, 85(9): 48–53.

Kerr, R. 1792. *The Animal Kingdom or Zoological System of the Celebrated Sir Charles Linnaeus*. London: Murray and Faulder.

Kershaw, W. 1963. "Vector-borne Diseases in Man: A General Review." *Bulletin of the World Health Organization*, 29 (Suppl.): 13–17.

King, J.C.H. 1996. "New Evidence for the Contents of the Leverian Museum." *Journal of the History of Collections*, 8(2): 167–86.

King-Hele, D. (ed), and C. R. Darwin. 2003. *Charles Darwin's The Life of Erasmus Darwin*. Cambridge, UK: Cambridge University Press.

Kingsley, C. 1890. *Glaucus or the Wonders of the Shore*. London: Macmillan.

———. 1917. *The Water-Babies: A Fairy Tale for a Land-Baby*. New York: The Macmillan Company.

Kingsley, M. H. 1893–1899. Macmillan Archive. Correspondence with Mary H. Kingsley, two volumes. British Library, London.

———. 1897. *Travels in West Africa: Congo Français, Corisco and Cameroons*. New York: The Macmillan Company.

Kingston, R. 2007. "A Not So Pacific Voyage: The 'Floating Laboratory' of Nicolas Baudin." *Endeavour*, 31(4): 145–51.

Klarer, M. 2005. "Humanitarian Pornography: John Gabriel Stedman's Narrative of a Five Year Expedition Against the Revolting Negroes of Surinam (1796)." *New Literary History*, 36(4): 559–87.

Koerner, L. 1999. *Linnaeus: Nature and Nation*. Cambridge, MA: Harvard University Press.

Kohler, R. E. 2006. *All Creatures: Naturalists, Collectors, and Biodiversity, 1850–1950*. Princeton, NJ: Princeton University Press.

Lecercle, J. 1994. *Philosophy of Nonsense: The Intuitions of Victorian Nonsense Literature*. New York: Routledge.

Lee, D. 2004. *Slavery and the Romantic Imagination*. Philadelphia, PA: University of Pennsylvania Press.

Lee, S. 1833. *Memoirs of Baron Cuvier*. London: Longman, Rees.

Lettsom, J. C. 1799. *The Naturalist's and Traveler's Companion*. London: C. Dilly.

Li, S.-J. 2002. "Natural History of Parasitic Disease: Patrick Manson's Philosophical Method." *Isis*, 93(2): 206–28.

Lightman, B. 1997. *Victorian Science in Context*. Chicago, IL: University of Chicago Press.

———. 2007. *Victorian Popularizers of Science: Designing Nature for New Audiences*. Chicago, IL: University of Chicago Press.

Lindroth, S. 1994. "The Two Faces of Linnaeus." In *Linnaeus: The Man and His Work*, ed. Tore Frängsmyr, rev. ed. Canton, MA: Science History Publications.

Linnaeus, C. "The Linnaean Correspondence," http://linnaeus.c18.net.

Lloyd, C. 1985. *The Travelling Naturalists*. Chicago, IL.: The Art Institute of Chicago.

Lucas, F. A. 1914. "The Story of Museum Groups." *The American Museum Journal*, xiv(1): 9, 10.

Lyell, C. 1842. *Principles of Geology*. Boston: Hilliard, Gray.

———. 1881. *Life, Letters and Journals of Sir Charles Lyell, Bart, Vol. II*, ed. K. M. Lyell. London: John Murray.

———. 1970. *Sir Charles Lyell's Scientific Journals on the Species Question*, ed. L. G. Wilson. New Haven, CT: Yale University Press.

Lyon, J., and P. Sloan. 1981. *From Natural History to the History of Nature: Readings from Buffon and His Critics*. Notre Dame, IN: University of Notre Dame Press.

MacGregor, A. (ed.) 1994. *Sir Hans Sloane: Collector, Scientist, Antiquary, Founding Father of the British Museum*. London: British Museum Press.

Magyar, L. 1994. "John Hunter and John Dolittle." *The Journal of Medical Humanities*, 15(4): 217–20.

Mandelstam, J. 1994. "Du Chaillu's Stuffed Gorillas and the Savants from the British Museum." *Notes and Records of the Royal Society of London*, 48: 227–45.

Manson P. 1894. "On the Nature and Significance of the Crescentic and Flagellated Bodies in Malarial Blood." *British Medical Journal*, 2:1306–08.

———. 1922. "A Short Autobiography." *Journal of Tropical Medicine and Hygiene*, 25:156–64.

Matthew, P. 1831. *On Naval Timber and Arboriculture*. London: Longman, Rees, Orme, Brown, and Green.

McCook, S. 1996. " 'It May Be Truth, But It Is Not Evidence': Paul du Chaillu and the Legitimation of Evidence in the Field Sciences." *Osiris* (2nd Ser.), 11: 177–97.

McMillan, N. 1996. "Robert Bruce Napoleon Walker, West African Trader, Explorer and Collector of Zoological Specimens." *Archives of Natural History*, 23(1): 125–41.

Mearns, B., and R. Mearns. 1998. *The Bird Collectors*. Toronto: Academic Press.

Meek, A. S. 1913. *A Naturalist in Cannibal Land*. London: T. Fisher Unwin.

Meigs, C. D. 1851. *A Memoir of Samuel George Morton, MD*. Philadelphia: P. G. Collins.

Meijer, M. C. 1999. *Race and Aesthetics in the Anthropology of Petrus Camper (1722–89)*. Amsterdam: Editions Rodopi B.V.

Michael, J. S. 1988. "A New Look at Morton's Craniological Research." *Current Anthropology*, 29(2).

Mills, E. 1984. "A View of Edward Forbes." *Archives of Natural History*, 11(3): 365–93.

Moore, W. 2005. *The Knife Man: The Extraordinary Life and Times of John Hunter, Father of Modern Surgery*. New York: Broadway Books.

Morris, P. A. 1993. "An Historical Review of Bird Taxidermy in Britain." *Archives of Natural History*, 20(2): 241–55.

Morton, S. G. 1839. *Crania Americana: or, a Comparative View of the Skulls of Various Aboriginal Nations of North and South America*. Philadelphia: J. Dobson.

———. 1847. "Hybridity in Animals, Considered in Reference to the Question of the Unity of the Human Species." *American Journal of Science and Arts*, 3: 39–50, 203–12.

Moyal, A. 2004. *Platypus: The Extraordinary Story of How a Curious Creature Baffled the World*. Washington, D.C.: Smithsonian Institution Press.

Munoz, P. 1999. "Rhinopithecus roxellana." Animal Diversity Web site: http://animal diversity.ummz.umich.edu, accessed October 29, 2009.

Nieuwenhuys, R., and E. M. Beekman. 1982. *Mirror of the Indies: A History of Dutch Colonial Literature*. Amherst, MA: University of Massachusetts Press.

Noakes, R. 2002. "Science in Mid-Victorian Punch." *Endeavour*, 26: 92–96.

Nobles, G. 2005. "A French Affair? Jean Audubon, Père, and the West Indian Origins of John James Audubon." Atlanta, GA: Georgia Institute of Technology. http://www.dssi.unimi.it/dipstoria/mg/papers/greg_nobles.pdf.

Nott, J. C. 1849. *Two Lectures on the Connection Between the Biblical and Physical History of Man*. New York: Bartlett & Welford.

O'Brian, P. 1997. *Joseph Banks: A Life*. Chicago, IL: University of Chicago Press.

Owen, R. 1845. "Recollections and Reflections of Gideon Shaddoe, Esq., No. IX." *Hood's Magazine and Comic Miscellany*.

———. 1848. "On a New Species of Chimpanzee." *Proceedings of the Zoological Society of London*, XVI: 27–35.

———. 1855. "On the Anthropoid Apes." *Proceedings of the Royal Institution of Great Britain*, 2:41.

Owen, R. 1894. *The Life of Richard Owen*. London: John Murray.

Paget, J. 1866. "On the Discovery of Trichina." *The Lancet*, 1: 269, 270.

Park, M. 1878. *Travels in the Interior of Africa*. Edinburgh: Adam and Charles Black.

Pascoe, J. 2005. *The Hummingbird Cabinet: A Rare and Curious History of Romantic Collectors*. Ithaca, NY: Cornell University Press.

Patterson, K. D. 1974. "Paul B. Du Chaillu and the Exploration of Gabon, 1855–1865." *The International Journal of African Historical Studies*, 7(4): 647–67.

Patterson, T. C. 2001. *Social History of Anthropology in the U.S.* New York: Berg Publishers.

Peterson, D. 2006. *Jane Goodall: The Woman Who Redefined Man*. New York: Houghton Mifflin Company.

Philbrick, N. 2004. *Sea of Glory: America's Voyage of Discovery, the U.S. Exploring Expedition, 1838–1842*. New York: Penguin.

Piccolino, M., and M. Bresadola. 2002. "Drawing a Spark from Darkness: John Walsh and Electric Fish." *Trends in Neurosciences*, 25(1): 51–57.

Pickering, C. 1854. *The Races of Man and Their Geographical Distribution*. London: H. G. Bohn.

Pickering, J. 1998. "William John Burchell's Travels in Brazil, 1825–1830, with Details of the Surviving Mammal and Bird Collections." *Archives of Natural History*, 25: 237–65.

Poe, E. A. 1840. *The Conchologist's First Book: A System of Testaceous Malacology*. Philadelphia, PA: Haswell, Barrington.

Porter, C. M. 1979. " 'Subsilentio': Discouraged Works of Early Nineteenth-Century American Natural History." *Journal of the Society for the Bibliography of Natural History*, 9(2): 109–19.

Porter, R. 1994. *London: A Social History*. London: Hamish Hamilton.

Prak, M. 2005. *The Dutch Republic in the Seventeenth Century: The Golden Age*. New York: Cambridge University Press.

Prince, S. A. (ed.) 2003. *Stuffing Birds, Pressing Plants, Shaping Knowledge: Natural History in North American, 1730–1860*. Philadelphia, PA: American Philosophical Society.

Raby, P. 1997. *Bright Paradise: Victorian Scientific Travellers*, Princeton, NJ: Princeton University Press.

———. 2002. *Alfred Russel Wallace: A Life*. Princeton, NJ: Princeton University Press.

Raverat, G. 1952. *Period Piece: A Cambridge Childhood*. London: Faber.

Recker, D. 1990. "There's More Than One Way to Recognize a Darwinian: Lyell's Darwinism." *Philosophy of Science*, 57(3): 459–78.

Reitter, E. (1960). *Beetles*. New York: G. P. Putnam's Sons.

Rhodes, R. 2004. *John James Audubon: The Making of an American*. New York: Alfred A. Knopf.

Ripley, D. 1970. *The Sacred Grove*. New York: Simon and Schuster.

Riviere, P. 1998. "From Science to Imperialism: Robert Schomburgk's Humanitarianism." *Archives of Natural History*, 25: 1–8.

Robley, H. G. 2003. *Moko or Maori Tattooing*. Mineola, NY: Dover Publications.

Roemer, B. 2004. "The Relation Between Art and Nature in a Dutch Cabinet of Curiosities from the Early Eighteenth Century." *History of Science*, 42: 47–84.

Roger, J. 1997. *Buffon: A Life in Natural History*. Trans. S. L. Bonnefoi. Ithaca, NY: Cornell University Press.

Rolt, L.T.C. 1957. *Isambard Kingdom Brunel*. London: Longman.

Rookmaaker, L. C. et al. 2006. "The Ornithological Cabinet of Jean-Baptiste Bécoeur and the Secret of the Arsenical Soap." *Archives of Natural History*, 33: 146–58.

Ross, R. 1923. *Memoirs with a Full Account of the Great Malaria Problem and its Solution*. London: John Murray.

Rotella, C. 1991. "Travels in a Subjective West: The Letters of Edwin James and Major Stephen Long's Scientific Expedition of 1819–1820." *Montana*, 41: 20–35.

Rothschild, H. 2009. "The Butterfly Effect." *Bonhams*, Spring 2009.

Rothschild, M. 1983. *Dear Lord Rothschild*. London: Hutchinson.

Rousseau, G. S. 1982. "Science Books and Their Readers in the Eighteenth Century." In *Books and Their Readers in Eighteenth Century England*, ed. I. Rivers et al. New York: St. Martin's Press.

Rudwick, M.J.S. 1972. *The Meaning of Fossils: Episodes in the History of Paleontology*. Chicago, IL: University of Chicago Press.

———. 1997. *George Cuvier, Fossil Bones, and Geological Catastrophes: New Translations and Interpretations of the Primary Texts*. Chicago, IL: University of Chicago Press.

———. 2005. *Bursting the Limits of Time: The Reconstruction of Geohistory in the Age of Revolution*. Chicago, IL: University of Chicago Press.

Rumpf, G. E. 1981. *The Poison Tree: Selected Writings of Rumphius on the Natural History of the Indies* (Trans. E. M. Beekman) Amherst, MA: University of Massachusetts Press.

Rumphius, G. E. 1999. *The Amboinese Curiosity Cabinet*. Trans. and ed. E. M. Beekman. New Haven, CT: Yale University Press.

Rupke, N. A. 2002. "Geology and Paleontology." In *Science and Religion: A Historical Introduction*, ed. G. B. Ferngren. Baltimore, MD: Johns Hopkins University Press.

Ruse, M. 2003. *Darwin and Design: Does Evolution Have a Purpose?* Cambridge, MA: Harvard University Press.

St. John, H. 1864. *Audubon, The Naturalist of the New World, His Adventures and Discoveries*. Boston: Crosby and Nichols.

Sarton, G. 1937. "Rumphius, Plinius Indicus (1628–1702)." *Isis*, 27(2): 242–57.

Savage, T. S. 1847. "On the Habits of the 'Drivers' or Visiting Ants of West Africa." *Transactions of the Royal Entomological Society London*, 5:1–15.

Savage, T. S., and J. Wyman 1844. "Observations on the external characters and habits of the *Troglodytes niger*, Geoff. and on its organization." *Boston Journal of Natural History*, 4:362–386.

———. 1847. "Notice of the External Characters and Habits of a New Species of Troglodytes Gorilla." *Boston Journal of Natural History*, 5: 245–47.

Savage, W. R. 1843. William R. Savage papers, Southern Historical Collection, Manuscripts Department, Wilson Library, University of North Carolina at Chapel Hill.

Schaller, G. B. 1994. *The Last Panda*. Chicago, IL: University of Chicago Press.

Schiebinger, L. 1993, *Nature's Body: Gender in the Making of Modern Science*. Boston, MA: Beacon Press.

Schwartz, J. H. (ed.). 1988. *Orang-utan Biology*. Oxford, UK: Oxford University Press.

Schwartz, J. S. 1999. "Robert Chambers and Thomas Henry Huxley, Science Correspondents: The Popularization and Disseminationation of Nineteenth Century Natural Science." *Journal of the History of Biology*, 32(2): 343–83.

Scott, E. L. 1974. "Edward Jenner, F.R.S., and the Cuckoo." *Notes and Records of the Royal Society of London*, 28(2): 235–40.

Sears, S. W. 1988. *George B. McClellan: The Young Napoleon*. New York: Ticknor & Fields.

Secord, A. 2003. "'Be What You Would Seem to Be': Samuel Smiles, Thomas Edward, and the Making of a Working-Class Scientific Hero." *Science in Context*, 16:147–73.

Secord, J. 2000. *Victorian Sensation: The Extraordinary Publication, Reception and Secret Authorship of Vestiges of the Natural History of Creation*. Chicago, IL: University of Chicago Press.

Sellers, C. C. 1951. "Charles Willson Peale and 'The Mammoth Picture,'" in *The Peale Museum Historical Series*, no. 7. Baltimore, MD: Municipal Museum of the City of Baltimore. http://www.lewis-clark.org/content/content-article.asp?ArticleID=2757.

———. 1969. *Charles Willson Peale*. New York: Charles Scribner's & Sons.

———. 1980, *Mr. Peale's Museum: Charles Willson Peale and the First Popular Museum of Natural Science and Art*. New York: W. W. Norton.

Semonin, P. 2000. *American Monster: How the Nation's First Prehistoric Creature Became a Symbol of National Identity*. New York: New York University Press.

Shaw, G. D. 2005. "'Moses Williams, Cutter of Profiles': Silhouettes and African

American Identity in the Early Republic." *Proceedings of the American Philosophical Society*, 149(1): 22–39.

Sheffield, S. L. 2001. *Revealing New Worlds: Three Victorian Women Naturalists*. London: Routledge.

———. 2004. *Women and Science: Social Impact and Interaction*. Santa Barbara, CA: ABC-CLIO.

Simpson, G. G. 1942. "The Beginnings of Vertebrate Paleontology in North America." *Proceedings of the American Philosophical Society*, 86(1): 130–88.

———. 1959. "Review." *Science*, 130(3368): 158.

Smallwood, W. M., and M.S.C. Smallwood. 1941. *Natural History and the American Mind*. New York: Columbia University Press.

Smiles, S. 1879. *Life of a Scotch Naturalist*. London: John Murray.

Smith, H. E. 1903. "Reminiscences of Paul Belloni Du Chaillu." *The Independent*, 55:1146–48.

Smith, R. F., T. E. Mittler, and C.N. Smith (eds.). 1973. *History of Entomology*. Palo Alto, CA: Annual Reviews Inc.

Smith, S. S. 1810. *An Essay on the Causes of the Variety of Complexion and Figure in the Human Species . . .* New Brunswick, NJ: J. Simpson and Co.

Smith, W. J. 1962. "Sir Ashton Lever of Alkrington and His Museum 1729–1788." *Transactions of the Lancashire and Cheshire Antiquarian Society*, 72: 61–92.

Snowden, F. 2006. *Conquest of Malaria: Italy, 1900–1962*. New Haven, CT: Yale University Press.

Somerset, R. 2002. "The Naturalist in Balzac: The Relative Influence of Cuvier and Geoffroy Saint-Hilaire." *French Forum*, 27(1): 81–111.

Spary, E. C. 2004. "Scientific Symmetries." *History of Science*, 62: 1–46.

Stanford, D. E. 1959. "The Giant Bones of Claverack, New York, 1705." *New York History* 40: 47–61.

Stanton, W. 1960. *The Leopard's Spots: Scientific Attitudes Toward Race In America, 1815–59*. Chicago, IL: University of Chicago Press.

Stedman, J. G. and S. Price (eds.). 1988. *Narrative of a Five Years Expedition Against the Revolted Negroes of Surinam*. Baltimore, MD: Johns Hopkins University Press.

———. *Stedman's Surinam: Life in Eighteenth-century Slave Society*. Baltimore, MD: Johns Hopkins University Press.

Stephens, L. D. 2000. *Science, Race, and Religion in the American South: John Bachman and the Charleston Circle of Naturalists*. Chapel Hill, NC: University of North Carolina Press.

Stott, R. 2003. *Darwin and the Barnacle*. New York: W. W. Norton.

Stratton, J. A., and L. H. Mannix. 2005. *Mind and Hand: The Birth of MIT*. Cambridge, MA: MIT Press.

Stresemann, E. 1975. *Ornithology: From Aristotle to the Present*. Cambridge, MA: Harvard University Press.

Stroud, P. T. 1992. *Thomas Say: New World Naturalist*. Philadelphia, PA: University of Pennsylvania Press.

———. 1995. "Forerunner of American Conservation: Naturalist Thomas Say." *Forest & Conservation History*, 39(4): 184–90.

————. 2000. *The Emperor of Nature: Charles-Lucien Bonaparte and His World*. Philadelphia, PA: University of Pennsylvania Press.

Stukely, W. 1719. "An Account of the Impression of the Almost Entire Sceleton of a Large Animal in a Very Hard Stone . . . " *Philosophical Transactions (1683–1775)*, 30: 963–68.

Swainson, W. 1834. *A Preliminary Discourse on the Study of Natural History*. London: Longman, Rees.

Thomson, C. W. 1873. *The Depths of the Sea*. London: Macmillan.

Thoreau, H. D. 1980. *Natural History Essays (Literature of the American Wilderness)*. Ed. R. Sattelmeyer. Salt Lake City, UT: Peregrine Smith Books.

Tyson, E. 1699. *Orang-outang, Sive, Homo Sylvestris, or, The Anatomy of a Pygmie*. London: Thomas Bennet.

Twain, M. 1963. *The Complete Essays of Mark Twain*, ed. C. Neider. New York: Doubleday.

Vanhaeren, M., et al. 2006. "Middle Paleolithic Shell Beads in Israel and Algeria." *Science*, 312(5781): 1785–88.

Van Seters, W. H. 1962. *Pierre Lyonet (1706–1789): sa vie, ses collections de coquillages et de tableaux, ses recherches*. La Haye: M. Nijhoff.

Vaucaire, M. 1930. *Paul Du Chaillu, Gorilla Hunter: Being the Extraordinary Life and Adventures of Paul Du Chaillu*. New York: Harper.

Walker, A. R. 1960. *Notes D'histoire du Gabon*. Libreville: Editions R. Walker.

Wallace, A. R. 1852. "On the Monkeys of the Amazon." *Proceedings of the Zoological Society of London*, 20: 107–10.

————. 1855. "On the Law Which Has Regulated the Introduction of New Species." *Annals and Magazine of Natural History* (2nd ser.), 16: 184–96.

————. 1886. *The Malay Archipelago: The Land of the Orang-utan, and the Bird of Paradise*. London: Macmillan & Co.

————. 1895. *A Narrative of Travels on the Amazon and Rio Negro*. London: Reeve.

————. 1905. *My Life: A Record of Events and Opinions*, 2 vols. London: George Bell & Son.

Wallace, A. R., and A. Berry. 2003. *Infinite Tropics: An Alfred Russel Wallace Collection*. New York: Verso.

Waller, M. 2002. *1700: Scenes from London Life*. New York: Four Walls Eight Windows.

Waterton, C. 1909. *Wanderings in South America*. New York: Sturgis & Walton.

Wayne, A. T. 1906. "The Date of Discovery of Swainson's Warbler (*Helinaia swainsonii*)." *The Auk*, 23(2): 231, 232.

Wells, L. A. 1974. " 'Why Not Try the Experiment?' The Scientific Education of Edward Jenner." *Proceedings of the American Philosophical Society*, 118(2): 135–45.

Westwood, J. O. 1847. "New Orang-outang." *The Annals and Magazine of Natural History*, 20: 286.

Wheeler, A. 1984. "Daniel Solander and the Zoology of Cook's Voyage." *Archives of Natural History*, 11(3): 505–15.

Wheeler, M. 1993. *The Lamp of Memory: Ruskin, Tradition and Architecture*. New York: Manchester University Press.

Wildt, D. et al. (eds.). 2006. *Giant Pandas: Biology, Veterinary Medicine and Management*. New York: Cambridge University Press.

Wilkins, G. L. 1957. "The Cracherode Shell Collection." *Bulletin of the British Museum (Natural History) Historical Series*, 1(4): 121–84.

Wilson, A. 1839. *American Ornithology*, Boston: Otis Broaders and Co.

Wilson, E. O. 1994. *Naturalist*. Washington, D.C.: Island Press.

———. 2003. "The Encyclopedia of Life," *Trends in Ecology and Evolution*, 18: 77–80.

Wilson, G., and A. Geikie. 1861. *Memoir of Edward Forbes*. London: Macmillan.

Wilson, L. G. 1996. "The Gorilla and the Question of Human Origins: The Brain Controversy." *Journal of the History of Medicine and Allied Sciences*, 51: 184–207.

Wood, R. G. 1966. *Stephen Harriman Long, 1784–1864: Army Engineer, Explorer, Inventor*, Glendale, CA: A. H. Clark Co.

Woodcock, G. 1969. *Henry Walter Bates: Naturalist of the Amazons*. London: Faber and Faber.

Wurtzburg, C. E. 1954. *Raffles of the Eastern Isles*. Singapore: Oxford University Press.

Wyman, J. 1866. Three journal articles bound with related manuscripts in the Ernst Mayr Library, Museum of Comparative Zoology, Harvard University, Cambridge, MA.

———. 1868. "Observations on Crania." *Proceedings of the Boston Society of Natural History*, XI.

ILLUSTRATION CREDITS

121   Library of Congress
135   American Philosophical Society
136   American Philosophical Society
140   Wellcome Library, London
144   Collection of Paul Farber
154   Wellcome Library, London
161   Wellcome Library, London
170   Wellcome Library, London
172   Collection of Richard Conniff
180   Wellcome Library, London
184   American Philosophical Society
194   Collection of Richard Conniff
202   Wellcome Library, London
203   Wellcome Library, London
205   Collection of Richard Conniff
214   Wellcome Library, London
221   Wellcome Library, London
228   Special Collections, University of Virginia
       Library
228   Courtesy of Harvard University Archives, call #HUP
       Wyman, Jeffries (6)
243   Collection of Richard Conniff
246   Collection of Richard Conniff
248   Wellcome Library, London
256   Wellcome Library, London
268   Collection of Richard Conniff
278   Wellcome Library, London
278   Wellcome Library, London
286   Collection of Richard Conniff
292   © Natural History Museum, London
294   Wellcome Library, London
310   Collection of Richard Conniff
313   Illustration by Clare Conniff
322   Collection of Richard Conniff
339   Collection of Richard Conniff
348   Wellcome Library, London
361   Wellcome Library, London
365   Wellcome Library, London
370   Wellcome Library, London

INDEX

Page numbers in *italics* refer to illustrations.

437